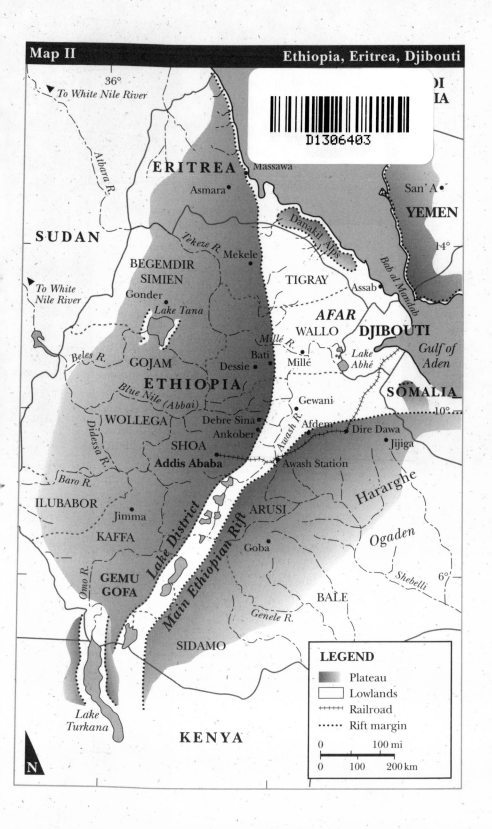

Map II   Ethiopia, Eritrea, Djibouti

36°
▼ To White Nile River

ERITREA
Massawa
Asmara •

San' A •
YEMEN

SUDAN

Athara R.

Danakil Alps

14°

Tekeze R.
Mekele •

Bab al Mandab

BEGEMDIR
SIMIEN
Gonder •

TIGRAY

Assab

▼ To White
Nile River

Lake Tana

AFAR
WALLO

DJIBOUTI

Gulf of
Aden

Beles R.

GOJAM

Millé R.
Bati •
Dessie •
Millé •

Lake
Abhé

ETHIOPIA

Blue Nile (Abbai)

Gewani •

SOMALIA

WOLLEGA

Debre Sina •
Ankober •

Afdem •

Dire Dawa •
Jijiga •

10°

Didessa R.

SHOA
Addis Ababa •

Awash Station •

Baro R.

ILUBABOR
Jimma •

ARUSI

Hararghe

KAFFA

Goba •

Ogaden

GEMU
GOFA

BALE

Shebelli

6°

Genele R.

SIDAMO

Omo R.

Lake District

Main Ethiopian Rift

Awash R.

KENYA

Lake
Turkana

LEGEND

Plateau
Lowlands
+++++ Railroad
••••• Rift margin

0   100 mi

0   100   200 km

N

3765300834056
Main Library: 1st floor
569.9 KALB
Adventures in the bone
trade : the race to
discover human ancestors in

APR - - 2001

CENTRAL ARKANSAS
LIBRARY SYSTEM
LITTLE ROCK PUBLIC LIBRARY
100 ROCK ST.
LITTLE ROCK, ARKANSAS

Adventures
in the
Bone Trade

Adventures
in the
Bone Trade

The Race to Discover Human
Ancestors in Ethiopia's
Afar Depression

JON KALB

COPERNICUS BOOKS
AN IMPRINT OF SPRINGER-VERLAG

© 2001 Jon E. Kalb

All rights reserved. No part of this publication may be reproduced, stored
in a retrieval system, or transmitted, in any form or by any means,
electronic, mechanical, photocopying, recording, or otherwise,
without the prior written permission of the publisher.

Published in the United States by Copernicus Books, an
imprint of Springer-Verlag New York, Inc., a member of
BertelsmannSpringer Science+Business Media GmbH.

Copernicus Books
37 East 7th Street
New York, NY 10003
www.copernicusbooks.com

Library of Congress Cataloging-in-Publication Data
Kalb, Jon E.
   Adventures in the bone trade: the race to discover human
ancestors in Ethiopia's Afar Depression/Jon E. Kalb.
      p.   cm.
   Includes bibliographical references and index.
   ISBN 0-387-98742-8 (hardcover : alk. paper)
   1. Fossil hominids—Ethiopia—Awash River Valley.
2. Human remains   (Archaeology)—Ethiopia—Awash River Valley.
3. Paleoanthropology—Ethiopia—Awash River Valley.   4. Excavations
(Archaeology)—Ethiopia—Awash River Valley.   5. Awash River Valley
(Ethiopia)—Antiquities.   I. Title.
GN282 K35 2000
599.9'0963'3—dc21   00-030837

Manufactured in the United States of America.
Printed on acid-free paper.

9 8 7 6 5 4 3 2 1

ISBN 0-387-98742-8   SPIN 10557407

For

Judy, Justine, Spring,

and

Sleshi

CENTRAL ARKANSAS LIBRARY SYSTEM
LITTLE ROCK PUBLIC LIBRARY
100 ROCK STREET
LITTLE ROCK, ARKANSAS 72201

Preface

I n the last few hundred years Africa has witnessed the slave trade, the ivory trade, the diamond trade, and the rubber trade. Each has represented a separate chapter of discovery and exploitation. Beginning in the 1920s, another type of "trade" burst onto the scene with the discovery of our oldest human ancestors: Fossil Man. Most notable was the headline-making Taung Child, *Australopithecus africanus,* found in South Africa in 1924. Then, beginning in the late 1950s, came the sensational discoveries in East Africa by Louis and Mary Leakey, *Zinjanthropus* and *Homo habilis;* and those by their son Richard Leakey, which, beginning in the 1960s, consisted of an unprecedented array of australopithecine and early *Homo* fossils. Each new discovery represented extraordinary wealth of information about our origins and instantly captured the public's attention.

Although trafficking in slaves and extracting minerals can hardly be equated with the pursuit of human origins, these diverse quests have followed a similar trajectory: exploration, discovery, territorial competition, exploitation, and personal gain or acclaim. In the early 1970s the search for early hominids (from the family Hominidae) shifted to the Afar Depression of Ethiopia, also known as the Danakil—one of the last major regions in eastern Africa to be scientifically explored. There the "bone trade" would reach its zenith amid the discovery of immense fossil- and artifact-bearing deposits, yielding Lucy, the First Family, Bodo Man, the Aramis skeleton, the Buri Skull, and some of the oldest and most extensive stone tool finds in Africa—what may prove to be the longest and most complete single record of hominid habitation known in the world. In the course of these events, intense competition developed among scientists and scientific teams, which resulted in treachery and bloodletting reminiscent of the exploration duels of the nineteenth century that involved such famous explorers as Richard Burton, John Speke, and Henry Stanley.

I know about these things because I was a co-founder of the expedition that discovered Lucy and later led the first surveys to most of the sites in the Afar that have since produced fossil hominids. I was a participant in the "bone wars" that accompanied these discoveries, and I have firsthand knowledge of the cutthroat competition and backstabbing that plague media-driven hominid hunts and the race for the *oldest* fossil or artifact. As a geologist I also know something of the extraordinary uniqueness of the Afar and of the reasons why it is both a geological marvel and one of the most punishing environments on earth. Having worked among the Afar nomads, I have come to understand their special relationship with their desert habitat and to identify with their life on the edge of survival.

Finally, as a resident of Ethiopia with my family from 1971 to 1978, I experienced the political upheavals that rocked the country with the overthrow of Emperor Haile Sellassie in 1974 and the subsequent rule by a brutal tyrant, who steered the country on a course of revolution and "red terror" amid a succession of tribal wars, civil wars, and invasions.

JON KALB
AUSTIN, TEXAS
AUGUST 17, 2000

Acknowledgments

At the conclusion of three decades of research that underlies the writing of this book, it is hard to say who is most to blame. Would it be Dick Williams, the fellow graduate student who dropped that article about the Afar Triangle in my lap in 1970? Or my publisher who dared, 30 years later, to publish a book that mixed science adventurism with totalitarianism and human origins? In between the two were scores of others who helped, or hindered, me along the way, and all bear some responsibility for this work.

First there is my wife Judy, who tolerated us, both the good guys and the bad guys, over the years. She also encouraged me to develop my skills in written English, as she has hundreds of school children over the years. I owe her so much for every step along the way. And our daughters, Justine and Spring, who were born at the beginning and middle of this story, respectively, yet grew up to be responsible citizens and are the delight of their parents. Then there is the rest of my family—my father BB, my mother, my brother Peter, and my sister Claire, who bore witness to my early years and who collectively believed that I would not live long enough to vote. Also, I am grateful to Ervin Kalb and to the matriarch of our family, Elva Kalb Dumas, who, at the age of 98, read and critiqued the entire final, unsanitized draft of this book.

I pay special tribute to Sleshi Tebedge, to whom this book is partially dedicated, who was a pioneer for his country, and for his continent, and was my close friend.

I thank other friends who have encouraged or helped me over the years—Kelati Abraham, Tsrha Adefris, Dejene Aneme, Mesfin Asnake, Robert Bell, Glen F. Brown, Karl Butzer, Doug Cramer, Eric Glitzenstein, Conrad Hersh Tewolde Berhan G. Igziabher, Clifford Jolly, Alfred Kelleher, Assefa Mebrate, Ted Morse, Herb Mosca, Liz Oswald, Dennis Peak, Paul Whitehead, and Craig Wood—and friends who offered assistance or inspiration—Glen Bailey, Dan Barton, Gloria Cook, Sutton Grant, Lynn Hughes, Ken and Jackie Jacobs, Bill and Mary Kaiser, Wann Langston, Ernie Lundelius Jr., Linda and David Maraniss, Don Reid, Will Reid, Larry Schwab, Naim Sipra, Dwight Schmidt, and Turu Workineh Tebedge.

I am indebted to those Ethiopian institutions that gave me their support, particularly the Ministry of Mines, the Antiquities Administration and later the Ministry of Culture, Addis Ababa University, the Commission for Science and Technology, and the Commission for Higher Education. Among those officials who were particularly helpful were Duncan Dow, Mahdi Shumburo, Shiferaw Demissie, Bekele Negussie, Soloman Tekalign, Aklilu Habte, Haile Wolde Michael, Haile Lul Tebecke, Yayehyirad Kitaw, Mamo Tessema, Daniel Tuafe, Wolde Senbet Abomsa, and, of course, Alemayehu Asfaw, who made more discoveries than all the rest of us put together.

My family and I are grateful for the assistance and friendship of many at the U.S. Embassy over the years. Richard Matheron stands out; with the assistance of Glen Brown at the U.S. Geological Survey, he helped me get much of my data out of the country. Likewise, my family and I are grateful for our friends at the British and Canadian Embassies, and those, foreign and Ethiopian, who shared so many experiences with us during both pleasant and dangerous times.

I thank my former colleagues with the International Afar Research Expedition, and the Rift Valley Research Mission in Ethiopia, all of whom are referred to by name in the pages that follow. I also thank Fred Simon and Mike Foose with the U.S. Geological Survey.

For special assistance in this country I owe much to the Vertebrate Palaeontology Laboratory with the Texas Memorial Museum of the University of Texas at Austin, which has provided me with a research base over the years. For legal and technical support, I am grateful to Public Citizen, Meyer & Glitzenstein, and Graybill & English, all of Washington D.C., and also to The Authors Guild of New York, Austin Lawyers and Accountants for the Arts, and the Austin Writers League.

Those individuals who kindly gave me support or valuable information include Bill Adler, Sally Baker, Daniel Barker, Robert Bell, Ray Bernor, Claud Bramblett, Lana Castle, Bill Cayce, H. B. S. Cooke, Tom Doyal, Jack Edwards, Veronica Evering, Mulugetta Fissaha, Tafara Ghedamu, Mike Hamilburg, Charlie Harrell, Clifford Jolly, John Kappelman, Edward Kimball, Ernest Lundelius, Jr., Vincent Maglio, N. O. Nelson, Workineh Wolde Rufael, Heather Page, Jean-Pierre Slakmon, Angela Smith, Louise Sperling, Turu Workineh Tebedge, Wonde Wossen Tebedge, John Van Couvering, and Carolyn Wylie.

For valuable technical assistance with the research, editing, or preparation of the book I thank Becky Ballou, Mary Ann Brickner, Sandra Bybee, Roxanne Bugucka, Connie Day, Jennifer Haas, Judy Hogan, Leah Linney, Valeria Liverini, Ran Moran, Susan Murphy, Gianluca Paganoni, Pierre Rico, Rebecca Sankey, and, with special gratitude for years of assistance, Sharon Robinson. Credit for the maps goes to Jana Robinson and Maria Saenz.

We thank Fort Knox Music Inc. for permission to use lyrics from "I Got You (I Feel Good)" by James Brown © 1966.

Many thanks to those who gave their valuable time in reading all or parts of the manuscript and supplying valuable comments: Richard Benson, LaVerle Berry, Loren Bliese, Claud Bramblett, Craig Feibel, Ron Girdler, Wulf Gosa, John Harbeson, Terry Harrison, Richard Hayward, Clifford Jolly, Ioan Lewis, Larry Martin, and Laura Wood.

Special gratitude goes to my publishers at Springer-Verlag, Jerry Lyons and Kevin Lippert, who made the publication of *Adventures* possible, and also to my editors, Bill Frucht for his encouragement, Jonathan Cobb for his guidance, and especially Paul Farrell, who steered the book to completion.

Finally, nothing can be said of the crusty flats of the Afar Depression, or of its deep valleys and many volcanoes, or of its mysteries or its mystique, without recognizing that its people, the Afar nomads, are its ultimate caretakers and that without them little is possible. To them go my warmth and deepest feelings. I especially remember my friend Selati Alemma Ali, companion of many delightful days, of river crossings, long treks, and evenings recounting the rewards of another day.

JON KALB
AUSTIN, TEXAS
AUGUST 17, 2000

Contents

Maps and Appendices

Maps

Appendices

With regard to the surviving mule, his having survived was entirely due to a series of very fortunate circumstances and to the pride eventually taken by all of us in seeing him through.

—E.M. NESBITT, *From North to South Through Danakil*

CHAPTER 1

An Oceanic Desert

BALTIMORE, MARYLAND
FEBRUARY 1970

One winter day I was eating lunch in the cafeteria of Johns Hopkins University, my brain still reeling from another excruciating lecture in crystallography. As the class ended, the professor had said, just one more time, "I don't ask you to believe, I *challenge* you to deny!" Ringing words those, but if you chose not to *listen* to his lectures, you were begging for disaster. As the professor told me at the end of the Christmas holidays, after I had worked two weeks on his take-home exam, "Mr. Kalb, how are you ever going to be a geologist if you don't learn crystallography?" What could I say? The man knew I was a crystallographic illiterate. To this day I could not tell you the difference between a rhombic dodecahedron and a bowl of pine nuts.

As I sat there sodden with remorse while eating my tuna fish sandwich, a fellow graduate student in geology walked by, dropped a magazine in my lap, and said, "Read the cover story. It sounds like your thing." Because my friend had tried in vain to help me through the intricacies of thermodynamics the previous semester, I wondered what he thought I would understand. But I trusted his judgment—he was just wrapping up the second volume of a thousand-page doctoral dissertation, a feat I thought dazzling.

So, picking up the magazine, I said, "Sure, why not give it a shot."

It was about Ethiopia. Ah yes, Africa had been on my mind since high school, when I first thought of being a geologist. Like many in the late 1950s and 1960s, I had been fascinated by articles in *The National Geographic* about the East African Rift Valley and "fossil man" discoveries made by Louis Leakey and family at Olduvai Gorge. But I always thought of anthropology as too limiting. My idea, even as a youngster, was that geology encompasses the entire earth, so that was the direction to go. Also, growing up in Houston, Texas, surrounded by people connected with the still-booming oil industry, made me think of geology as a profession.

My interest in Ethiopia was aroused in 1962 when I read Alan Moorehead's book *The Blue Nile*, which described the exploration of this mighty river from its origins in the Ethiopian highlands to its confluence with the White Nile in the central Sudan. To this day I recall Moorehead's description of a mysterious man-made earthwork found by geologists that stretches across the Ethiopian highlands—"an immense ditch too wide for a horse to jump across has been discovered, and it winds away over valleys and hills for hundreds of miles."[1] The author speculated that the ditch was an ancient boundary between two tribes, or perhaps the work of some past emperor. I was certain that if given the chance, I could find out who made that ditch and why.

The article my friend gave me to read appeared in the February 1970 issue of *Scientific American* and described "one of the world's most forbidding regions," an area in northeastern Ethiopia called the Afar Triangle.[2] Named for its inhabitants, the Afar nomads, and its isosceles shape, the area was further described as one where a "new ocean" is being created in the middle of a "nightmarish desert landscape."

Intriguing.

Because of its isolation and inhospitality, the Afar had been explored by few, and much of it was still *terra incognita*. Several expeditions that ventured into the area in the nineteenth century had been slaughtered and the men mutilated by fierce tribesmen. Their descendants, a nominally Muslim semi-nomadic people inhabiting the region today, reportedly still emasculated their male victims for trophies to be offered to their womenfolk.

Not so intriguing.

The author of the article, Haroun Tazieff, a Polish-born volcanologist and adventurer, described the Afar as a wild and rugged lowland containing active volcanoes, a below-sea-level desert, blistering salt flats, boiling hot springs, and temperatures soaring to 56°C (133°F). Temperatures recorded on a salt pan in the northern Afar, 35°C (94°F), are the highest average mean temperatures in the world. An accompanying map showed the triangle encompassing parts of Ethiopia (including what is now Eritrea) and the neighboring French Territory of Afars and Issas (now Djibouti) (see Maps I and II). The Issa are ethnic Somalis and are the second most predominant tribe in the area. Altogether, the region comprises some 140,000 square kilometers, the size of England. To the west, the Afar is bounded by a towering escarpment rising to 3500 meters above sea level, described by Tazieff as the highest cliff top in the world. The Afar is bounded to the east by the low-lying Danakil Alps, which straddle the southern end of the Red Sea, and to the south by a 3000-meter escarpment that merges with the margins of the East African Rift Valley and the Gulf of Aden.

In the far north, the floor of the Afar is covered by a salt plain, which lies 122 meters below sea level; to the east, in the Asal basin in Djibouti, the elevation drops to 156 meters below sea level, the lowest in Africa. Some 500 kilometers to the south, the Afar floor rises to 800 meters above sea level. Because of its low elevation, accentuated by surrounding escarpments and mountains, the Afar region is widely referred to by geologists as the Afar Depression, although historically the term *depression* has been applied just to the northern lowlands.

The Afar lies at the intersection of three major structures: the East African Rift, the Red Sea, and the Gulf of Aden. All three are giant troughs—rifts—created by separation of the earth's crust. The rifting in the Red Sea and the Gulf of Aden is so wide and so deep, however, that not only are these inlets filled with sea waters, but their floors are made of particularly dense volcanic rocks like those found in the deepest oceans. When this and other oceanic features of these inland waterways were recognized during marine surveys in the early 1960s, it was realized that the area at the junction of the Red Sea and the Gulf of Aden, the Afar Depression, must be subject to ocean-forming processes that are ongoing today and that should be observable on dry land. For many years the Red Sea was the focus of theories about the early stages of continental drift—the idea that the continents were once assembled like a giant jigsaw puzzle—but now, Tazieff was saying in his article, the Afar could answer even more fundamental questions about the embryonic history of the oceans.

To observe the "birth of an ocean" firsthand, in 1967 Tazieff and his team of French and Italian geologists began exploring and mapping the many volcanoes and lava flows in the Afar. After several field seasons, the team concluded that the northern Afar represents the landward extension of the Red Sea floor and that the lofty western escarpment is simply an extension of the western coastal scarp. Furthermore, ocean-forming processes *are* active today in the Afar, as revealed by deep rifting of the outer layers of the earth, called continental crust, and the extrusion of dense oceanic-like basalts. Tazieff's team also discovered that much of the northern Afar had been covered by sea water as recently as 10,000 years ago, as indicated by fossil coral reefs and the discovery of stone tools encrusted with seashells along ancient shorelines. When the waters later receded, thick salt deposits were formed in the landlocked Depression, and they continue to accumulate today, as waters flowing from the adjacent highlands evaporate in the cauldron-like lowlands.

With renewed energy, I gobbled down my lunch and resolved to learn more about this "oceanic desert." Over the coming weeks—when I could tear myself away from trigonal bipyramids and hexoctahedrons—I dug through the Johns Hopkins library. A good start was Alfred Wegener's classic book *The Origin of Continents and Oceans,*[3] originally published in German in 1915.[4] Wegener, born in Berlin in 1880, is the most famous early proponent of the theory of continental drift. Trained in astronomy and meteorology, he wrote his book while convalescing after being shot in the neck during World War I.[5] In the 15 years before he died, on an icecap during his third expedition to Greenland, Wegener refined his ideas of drifting continents. Basing his conclusions largely on paleoclimate and fossil evidence, he envisioned that 300 million years ago, only a single supercontinent had existed, which eventually split into the smaller continents of today.

Whoever was the first to draw a reasonably accurate map of the Red Sea was probably also the first to ponder whether some mighty force, biblical or otherwise, pulled apart its coastlines. Certainly the striking fit of the opposing shores was crucial evidence for Wegener's theory. And with the increasing accuracy of each new Red Sea map over the years, the fit became more and more evident. Wegener believed that Arabia had drifted away from Africa, leaving the bottom of the Red Sea filled with volcanic rock, along with the Afar region, in its wake. He proposed that "if one cuts out this triangle, the opposite corner of Arabia fits perfectly into the gap."[6] Wegener also believed that the East African Rift represented an early stage of continental splitting and that the Red Sea and the Gulf of Aden represented advanced stages of actual crustal separation.

A colleague of Wegener's at the University of Marburg in Germany, where they both taught in the early 1900s, took issue with the idea that continental drift formed the Red Sea. At age 24, in 1909, Hans Cloos had seen the rising Red Sea coastal scarps firsthand while traveling by ship to Southwest Africa, where he was to begin his career as a geologist.[7] Cloos was to become an internationally renowned geologist who specialized in simulating deformations of the earth's crust in the laboratory. He did so by using blocks of moist clay, subjecting them to varying stresses on a movable platform. One of his most famous re-creations was to model the formation of a rift valley by extension, by slowly pulling apart a thick clay "cake." Left in the middle of the clay slab was a trough—*graben* in German—bounded on either side by elevated margins (*horsts*) like those on either side of the Rhine Valley or the Red Sea (see figure on page 7).[8]

In his memoir, *Conversation with the Earth,* Cloos described the day in 1913 when Wegener, a man with "penetrating gray-blue eyes," first came to see him in his office.[9] Wegener lost no time launching into his still-developing theory of continental drift:

> "Just look at Arabia!" Wegener cried heatedly, and let his pencil fly over the map. "Is that not a clear example? Does the peninsula not turn on Sinai to the northeast like a door on a hinge, pushing the Persian mountain chains in front of it . . . ! In the rear, the Arabian table has been torn off Africa. It has moved away from the [southern Afar], opening a rift 200–250 miles wide, exactly the amount of narrowing suffered by the Persian mountain chains."

> "But the triangle of Danakil in the southeast corner of the rift [Cloos replied], how does it fit into your movement?"[10]

Wegener believed that Arabia rotated counterclockwise, pivoting on the Sinai peninsula, while compressing the Persian, or Zagros, Mountains in Iran. Lava filled the Danakil (Afar) interior, as it did the Red Sea.

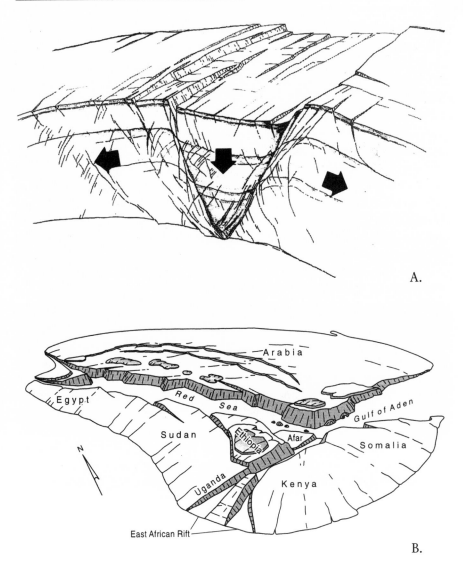

FIGURE 1. (A) HANS CLOOS'S "CLAY CAKE" MODEL of a rift valley, or graben, in the form of a wedge-shaped trough created by uplift and stretching.
(B) Cloos's view in 1953 of the formation of the Afro-Arabian shield as a
result of domal uplift and extension. The Afar lies in the center at the junction of the Red Sea, the Gulf of Aden, and the East African rifts.

Cloos was skeptical, and over the years he gave the Red Sea and nearby landforms much thought. In 1929 he created a model for the formation of the Red Sea
that convinced him the waterway was underlain by elongate, collapsed blocks of
continental rock, not oceanic basalt. During a second trip through the Red Sea and
the Gulf of Aden in 1933, Cloos was impressed by evidence of the great graben-like

vertical movements—faults—that seemed responsible for much of the topography of the region. Back in Germany he conducted more clay cake experiments that convinced him the Red Sea was part of "a gigantic structural dome" formed by a vast underground "swelling of the earth's skin, which split [in graben fashion] where too tautly stretched."[10]

To simulate this movement, Cloos placed layers of clay on the surface of a hot-water bottle and observed their deformation as the bottle was expanded. From this he then drew an extraordinarily accurate reconstruction of the Afro-Arabian shield (see figure on p. 7). He demonstrated that the Red Sea, the Gulf of Aden, and the East African Rift are all immense tensional cracks created by stretching and that all three intersect in a triple junction in the Afar Depression, across the crest of a large regional dome. Cloos proclaimed, "The piece of land called Danakil, which had been a stumbling block to Wegener's drift theory, now had its organic place in the dynamic process which formed the dome."[10] Cloos concluded that continental drift was unnecessary to form the Afar.

Cloos's experiments were convincing to later generations of geologists, who would invoke "crustal swelling" or "domal uplift" caused by upwelling magmas to explain how the Afar triple rift had formed. But Cloos was to find out that he still did not have all the answers, by far.

After the outbreak of World War II, while at the University of Bonn where he was by then a distinguished professor, Cloos obtained a newly completed chart of the bathymetry of the Gulf of Aden sea floor.[11] He fully expected to see a series of faults more or less parallel to the coastlines of the Gulf, which would be indicative of graben formation and crustal swelling—and a splendid confirmation of his methodical laboratory experiments. Instead, he saw a series of enormous *transverse* ridges and troughs running northeast–southwest—that is, *obliquely*—across the Gulf. Cloos was dumbfounded. As he related in his memoir, so perplexed was he over these features that even when fleeing Bonn with his family to a remote farmhouse in the midst of an intensified war, he took with him his chart of the Gulf of Aden and clay cake apparatus.[10] From morning until night, with war planes droning overhead, Cloos conducted more experiments while poring over the chart, trying to figure out how those structures had formed. He had no idea that it would take 25 years and many millions of dollars worth of basic mapping of the ocean floors before the origin of those faults was known. But to his great credit, Hans Cloos recognized that these faults were a major discovery.

My favorite course at Johns Hopkins was field geology. A dozen or so students would go out every weekend to visit one famous geological outcrop or another. On several occasions the group was led by Professor Ernst Cloos, who had come to the United States from Germany in 1935. It is said that Ernst emigrated partly to escape the shadow of his acclaimed older brother, Hans. Certainly he succeeded. In

1954 he was elected president of the Geological Society of America.[12] Nevertheless, Hans had clearly influenced Ernst, who not only became a geologist like his
brother but also used clay models to replicate earth movements. Whereas much of
Hans's research concentrated on extensional earth movements, such as graben formation, Ernst made compressional tectonics his life's work. He specialized in the
geology of the Appalachian Mountains, which are squeezed together much as are
the Zagros Mountains in Iran.

Unfortunately for me, Professor Cloos retired the year before I went to Baltimore, but he continued to lead excursions with students. I never heard him say
what he thought about continental drift, or plate tectonics, an idea then rev-
olutionizing the world of classical geology. However, I do recall him standing
before a roadcut one clear, crisp winter day high in the Appalachians, as he lec-
tured to my field geology class. The expanse of the folded crystalline mountains
lay in the background. As he picked through the outcrop, the erect, white-haired
professor spoke in his German accent, devoting much time to explaining the
origin of deformed pebbles and squashed grains of sand. Bored with all this, I
thought *But how in the hell were these mountains formed?* So I blurted out, "But
Professor Cloos, how were these *mountains* formed?" All of a sudden you could
hear the birds chirping on either side of the highway. Professor Cloos looked
straight at me and said, "Mr. Kalb, how are you going to learn anything if you
do not *listen!*"

Gulp.

Back in the library late at night, I looked for more information about the Afar
Depression. Students were here and there, a few asleep. With the neon light
above my desk humming over the silence, I paged through the *magnum opus* of
Arthur Holmes, an Englishman and one of the great geologists of the century.[13]
His *Principles of Physical Geology,* published in 1944, has since gone through 4
editions and 31 printings.[14] Like Hans Cloos, Holmes began his career in Africa.
After graduating from the Imperial College of London in 1911, he went on a geological expedition to Mozambique, pursuing what was to be a lifelong interest in
Africa. Soon after his return, he published *The Age of the Earth,* which established him as a pioneer in the use of radioactivity for dating minerals.[15] In 1929
Holmes proposed that deep-seated heat in the earth generated by radioactivity
has created immense convection cells that have helped shape the earth's surface.[16] In *Principles* he speculated that the drag of these cells moving in opposite
directions on the underside of continents could break and pull them apart and
eventually form ocean basins. Holmes's radical idea proved to be the first cred-
ible mechanism proposed for continental drift.[17] Over the years, Holmes
wrongly favored the view that the Afro-Arabian rift valleys resulted from convection cells moving toward each other in the form of compression.[18] By the time
he published his monumental 1288-page second edition of *Principles* in 1965,
however, he had embraced the meticulous work of Hans Cloos and referred to
the Afar as part of the vast swell trisected by the Red Sea, the Gulf of Aden, and

the East African rifts. "Tension clefts in at least three directions, which box the compass, are compatible only with crustal extension."[19]

Harry Hess of Princeton University and Arthur Holmes had much in common. Hess also began his career in Africa (in Rhodesia), and the research of both spanned the breadth of geology, from petrogenesis to global tectonics.[20] In a seminal paper written in 1960, Hess advanced the ideas of Holmes and accounted for the massive data collected from the ocean floors through the 1950s.[21] Submarine mapping had revealed an elaborate global system of midoceanic ridges that are commonly split at their crests, forming oceanic rift valleys, or "rift-ridges." In contrast, deep marine trenches line many continental margins at the front of deformed mountain ranges. Hess put this and other information together and believed, like Holmes, that the formation of ocean basins begins with opposing convection cells of magma splitting continents apart. However, Hess concluded that oceanic rift-ridges are active sources of basaltic lavas that spread out from ridge crests forming the sea floor. With time the outwardly moving ocean floor collides with continents and plunges back into the upper mantle, forming the deep marine trenches. Hess emphasized the "ephemeral" nature of oceanic ridges and the youthfulness of the sea floor. Both are newly created by the rising limbs of convection cells and are ultimately destroyed by descending limbs that drag oceanic crust into trenches like a "jaw crusher."[22] Eventually the crust is regurgitated at midoceanic ridges. Hess estimated that the sea floor is recycled in this way every 200 to 300 million years.

Mountains at the bottom of the oceans are "ephemeral"? The sea floor is consumed by a "jaw crusher"? As I finished Hess's paper, I thought that if the creationists have a problem with Darwin, they must be going apoplectic with Hess. And Hess had an Alfred Russel Wallace and a Thomas Huxley rolled into one: Robert Dietz, an oceanographer with the U.S. Navy, who coined the phrase *sea-floor spreading*. Like Wallace, who had independently conceived the theory of natural selection, Dietz claimed ownership of the theory of a spreading sea floor. In 1961 he presented his ideas in a brief but cogent article in the British journal *Nature*.[23] Like Huxley, who was an outspoken supporter of evolution, Dietz then became his own best advocate for his theory by producing a plethora of scientific papers on spreading over the next few years. In a *Saturday Evening Post* article, he described the mobile ocean floors as "gaping wounds in the earth's skin, exposing the raw moving flesh of the earth."[24]

Dietz envisioned, as had Hess before him, that the sea floor represented the cooled and hardened outer surface of convection cells that lay revealed between oceanic ridges and trenches. As Dietz described the phenomenon, the mantle, fueled by radioactivity, "turns over slowly and inexorably like a vat of asphalt over a low fire."[24] In his version of the theory, however, he went one giant step further. He proposed that parallel ridges of volcanic rock on either side of the Mid-Atlantic Ridge and linear bands of magnetized rock that had been mapped along ridges in the

Pacific were the same thing: newly created sea floor. These magnetic lineations soon provided proof to the theory of sea-floor spreading.

In 1963 Lawrence Morley, a Canadian geologist, and Fred Vine and Drummond Matthews of Cambridge University suggested independently that such linear bands of rock on either side of midoceanic ridges represent changes in polarity of magnetic minerals contained in basaltic rocks.[25] When magnetic minerals in lavas are cooling, they acquire magnetization parallel to the earth's magnetic field and align themselves like a compass. Because the magnetic field periodically reverses, if seafloor spreading is real, then stripes of alternately normal and reversed magnetized rock should be present symmetrically on either side of oceanic ridges. Vine and Matthews were able to show that such alternating magnetic bands exist on both sides of the Carlsberg Ridge, a midoceanic ridge in the Indian Ocean that continues into the Gulf of Aden. Other scientists discovered the presence of symmetrical anomalies in the Gulf itself and in the Red Sea.[26] Altogether, these efforts were part of a British oceanographic assault on the region in the 1960s that would profoundly influence ideas on the origins of oceans.[27]

This work captured the attention of the well-known Canadian geologist J. Tuzo Wilson, who while studying the Cambridge data recognized a totally new class of faults associated with spreading zones.[28] These were the same type of "mystery faults" as those cutting transversely across the Gulf of Aden that had so astonished Hans Cloos 25 years earlier.

Wilson concluded that the faults represent fractures connecting offset segments of oceanic ridges undergoing sea-floor spreading. He called them *transform* faults because movement along spreading zones is "transformed" into movement along fault zones. Thus these transverse faults are expressions of horizontal movement of newly created oceanic crust; they are not artifacts of vertical movements of ancient continental crust associated with graben formation, which Cloos was so familiar with. In his paper on transforms—published in *Nature* in 1965, just three weeks after it had been accepted[29]—Wilson described a special type of transform fault, mapped just east of the Gulf of Aden.[28] He described the fault as having been created by the northeast movement of the sea floor from the Carlsberg Ridge spreading center (see figure on page 12). This movement in turn forced the collision of India into Eurasia, creating the folded Zagros and Himalayan Mountains. He described a similar fault at the northern end of the Red Sea that passed through the Gulf of Aqaba and the Dead Sea. The movement along both of these transform faults resulted in the northeast movement of Arabia and the creation of the Red Sea, the Gulf of Aden—and the Afar Depression.

Like Dietz and others before him,[30] Wilson believed that the East African Rift is the westward extension of the Carlsberg Ridge and that the Red Sea is its northern extension. By the mid-1960s, Vine and others had confirmed that oceanic ridges are emerging beneath the Gulf of Aden and the Red Sea and that both inlets are embryonic oceans.[31] In support of this idea, earthquake data revealed a line of epi-

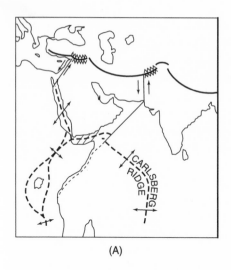
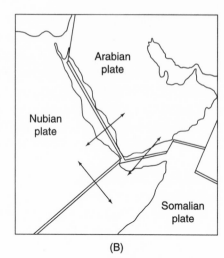

(A)   (B)

FIGURE 2.   (A) J. TUZO WILSON'S SKETCH in 1965 depicting the offset of the
Carlsberg midoceanic ridge and the northeast drift of Arabia along transform faults. Note the collision of Arabia with Eurasia in the area of Turkey
and the Zagros Mountains. (B) The boundaries between the Nubian (African), Arabian, and Somalian (East African) plates, the relative directions of
plate movements, and the formation of a triple junction, as viewed by D. P.
McKenzie et al. in 1970.

centers that extend westward from the Carlsberg Ridge into the Afar and up the
Red Sea, indicating instability along the area separating Arabia and Africa.[32] Soon
seismologists with Columbia University determined that this "earthquake belt" is
just one piece of a worldwide rift-ridge network that courses throughout the
oceans, patterned like the seams on a baseball.[33] They would also learn that there
are only three places in the world where oceanic ridges plunge into continents. One
is in the Gulf of California, the second is in northern Siberia, and the third lies at
the junction of the Red Sea and the Gulf of Aden—in the Afar Depression.

By 1966 scientific attention was firmly focused on the Afar as a region where "an
ocean basin is being born." Several international symposia featuring global tectonics recommended wide-ranging studies in the Afar.[34] Two major programs of
exploration were adopted under the sponsorship of the International Upper Mantle Project: one calling for geological studies to begin in 1967 by Haroun Tazieff's
French–Italian team, the other for predominantly geophysical studies to begin in
1969 by a large West German team.

    While these groups were gearing up for a scientific campaign in the Afar, another
major breakthrough resulted in the "unified field theory" of the earth sciences and further focused attention on the Depression. In 1967, at a sparsely attended lecture

held just before lunch break at the annual meeting of the American Geophysical Union in Washington, D.C.,[35] W. Jason Morgan of Princeton University presented a master plan to explain the interrelationships among continental drift, sea-floor spreading, and transform faults.[36] He divided the earth's surface into a network of "plates" that form its outer shell—the continents, the oceanic crust, and the semi-solid upper mantle. He showed that oceanic plates are created at oceanic ridges by sea-floor spreading, are destroyed at marine trenches and in folded mountain belts, and slide past one another along transform faults. The example Morgan used to describe these relationships was the Arabian plate, which is surrounded by oceanic ridges, transform faults, and the folded Zagros Mountains.

The theory of "plate tectonics" was immediately applied by researchers from Cambridge, who demonstrated that earthquakes reveal the direction of plate movements.[37] Scientists at Columbia University then used both seismicity and magnetic anomalies to describe the geometry and movements of plates.[33] The Columbia team used the Gulf of Aden as one example to illustrate that earthquake epicenters mark the boundaries and direction of movement of the Arabian and East African plates along transform faults. These observations made it possible to reassemble the two plates into their pre-drift position by pulling together the opposing shorelines of the gulf. Such a confirmation of continental drift had broad implications for the worldwide rift system and for the concept of plate tectonics generally. And certainly for the formation of the Afar Depression.

CHAPTER 2
Afar Gold

By the time spring had rolled around in Baltimore and the black soot resting on top of the last snowfall had long since completed its final journey to the earth, I was once again looking for diversions from my course work. In the April 1970 issue of *Nature,* an article described the Afar as a triple junction created by the Red Sea, Gulf of Aden, and East African rifts following the separation of the African, Arabian, and East African plates (see figure on p. 12).[1] The authors concluded that the voids created by the separation of Africa from Arabia should be partially filled with rock rising from the upper mantle. Thus the Afar should possess the only emerging oceanic ridge in the world above sea level and, by implication, should be the only place in the world where active sea-floor spreading could be observed on dry land.

With my interest in the Afar heightened, I traveled the 40 kilometers to Washington, D.C., to hear two lectures on Afar geology given at the annual meeting of the American Geophysical Union. The last AGU meeting I had attended had begun on April 2, 1968, just two days before the assassination of Martin Luther King, Jr. In the middle of the proceedings, widespread rioting had erupted in the ghetto area of the city, and thick clouds of black smoke from burning buildings hung above the nation's capital. Those attending the meeting, held at the Sheraton-Park Hotel, had to observe the same 4:00 P.M. curfew as the rest of the city. (That was a bad year for geophysicists. The following April many participants in that ill-fated AGU meeting attended the International Geological Congress in Czechoslovakia: the Prague Spring. Friends of mine came back with stories about how they had dodged armored tanks while frantically trying to get out of the city during the Soviet invasion.)

The first lecture I heard at that 1970 meeting was by Paul Mohr of the Smithsonian Institution, who had worked in Ethiopia for some years and had written a book on Ethiopian geology.[2] Mohr had recently published a paper contesting much that had been learned about the Afar.[3] He argued that the northern extension of the East African Rift plowed across the Danakil Alps, and he disputed the idea that Red Sea–related rifting accounted for much of the geology of the northern Afar. Instead, he proposed that the depression was simply an ancient sedimentary basin cut by faults associated with the East African Rift. Mohr's lecture, equally provocative, argued that drifting plates were simply unnecessary to explain Afar geology.[4]

The next lecture was by a member of Tazieff's team, Enrico Bonatti of the University of Miami. He offered convincing evidence that (1) the northern Afar was indeed a tectonic depression associated with Red Sea rifting and (2) the East African Rift came to a halt at the triple junction well short of the Danakil Alps. He described the Alps as part of an elongate splinter of ancient crustal rock that had been ripped off the African mainland with drifting and was left straddling the southern end of the Red Sea. Bonatti's slides of the northern Afar showed deep tensional fissures and associated volcanism clearly aligned with the Red Sea rift. He also described submarine volcanoes in the Afar, called *guyots,* that were later uplifted to well above sea level.[5] As evidence for their marine origin, Bonatti described finding marine shells and coral reefs at their summits. Even more graphic was his description of the heat and working conditions in the Afar, which he described in his Italian accent as "veri painfeel."

When summer arrived in Baltimore, and students had exchanged their jeans for shorts and were spread out lounging on the campus lawn, I began exchanging letters with geologists in Ethiopia and elsewhere who had knowledge of the Afar. My original idea was to select a project on Afar geology for a Ph.D. dissertation, but that idea—at Johns Hopkins at least—had grown more and more remote as the second semester progressed. I had never been cut out for the classroom, and I had spent my school years proving it. If there was ever an alternative to sitting through a class, that was for me.

On the other hand, my interest in exploration and in nature knew no end. In early childhood, I had regularly escaped from the backyard of my home with my older brother and sister to explore a large forest and pond belonging to a wealthy neighbor. There we followed animal trails, chased armadillos, and caught poison ivy, while building forts and making secret hiding places for our childhood treasures. As a youngster, I had filled my room with collections of rocks, fossils, arrowheads, and live snakes. By the time I reached high school, there was scarcely a bayou in Houston that my buddies and I had not explored, and later, as we grew older, scarcely a field outside the city that we had not plowed across in our beat-up automobiles. We undertook expedition after expedition into the hinterlands outside the city, exploring farmlands while shooting rabbits, ducks, and geese. We rafted down the Brazos River, trekked up dry river beds, and survived clouds of mosquitoes during campouts. My outdoor skills were honed during summers on my grandfather's ranch west of Houston exploring pasturelands, gravel pits, and the graveyards of early settlers.

By the time I graduated from high school in 1959, I had crisscrossed Texas, hiked through the Colorado Rocky Mountains, and traversed much of New Mexico on foot. And by the time I arrived at Johns Hopkins ten years later, I had worked in Mexico, Central and South America, the Caribbean, and the South Pacific, mostly with expeditions of one kind or another. I had managed to avoid going to Vietnam by blowing out my eardrums in an underwater accident while working on a Spanish shipwreck off the coast of Yucatan. My interest in geology, archeology, and paleontology eventually landed me in Washington, D.C. There I worked at the Smithsonian Institution for several years as a research assistant in paleontology and later at the Geophysical Laboratory of the Carnegie Institution as a "predoctoral research follow." During all this time I pecked away at an undergraduate degree in geology, first in Texas, then at American University in the national capital. I was a terrible student, but my experience won me a graduate scholarship at Johns Hopkins in geology.

The year before I made the trip to Baltimore, I married my girlfriend from Washington State, who worked in Washington, D.C., for her home-state senator. During a three-year courtship we had gone on long hikes in the Appalachians, stayed in youth hostels from Maryland to West Virginia, and become immersed in the charged climate of civil rights and anti-war marches that raged across the nation in the mid-1960s. The summer before we married, Judy had visited me in Panama, where I was on leave from the U.S. Army Corps of Engineers, with whom I was working as a geologist in the Chaco rainforest of Colombia. In the middle of our four-day rendezvous, I became violently ill with paratyphoid from drinking untreated water some idiot had put into the camp water supply. Judy nursed me for two days in our hotel room before I was admitted to the Gorgas Hospital in the Canal Zone for a two-week recovery. I proposed that winter, back in Washington, D.C., and we enjoyed a two-day honeymoon at an old inn located in the foothills of the Appalachians.

Then at 7:30 A.M. on January 16, 1970, the alarm went off in our roach-infested apartment in the graduate student housing of Johns Hopkins, and Judy promptly announced that she was having labor pains. Exactly one hour later she gave birth to a gorgeous baby girl in the Baltimore Union Hospital. We called her Justine, after the sultry character in Lawrence Durrell's book of the same name. I insisted that her middle name be Demerara, because I was convinced that she had been conceived up the Demerara River in Guyana the previous summer while Judy was visiting me in a bush camp for geologists. As I sat in the hospital anxiously awaiting Justine's imminent appearance in the world, it occurred to me that I was missing another lecture on crystallography. *Damn!*

Although I was to receive no advanced degrees from Johns Hopkins, I remained in Baltimore throughout the summer, giving desultory attention to a research project in petrology while continuing my readings on the Afar and contacting geologists with knowledge of Ethiopia. From Paul Mohr I learned about a graduate student at the University of Paris named Maurice Taieb, who was studying the sedimentary history of the Awash River Valley, which was depicted on the map that accompanied Tazieff's article in *Scientific American*. The map showed that the river flowed out of the Ethiopian sector of the East African Rift onto the Afar plain, then followed a sigmoidal northern course, and ultimately drained east into Lake Abhé on Ethiopia's border with the French Territory of Afars and Issas (Djibouti) (see Map II). At this same time I also began reading reports about a U.S.–French team, the Omo Research Expedition, that was exploring vertebrate fossil deposits in Ethiopia's lower Omo River Valley.[6] This river drains south from the Ethiopian highlands into the East African Rift at Lake Turkana (formerly Lake Rudolf) on Kenya's northern border. The team had found "fossil man" remains of *Australopithecus* that were several million years old. Even more spectacular hominid finds were being made at Turkana by Richard Leakey.[7] Perhaps similar remains could be found in the Afar, I thought. Maybe our earliest ancestors camped on the flanks of a spreading center? Or astride a transform fault? Or made their stone tools out of oceanic basalts? I could just see the headlines: "Fossil Man Found Ripped Apart by Sea-floor Spreading."

Rift valleys are famous as reservoirs containing thick sedimentary deposits that preserve prehistoric remains. Rifts serve as structural "traps" simply because rivers flowing into these deep linear depressions from adjacent highlands carry with them sediments that rapidly accumulate on the rift floor or on the bottom of rift valley lakes. Any remains of animals or stone tools left by humans are quickly buried. Subsequent erosion by modern rivers then exposes these remains to the surface, as does the faulting of strata associated with ongoing rifting.

While I was reading about the Afar, it occurred to me that if a rift valley is an ideal setting for preserving fossils and artifacts, then what about *three* rift valleys intersecting in one place? What about a depressed lowland that comprises one of the largest land-locked, multiple-rift-valley structures in the world?

A geological map of Ethiopia compiled by Paul Mohr in 1963 depicted the entire Afar portion of the Awash Valley as lake deposits of Pliocene age and younger—that is, just over some 5 million years old.[8*] In his book about Ethiopian geology, he referred to sketchy evidence of ancient lake basins throughout the lower reaches of the Awash Valley in the Afar.[10] In his office at the Smithsonian Astrophysical Laboratory in Cambridge, Massachusetts, Mohr showed me a recently published article by Taieb describing sedimentary deposits in the upper Awash Valley that were of Pleistocene age,[11] less than about 1.5 million years old. The upper Awash extends from the source of the river near the capital of Ethiopia, Addis Ababa, to Awash Station, a small train depot located at the point where the river flows into the extreme southwestern Afar lowlands (see Map II). Taieb's article also mentioned stone artifacts and elephant fos-

---

*The age boundaries given to different geological epochs have varied widely over the years as fossil and radiometric dating have improved. The ages used here are based on currently accepted dates.[9]

sils near Awash Station, but neither his work nor anything else I read provided similar descriptions for the rest of the Awash Valley to the north in the Afar. From what I could gather, the geology of that part of the Afar was virtually unexplored.

By midsummer my mind was made up. Clearly the Afar Depression was on the cutting edge of geological research and perhaps more. At this time I began corresponding with the director of the Ethiopian Geological Survey, an Australian named Duncan Dow. I learned that the Survey had been established only a few years earlier, largely with United Nations funds, and that it had no money to employ another geologist. Dow told me, however, that if I could come up with adequate grant funds to work in Ethiopia, the Survey would sponsor my stay and provide what logistical support it could for my field work in the Afar. Following more letters to Ethiopia and a telephone call, I requested that he send me a formal research proposal to conduct geological mapping in the Awash–Afar region that I could then use to secure funds. Dow sent me a one-page proposal with an accompanying sketch map outlining a field project in the middle Awash Valley, an area that covered the distance from Awash Station to the village of Gewani, 150 kilometers to the north.[12] During another telephone conversation he explained that the project would complement other studies under way to develop the agricultural potential of the region. The proposal required that I map this area, some 12,500 square kilometers, in six months. *No problem.* However, Dow said I would need $5700 in U.S. dollars to do the project. *Gag.* He broke the budget down as follows:

Hire of Land Rover	$2000
Wages for guides	400
Hire of carriers and/or donkeys	600
Living expenses in Addis Ababa	2400
Sustenance for field party	200
Cartographic supplies	100

"Sustenance for field party" for six months—$200! That came to about a dollar a day for what—three, four people? What did those people *eat* over there in Ethiopia? Dow also said I would have to provide my own transportation to Addis Ababa and show proof of return travel for myself and my wife and daughter (I had told him that my family would be joining me).

To raise money, I worked in Corpus Christi, Texas, on crews repairing hurricane-damaged homes and then later, at night, on a tugboat servicing offshore oil wells. By February I had $4500. Then I learned about a geologist in New York named Robert King, who had an interest in the Afar. King, recently retired as chief geologist with a major oil company, had just spent a year at the Woods Hole Oceanographic Institution, shortly after its research ships had discovered the now famous metal-rich hot muds in deep depressions in the Red Sea.[13] High bottom temperatures in the Red Sea had been known since the 1880s, but it was not until the

1960s that they were associated with a spreading center. Woods Hole scientists were amazed to discover ultrasaline brine pools overlying metal-rich oozes that contained high concentrations of manganese, zinc, copper, silver, and gold. The deposits were believed to be formed by hot solutions, percolating through fracture zones, that dissolved base metals out of basalts and overlying sediments. It was estimated that the metals were worth 2 to 3 billion dollars! Unfortunately, the Red Sea ores lay below 3000 meters of water. By 1970 various schemes had been proposed to pump the rich muds from the sea floor, but there were enormous technological and environmental problems to overcome, as well as major political problems, given that the Red Sea is bounded by seven countries that often seem to hate each other.

King reasoned that because the northern Afar is geologically the southern extension of the Red Sea basin, deposits with origins similar to those of the metalliferous Red Sea muds might be found in the Depression—on dry land. Support for the idea came from Woods Hole data suggesting that the Red Sea metals originated somewhere near the Eritrean coast or, as King thought, perhaps in the Afar itself. King offered to pay me a $50-per-day consulting fee once I was in Ethiopia to find out what I could about the prospects of finding precious metals in the northern Afar.

Done.

Thus gold was added to the list of reasons to work in the Afar. I would spend my own time on pursuits in the Awash Valley, a dual arrangement to which the Geological Survey agreed. If my financial situation took a turn for the worse while I was in the country, I could rely on backup support from family holdings in Texas that gave me a modest income.

In late February 1971, I kissed my wife and cherubic daughter good-bye with plans to send for them if all looked well in Ethiopia. En route to Addis Ababa, I met with King in New York, flew to London to meet more geologists and pick up a volume on Red Sea geology just published by the Royal Society, and then flew to Paris. There I met with the director of the French contingent of the Omo Research Expedition, Yves Coppens of the Musée de l'Homme, and with geologist Maurice Taieb. Both were friendly and agreeable to some sort of collaboration in Ethiopia. Coppens was courtly and mild-mannered and struck me as having the demeanor more of a poet than of a leader of an African expedition. Taieb was Mediterranean in appearance and ebullient about his work, which he discussed excitedly in a heavy French accent while waving his arms about for emphasis. I showed him the proposal that Duncan Dow had sent me for work in the middle Awash Valley, and after talking at length, we agreed to conduct joint field work during his planned trip to Ethiopia that November and December. He would provide a Land Rover, and we would share field costs.

After agreeing to help Maurice round up equipment and supplies for our pending field work, I jumped on an Ethiopian Airlines jet at the packed Orly Airport. Once we were off the ground en route to Addis Ababa, I began poring over the

Royal Society volume about the Red Sea.[14] I recalled a few days earlier charging up the steps of the Royal Society building at 6 Carlton House Terrace. In the entry hall, I had been stopped dead in my tracks by the mournful gaze of Charles Darwin. It was the famous portrait of him painted the year before he died, with a long, white beard and a black cape draped over his shoulders. Oh my gosh, I thought, as I caught my breath, what would he think of all this?

CHAPTER 3
The Awash Valley

Maurice was driving the Land Rover pickup. We had just left the village of Millé, where we were told that the Afar guide we were looking for, Ali Axinum, was out in the bush somewhere southeast of the village. We were zooming along on a flat, open plain, leaving a billowing cloud of dust behind us, when we spotted a dozen or so ostriches. Suddenly Maurice swerved the vehicle and began chasing one. Amusing. I guess this is what people do in Africa, I thought. Ostriches are fast, with a top speed of about 50 kilometers per hour. But Maurice showed no signs of letting up on the gas pedal. My amusement quickly faded.

"Maurice, what the hell are you doing?" I yelled. "Leave the goddamned ostrich alone."

Maurice laughed and gunned the Land Rover even faster. The ostrich was tiring and we got closer and closer. All of the sudden, *wham!* With great thuds and thump-thumps we ran over the ostrich, as I jerked my head around to see a sprawling mass of feathers behind us.

"Jesus, Maurice, you ran over the fucking bird!"

"I know, I know!" he shouted, wheeling the vehicle around to look at his handiwork. "We tell Ali's people and they will like us. They will have lots of meat. One ostrich weighs 150 kilos."

The ostrich was twitching around in its final throes. "Damn, Maurice, couldn't we just give them a sandwich?"

Maurice had been in Ethiopia for a week and I had come to expect such things from him. He was impetuous, excitable, often impatient, chain-smoking, and just slightly crazy. Yet Judy later found him charming, as well as handsome and studious-looking with his heavy-framed glasses. His olive complexion, brown eyes, and longish dark brown hair reflected what I guessed to be his mixed Arab and French background, Taieb being an Arab name. Whatever his origins, he also had a tendency to be quick to anger, often yelling at Ethiopians for not reading his mind when he wanted something, or arguing with me about how I drove the pickup. "Do not run over the Ethiopian!" he would scream in his heavy French accent. "You do that and we must pay the family one thousand *birr*." *Birr* is the Ethiopian equivalent of two U.S. dollars. Just as quickly, however, Maurice's anger would subside, and in the next sentence he would be asking me about some American movie. With time I came to ignore the angry outbursts altogether.

Satisfied that the ostrich was not going anywhere, Maurice jammed his foot onto the gas pedal and we continued on. A few minutes later we saw three or four low, oval huts made of woven grass mats and used by the Afar nomads. They were huddled together in the midst of some acacia trees. As we pulled up and jumped out, about a dozen Afar wandered over, mostly old men and women and some kids. The younger men and women were off tending their goats and camels.

Maurice said, "Ali Axinum *bet?*" *Bet* means "house" in Amharic, the dominant language of Ethiopia. No one answered, apparently not understanding, so Maurice repeated, louder, "ALI AXINUM bet?"

"*Ow*," yes, one replied, apparently catching Ali's name. "Ali *metfo!*" he said, Ali was sick. "*Doktor? Quinine?*" *Quinine*, which the Afar pronounce "keyneen," the traditional treatment for malaria, is their word for "medicine."

We squeezed into one of the huts, which was covered with grass mats and supported by bent sticks lashed together with leather strips. Ali was lying with a cloth wrapped around his middle. With the few Amharic words he knew, and his own brand of guttural sounds that somehow filled in the gaps, Maurice explained that we wanted Ali to guide us to Lake Abhé. The lake lay to the east, on the other side of a high volcanic plateau. Maurice wanted to study the geology of the lake basin— the final drainage point of the Awash River. This was to be the last part of his doctoral dissertation. He knew that the previous year Ali had led a UN team to the lake area as part of a survey of the hot springs in the Afar, which were being studied for possible geothermal uses.[1] The springs are abundant throughout the Depression and are created when ground waters are heated by underlying magma sources.

Ali indicated that he was willing to help us but was too ill. He pointed to his groin area, wrapped in a grimy white cloth. With his permission, I gingerly unwrapped the crude dressing, clearly a painful exercise for Ali. His genital area was covered with dried and wet blood and with open, pus-filled lesions.

"Jesus, Maurice, this guy has venereal disease! He's not going to work for anyone."

In fact, I thought Ali looked terminal. He was skin and bones. But to my surprise Maurice said, "Okay, we take him to clinic, get some medicine, then go to Lake Abhé."

That we did. We half carried Ali to the Land Rover, propped him up between us, and took off. *God, is this man contagious or what?* I thought. As we left, Maurice told one of the elders that we had "much food" for them nearby. We loaded a couple of men into the back of the pickup and bounced along to the dead ostrich. It was already covered with vultures, which flapped off as we approached. The Afar got out of the truck and looked at the tangled heap. After a few seconds, one of them said, "It's Ramadan." That is the Muslim holy time, and the Afar were fasting—not that they would eat the mess now anyway.

So much for French "AID."

The path Maurice and I took to that ostrich had begun at our meeting in Paris, nine months earlier. I had been eager, to say the last, to get to the Afar. After arriving in Ethiopia in early March of 1971, I had established myself at the Geological Survey, where I was given a small office. I had begun studying all the maps and reports I could find that might bear on base metal deposits in the northern Afar. Meanwhile, I had sent for Judy and Justine in late March, and we settled into a small apartment across the street from the soccer stadium. For 40 birr per month we hired a housekeeper, who helped Judy through the intricacies of running a low-budget household.

As opportunities arose, I traveled around Ethiopia on field trips with geologists from the Survey: to the Adola goldfields in Sidamo Province in the south, believed by some to be the fabled King Solomon's mines; to the Tigray highlands in the north, where in the town of Mekele I had watched blocks of salt being unloaded from camel caravans brought up from the northern Afar Depression; to the rugged bottomless Blue Nile basin north of Addis Ababa. With family and newly acquired friends, I had also visited the scenic Main Ethiopian Rift south of Addis Ababa.* Spotted with lakes and filled with birdlife, it would become a favorite vacation spot.

The field trip with Maurice would take us the length of the Awash Valley to the central Afar. En route we would pass through the middle Awash, the area of my proposed mapping project, giving me a chance to learn something of the geology and logistics of the region. Maurice had said we would briefly join Haroun Tazieff's team, which was then mapping the volcanic terrain in the area of the triple junction west of Lake Abhé, where the East African, Red Sea, and Gulf of Aden rifts come together. That had greatly excited me, because this was to be the team's last field season in Ethiopia before moving on to the nearby French Territory. Maurice had specifically wanted to collect some Holocene (the last 10,000 years) lake sediments in the Abhé area.[2] As he put it, he needed to tie up some loose ends for his dissertation. As I was to learn over the coming weeks and months, however, there were a great many such loose ends with his research. And small wonder: The region that was the subject of his dissertation project, the Awash Valley, comprised

---

*The northern segment of the East African Rift that passes through central Ethiopia is locally called the Main Ethiopian Rift.

some 50,000 square kilometers, equal in area to the volcanic terrain that Tazieff's entire team had mapped to date using helicopters.

Our excursion began in the uppermost drainages of the Awash River surrounding Addis Ababa. The basin is ringed by volcanoes rising more than 3000 meters and is breached by the river in the southeast, near a French archeological site called Melka Kontouré. Maurice had begun his work in Ethiopia in 1966 by studying the geological context of the site.[2] On this occasion he simply visited the director of the excavation, French archeologist Jean Chavaillon, one of his doctoral supervisors, and I was given a tour.

Although Melka Kontouré was a noteworthy Early Stone Age site where multiple artifact levels had been excavated,[3] only modest numbers of fossils had been recovered and, thus far, no human remains.* According to Chavaillon, the earliest and lowest levels of the site contained crudely made stone choppers and scrapers recovered from "australopithecine encampments."[5] He described the artifacts as similar to the nearly 2-million-year-old "pebble tools" recovered from the lowest levels of Olduvai Gorge, known as the Oldowan tradition. These tools, commonly made from stream cobbles, were believed by the Leakeys to have been manufactured by *Homo habilis,* the earliest known species of *Homo,* not by *Australopithecus,* its more ape-like relative.

Higher levels at Melka Kontouré contained abundant handaxes and cleavers in the Acheulean tradition (which takes its name from the archeological site of St. Acheul, in France). Both types of tools are "bifacial," or flaked on two sides. Handaxes are typically teardrop-shaped, whereas cleavers have a transverse cutting edge. On Sunday afternoons, tourists could make the short drive to Melka Kontouré from Addis Ababa and buy a sack of these stone tools from farmers, who hawked them on the roadside, much to the dismay of the French archeologists. Current thinking gave *Homo erectus,* the successor of *H. habilis* and the precursor of *H. sapiens,* credit for the Acheulean tradition.

From Melka Kontouré, we traveled east by road following the Awash River, whose total drainage basin covers 125,000 square kilometers. Fed by 12 major tributaries from the adjacent highlands, the river wends its way for 700 kilometers to Lake Abhé. I have friends who have wanted to raft the entire Awash, and with them Judy and I shot some of its rapids, but I know of no one who has navigated the entire river. Whoever does so should have ample time to spend, given the river's many meanders in the lowlands, and should be prepared to socialize with hippos and crocodiles.

The drive through the Main Ethiopian Rift to the train depot at Awash Station on the southwestern margin of the Depression is a geological marvel of volcanoes, calderas, open fissures, jet black lava flows, and fault escarpments that direct the path of the river. In places, broad strips of deep, fertile soils lie nearest the Awash, where

---

*In 1974, a single skull fragment believed to be *Homo erectus* was recovered from a stone tool level.[4]

several large Dutch- and Ethiopian-operated plantations were growing sugar cane, fruits, and vegetables. Near the point where the river turns north into the Afar Depression, about 150 kilometers east of Addis Ababa, the Awash cascades over 100 meters of basalt flows that create a spectacular waterfall in Awash National Park. From there we crossed a wide plain marked with 2-meter-high termite mounds and herds of oryx, kudu, gazelle, and bush buck. Geologists call this area the Afar "funnel," because here the rift opens dramatically into the Depression at the point where the Ethiopian escarpment turns sharply north toward the Red Sea and the southern Hararghe escarpment turns east toward the Gulf of Aden. Geologically, the escarpments represent the raised margins of the African and East African plates. The word *funnel* is misleading with regard to the Awash, however, because the river pours out of the top of the funnel, not its bottom (see Maps III and IV).

After traveling north a short distance from Awash Station, we spent our first night in the Afar in a German road construction camp. The Trapp Company was building a 300-kilometer asphalt road to the north through the western Afar. When finished, it would intersect an east–west road completed in the late 1930s during the Italian Occupation. This road connects the port of Assab on the Red Sea coast with the market towns of Bati and Dessie on the edge of the Ethiopian plateau. The Trapp camp, called Camp Arba for the nearby Arba River, housed a hundred or more people and served as the base of operations for the road project, which included a series of smaller line camps along the construction route. Until the road was completed a few years later, the camps would be a boon to anyone working in the region, because they were equipped with medical supplies, mechanics, Land Rover spare parts, water, and gasoline. Also, the German engineers and camp managers were always hospitable with a ready supply of cold beer.

Traveling north from Camp Arba on the freshly bulldozed dirt track, I could see towering volcanoes to the west on the Ethiopian escarpment. These peaks rise to 3700 meters and are separated from one another by chasms formed by rivers tumbling off the highlands. Similar peaks loom to the southeast on the Hararghe escarpment. Certainly, during my first venture into the Afar that November, I sensed nothing of the "hell-hole of creation" that I had read about. The Afar nomads that we encountered were friendly, and the daytime "winter" temperatures reached only about 35°C (95°F), while the evenings were cool and breezy. The landscape was a subdesert steppe of acacia scrub brush and rolling grassy savanna with intervening broken terrain of gravel-capped plateaus, dry streambeds, and craggy basalt flows. Volcanoes of every size lay to the north. Animal life was spotty: an oryx here, a gazelle or gerenuk there, no doubt more of everything deeper in the bush. The Germans spoke of herds of zebras, prides of lions, the occasional leopard, and a long list of animals they had hunted.

Along the Awash, the acacia brush gave way to grassy strips, riverside thickets, and a dense gallery forest filled with colobus and vervet monkeys, visiting troops of baboons, warthogs, small species of antelopes, and a thriving birdlife. Hot springs lay along the axis of the rift at places called Bilen, Hertale, and Mataka, all of which were being carefully mapped by geologists with the UN Geothermal Project.

At Mataka, 110 kilometers north of Camp Arba, Afar men, women, and children were bathing in the hot springs, their washed linen spread around them on rocks or draped on willow reeds growing along the margins of the mineralized waters. In this same area, the road rapidly dropped off a low basalt plateau through a gap in the rocks onto a broad, periodically flooded marshy plain that extends for 60 kilometers to the north. In 1934 the English explorer Wilfred Thesiger traveled this route and referred to the "ill-famed pass of Mataka," where "the tracks skirt an impossible swamp at the foot of a rugged chain of hills."[6] He said he felt "relieved" after his camel caravan had moved safely through the pass onto the plain beyond. The area just to the north was well inhabited by Afar then, as it is today, and evidently Mataka—which means "warning" in Afar—was a potentially hazardous checkpoint for out-of-town visitors. A remarkably clear photograph in Thesiger's report to the Royal Geographical Society shows his caravan stretching into the distance with the marshlands filling the background.

Maurice had found evidence that in the late Holocene (8000 to 10,000 years ago), during a wetter period, much of this lowland was filled with a shallow lake, indicated by thick deposits of stromatolites and diatomite.[7] Both are made of fossil algae; stromatolites typically form calcareous mats or crusts on rock surfaces around lake margins, and diatomite consists of siliceous skeletons of planktonic organisms that accumulate on lake bottoms, sometimes to great thicknesses. Both deposits are white and highly distinctive for purposes of geological mapping. I later saw chunks of the light, chalky diatomite in abandoned Afar encampments. I never learned what it was used for.

Gewani village lies near the base of the 2000-meter Ayelu volcano, which according to legend has spiritual powers.[8] One report claims that Christian hermits once lived on the mountain.[9] The village, with perhaps 700 to 800 people, consisted of a small military garrison, some mud-walled government buildings plastered with lime, a Red Cross clinic, a small school, a number of thatch-roofed dwellings, and a newly installed AGIP station* and rustic motel boasting a dozen or so spartan rooms and a small cafe. Also present were a handful of *souks* (small shops) where one could buy cooking oil, matches, soap, sugar, and the like. Gewani was a trading center for a thriving Afar population that lived in encampments at Beadu, to the west, between the village and the Awash, and seasonally along some sandstone hills across the river on the margin of a small lake called Yardi. The lake is also called Caddabasha by some and Hafya by others. Geographical features in the Afar frequently are known by several names given to them by different tribes at different times of occupation, and they often vary in name from map to map, depending on who first mapped the area and on when and how accurately the name was recorded. This is surely a common occurrence all over Africa.

---

*The state-owned petroleum company of Italy, Azienda Generale Italiana Petroli (AGIP).

Maurice and I were about to walk into one of the thatched huts selling soft drinks when a guy came blasting up on a motorcycle. He wheeled to a halt in a cloud of dust and introduced himself as Jim Ryan with the Peace Corps. Sandy-haired and friendly, he wore shorts and the cheap rubber sandals that the local merchants sold to the Afar. Inside the hut we sat down on wooden stools and ordered our drinks, which were kept cool in a breezeway under burlap sacks dampened with water. As our eyes adjusted to the dim light after the glare of the sun, Jim told us that he was an architect by training and that he and several others were building an agricultural school that would be used to teach the local Afar how to farm. He explained that the project was coordinated by the Awash Valley Authority, which was in charge of managing development of the region. The AVA had apparently decided that the future of the Afar in the Awash, historically a pastoral people, was to grow cotton and other cash crops along the fertile areas of the river, where the nomads traditionally grazed and watered their livestock. On another level, Jim explained that wealthy members of Emperor Haile Sellassie's extended family had taken over much of the Awash Valley and were courting foreign capital for venture-sharing development. Although the Afar have a reputation for independence, the highlanders were increasingly exercising their power in the area. However this was to play out, Jim recognized that the survival of the local Afar depended on their ability to adapt to the agricultural development of the region. In its way, the modest school would help. Jim also told us he was planning to volunteer for another year in the Peace Corps to see the construction of the school completed.

He invited us to his house to have some barbecued warthog. After buying a few supplies, we showed up at Jim's place on the edge of the village, a small, two-room house with a tin roof that he shared with another Peace Corps member. We sat down on a battered couch, and shortly Jim brought us a half-empty gallon jug of red wine, which we passed around. In front of us on a low table was a platter with a towel on it, which Jim flipped off, revealing the fire-blackened corpse of a roast pig. Jim noticed our skeptical looks and then said something like "Well, would you like a side dish to go with this?" With that, he hauled out a large plastic bag of cannabis, and after a smoke had been passed around, the roast pig looked better and better.

After a while I said, "Okay, Jim, why would anyone *really* want to re-up with the Peace Corps for another year in this place?" Before he replied, a tall, beautiful young Afar woman appeared at the open front door and leaned against the door frame, partly silhouetted in the streaming sunlight. She was bare from the waist up, as are all Afar women traditionally, with skin the color of mahogany and long braided hair that glistened like wet coal. Red and silver beads fell from her neck across her full breasts. *Stunning.* After a brief exchange with Jim in Afar, she remained standing there looking at us. Jim gazed at her for a long moment and then said, "Now, Jon, what was your question?"

I later learned that the woman was the wife of the local Afar chief.

In due course, Maurice and I thanked Jim for his hospitality, stumbled out into the afternoon sun, and climbed into the Land Rover. I got behind the wheel, and Maurice immediately nodded off. My mind was reeling as we rumbled north on the track. I would see Jim several more times over the coming months. On one occasion he was skinning a python that he had shot. The last time was in Addis Ababa when he went to the U.S. embassy to re-enlist for another year with the Peace Corps. But that never happened: He was bitten by a dog in the capital, and before he was properly diagnosed and treated, he fell deathly ill—it was rabies. The U.S. embassy flew him to an American military hospital in Germany for emergency treatment, but it was too late.

Fifteen kilometers north of Gewani, Maurice and I stopped at a roadcut where Trapp bulldozers were carving a path across a dry streambed. Still light-headed from "lunch," we went to see some vertebrate fossils scattered nearby that Maurice knew about. There were elephant teeth, a hippopotamus jaw, and cone-shaped crocodile teeth. It was the first time I had seen such fossils in the field.

"Christ almighty, how *old* are these things?"

Maurice thought they were about 3 million years old.

"Three *million* years old." I was staggered. "That's *wonderful!*"

Back in the Land Rover, we continued north until we came upon the strange sight of two grown men next to the road banging on rocks with little pointed hammers. But of course, they were *geologists*. They introduced themselves as Markward Schönfeld and Hans Schaefer, members of the German contingent of the International Upper Mantle Project. They were part of a larger team studying various aspects of the geology of the southern Afar. This struck me as odd, because their work surely overlapped that of Tazieff's group, and even Taieb's. I would learn that there was much competition among the French, Italian, and German teams working in the Afar. For the most part, however, it all seemed to sort itself out, because each team had different research priorities and goals.

Schönfeld told us he had just finished mapping paleolake sediments along the foothills of the Hararghe escarpment, where he had found abundant diatomites and some vertebrate fossils. He had determined that the lake was nearly the size of Lake Turkana and, from the age of the fossils, that it existed about 9 million years ago.[10]* The fossils came from near the Chorora area, southeast of Awash Station (see Map III). Schönfeld believed another large lake, far older than the Holocene lake described by Maurice, had once existed in the Gewani area. He estimated this lake to be at least 4 million years old on the basis of a radiometrically dated basalt.[12]† If

---

*Schönfeld's colleagues later dated the fossils radiometrically between 9 and 10.5 million years;[10] still later, French geologists redated them between 10.5 and 10.7 million years old.[11]

†A dating method based on measuring the decay of radioactive isotopes, such as potassium–argon and carbon-14.

that was true, then the fossils Maurice and I saw could be of a similar age, not 3 million years old as he believed.

The more Schönfeld talked about the geology of the Gewani area, the more agitated Maurice became, although I doubted our German friends realized this. As a heavy smoker, Maurice had a habit of fidgeting with his cigarette with his fingertips when something bothered him. In this case, it was obvious that he felt Schönfeld's work was infringing on his own.

We spent that night at another Trapp camp, named Camp 270 for the distance in kilometers from the start of the new road near Camp Arba (see Map IV). Here the Germans were building a large bridge across the Awash at the point where the river completes a major swing to the east. The next morning we drove north to the Bati-Assab road, then east to the village of Millé. From there we drove into the bush to pick up the guide, Ali Axinum. From the spot where Maurice had flattened the ostrich, we trundled north until we bounced back onto the road and came to a narrow gap in the mountains called Tendaho Pass.

———

At Tendaho the Awash River squeezes through a break in a basalt plateau before it spills out onto an expansive desert plain, the area where we intended to do field work over the next week (see Map V on page 30). Here the Awash takes a sharp turn to the southeast, following the trend of the Red Sea rift down a long graben before it turns east and drops into Lake Abhé, which partially fills another graben lying within the Gulf of Aden rift. The landscape beyond the pass changes strikingly from rolling scrub brush west of the gap to a nearly barren flat plain to the east with sparse grasses, occasional acacias, and isolated splinters of rock. The explorer L. M. Nesbitt described crossing the plain with camels in early 1928: "We were soon in motion on the burning desert, which was here intersected by walls of basalt. These stood out in sharp lines of metallic blackness against the white dazzling plain. Not a sign of life was visible; there was no animate thing besides ourselves in the whole glistening wilderness."[13]

Soon Maurice and I came to the village of Dubte, where we took Ali to a clinic. It was operated by an adjacent plantation, carved out of the desert in the early 1960s by the enterprising Mitchel Cotts Ltd. of London.[14] Some 4000 hectares of cotton were under cultivation, watered by a sophisticated irrigation system drawing on the Awash, which created a sprawling, spectacular oasis in the desert. Most of the labor force were highlanders, because the Afar were apparently scornful of picking cotton. Instead, the Afar men preferred to work as guards at checkpoints around the perimeter of the plantation, often stationed beneath tall tamarisk trees, carving on sticks or picking their teeth with their long, scimitar-shaped knives. They appeared to be looking longingly at the plantation managers walking around in white shorts, perhaps thinking what novel trophies the Englishmen would make for their womenfolk. Incongruously, the managers drove up and down the narrow

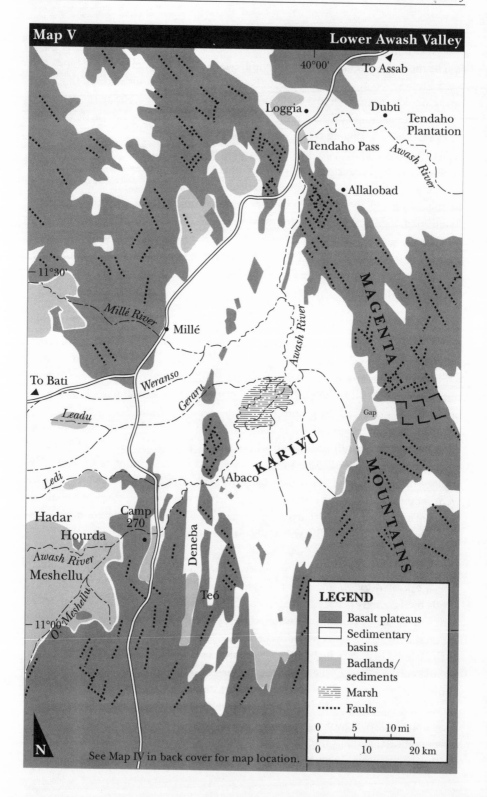

Map V   **Lower Awash Valley**

40°00'
To Assab

Loggia   Dubti
                        Tendaho
Tendaho Pass   Plantation
                   Awash River
• Allalobad

11°30'

Millé River

Millé

MAGENTA

To Bati
         Weranso
                   Gerara   Awash River
Leadu

Ledi   Gap
                        KARIYU
                  Abaco

Hadar   Camp
         270   MOUNTAINS
Hourda
Awash River   Deneba
Meshellu

11°00'   O. Meshellu
                        Teo

LEGEND

- Basalt plateaus
- Sedimentary basins
- Badlands/ sediments
- Marsh
- Faults

0   5   10 mi
0   10   20 km

N

See Map IV in back cover for map location.

levees between cotton fields in golf carts imported for that purpose. This and their dress gave the place the air of a country club in the wilderness, or in colonial Africa. However, Maurice and I would come to know the managers well and to appreciate their hospitality, as well as their fortitude for sticking it out year after year in the desert.

The clinic served both the plantation employees and the local Afar. Its one doctor told us that Ali did not have venereal disease—that was rare among the Afar. Rather, he had glandular tuberculosis, which had particularly attacked the lymph glands around his genital area. He explained that tuberculosis was common among the Afar, fostered by malnourishment and the close living quarters of the nomad dwellings. However, because the Afar were never, or rarely, exposed to antibiotics, they responded amazingly well to treatment. The doctor cleaned up Ali's bloodied midsection, gave him a shot of penicillin and some tetracycline for secondary infection, and told us that after some rest and a month on medication, he should fully recover.

We spent that night in the guest quarters of the plantation, sitting around a bona fide swimming pool sipping drinks while exchanging pleasantries with English agricultural experts, entomologists, and maintenance staff.

The next day, we left Ali in the hands of the clinic doctor and told the two of them that we would return in three days. We then drove southeast along a sandy track in the direction of Tazieff's camp and the ancient desert town of Aysaita.

With a population of several thousand people or more, Aysaita was the seat of the Afar sultanate of Awsa, the heart of which covered an area of some 7500 square kilometers. The town itself sits near the point where the Awash splits into two major branches (see Map IV). One takes a southern course through the graben before it turns east around an immense shield volcano and flows into Lake Abhé. The other flows due east from Aysaita and immediately splits into a half-dozen smaller branches, several flowing into nearby Lake Gemeri. These streams run through the middle of Awsa and collectively form a broad, deltaic bottomland made up of riverine forests with intervening marshes, swamps, and seasonally dry floodplains containing date groves. Much of this area was cultivated by the Afar, using ancient irrigation techniques.[15]

Along the eastern margin of Awsa are eight lakes of varying sizes. Two of the largest, Gemeri and Adobed, are interconnected with the eastern branch of the Awash before the two branches of the river rejoin and flow into Lake Abhé. The drainage and sedimentary history of Awsa and of the entire Tendaho graben can be envisioned as an enormous bathtub partially filled with sediment that has been lowered at its southern end and tilted to the east. Hence the dominant direction of water flow in the graben is to the southeast, where the buildup of sediment carried by the Awash is almost certainly the thickest. Awsa is very fertile and historically has been fiercely coveted by the reigning Afar sultans for pasturage, teeming wildlife, and agricultural use. The area has been protected from outsiders by its isolation at the center of the Afar desert and is made even more remote by the enormous piles of volcanic rock that almost surround it. The largest of these forms a nearly vertical wall of stratoid basalt that rises a kilometer above the floor of the Tendaho graben and constitutes the eastern boundary of the sultanate. At the southern end of Awsa is a shield volcano that rises 700 meters above the graben floor and is 20 kilometers across, almost the same width as Lake Abhé lying at its eastern flank.

Despite its isolation, over the last century Awsa has had its share of adventuresome visitors. High on the list was Werner Munzinger, a Swiss who had lived for some years in northern Eritrea and in the 1870s was an architect of an Egyptian scheme to conquer Ethiopia, then called Abyssinia.[16] As part of a multi-pronged invasion force, in late 1875 Munzinger led some 400 armed men, mostly Egyptians, from Tadjoura at the western end of the Gulf of Aden to Lake Abhé. After crossing through Awsa, he planned to proceed through the Awash Valley to the Ethiopian highlands, securing caravan routes along the way. Munzinger was up to his eyeballs in plots and hoped to obtain the support of different chieftains, as well as that of Menelik, then king of the vassal state of Shoa in the central highlands of Ethiopia, surrounding present-day Addis Ababa.[17] However, from several accounts it sounds as though Munzinger walked into a well-planned ambush.[18] He and most of his men were slaughtered on the outskirts of Awsa in the early hours of November 14, 1875, reportedly by a combined force of Afar and Issa. For unknown reasons, Munzinger was accompanied by his Eritrean wife, said to be a princess, and their child, both of whom were also killed. Overall, his failed mission was part of a fantasy of the Khedive of Egypt to extend the territories of the crumbling Ottoman Empire to the shores of the Indian Ocean. Apparently, King Menelik was in collusion with the Khedive and had ambitions of his own to use the Egyptian advance to help him win territorial advantages over his archival, Yohannes, the Emperor of Ethiopia.[17]

Munzinger had convinced the Egyptians to attempt not only to conquer Awsa but also to secure the valuable salt deposits of the northern Depression, which he knew were indispensable to the Ethiopians. For centuries salt bricks quarried from the salt plain were used as a trading currency. It was said that "Abyssinia could belong to whomever the salt belonged to."[19] Munzinger knew all about this because he was not only a political opportunist but a first-rate scholar and linguist as well. In 1867 he explored parts of the northern "triangle of the Afar" and made important ethnographical and geological observations.[20] He astutely recognized that the northern Afar was once part of the Red Sea, believing that the salt plain had been connected to the sea via channels that ran north on either side of Buri Peninsula. During his explorations he took lengthy notes on the "salt commerce" of the northern Afar, and he presented his findings to the Royal Geographical Society of London in 1869, six years before his abortive invasion of the Afar. He mistakenly informed the Society that the Awash River drained into the Awsa desert, forming a salt pan similar to the valuable deposits to the north.

Wilfred Thesiger, born in Addis Ababa and the son of a British diplomat, is widely credited with solving the puzzle of the Awash's final drainage into Lake Abhé with his travels through Awsa in 1934. In fact, however, this information is depicted on a map published nearly a century before. Charles Johnson, an English explorer who traversed the southern Afar in 1842, camped near the eastern margin of the lake and drew a reasonably accurate map of its position and of two of its smaller sister lakes to the north.[21] He called the terminal lake "Abhibhad," essentially identical to Abhé Bad (*bad* means "lake" in Afar). It would be interesting to know how well Johnson's map making and subsequent memoirs were received by Victorian England: For much of his journey he was accompanied by the "richest slave merchant of the southern Dankalli [Afar] tribes," a man he openly admired.[22] This slave merchant was widely traveled and helped Johnson draw a remarkably accurate sketch map of the entire Awash drainage basin.

The basic geography of the Awash was probably common knowledge to many of those who routinely traveled the trade routes, and it would not surprise me if similar maps were found on centuries-old Ethiopian and Arab parchments. Certainly we can assume that the ubiquitous Arab traders and slavers did business with Awsa for centuries. Both Nesbitt and Thesiger report that Arab-built bridges and a primitive irrigation system in Awsa date from "ancient times" or "some generations ago."[23] Interestingly, both Johnson and Henry Salt, another English explorer who traveled throughout northern Ethiopia from 1809 to 1810,[24] refer to man-made canals in Awsa. Salt's map of the "Hawush River" (based on secondhand information) shows the Awash draining into a single lake shaped like a hand with outstretched fingers, with the annotation "Here the River is lost in the sands, or led off in Canals by Natives."[25] Johnson's map shows a large canal diverting the Awash to the south of Awsa in a fashion like the branch of the river that flows west of Aysaita. Although some of the waterworks and geography depicted on these early maps are certainly fanciful, the various descriptions nevertheless suggest a knowledge of, and efforts to control the flow of, the Awash for at least 150 years before Europeans arrived on the scene in their golf carts. An examination of aerial photographs of the Awash delta by geographer Karl Butzer of the University of Texas suggests that much of the area east of Aysaita has been cultivated by the Afar using techniques dating back at least 500 years.[15] Butzer even suggests that agricultural origins in Awsa and trade with the coast may have a distant connection with the fabled Land of Punt, known in Pharaonic times.[15]

When Nesbitt, an Italian-born mining engineer of British ancestry, visited Awsa in 1928, he appears to have lacked any real interest in following the Awash to its end to answer the still-debated question of whether the river evaporated into the desert. Nesbitt did reach Aysaita and was cordially received, but from there he journeyed north, despite the warnings of the sultan who cited danger from another Afar sultanate in that area. Having already crossed the southern Afar, Nesbitt and his two Italian companions were intent on traversing the entire Depression from south to north, a task they successfully completed during a gutsy three-month trek. In his memoir, *Hell-Hole of Creation,* Nesbitt correctly described the eastward flow of the Awash from Aysaita, but from there he said the river "begins to lose itself in the sands" and that "its waters . . . eventually are evaporated, or absorbed by the earth, and so entirely disappear."[26]

Reading successive accounts of geographical exploration is like watching a fog lift off a mountain. The first person who comes along sees only the bottom third of the mountain from far away and describes it in his journal as a "long, low ridge." The next person moves a little closer when the fog is breaking up and describes it as a "series of moderately high hills." The next describes a "mountain with a flat-topped summit." Finally, someone actually climbs the mountain on a clear day and sees that in fact it is only the first of a chain of mountains. Until this is done and an accurate map is made of the mountain, any observations will remain suspect, and until the map is widely known or published, the explorer's efforts will remain largely unrecognized.

It was not until 1934 that Thesiger became the first European to confirm first-hand the fate of the Awash. Like Nesbitt, he was well received by the sultan of Awsa. From Aysaita, however, Thesiger was determined to follow the main course of the Awash to its end. He first reached Lake Gemeri (or Adobada as he called it), followed its narrow eastern shore, and then trekked south until he reached Lake Abhé.[27] In doing so, he circled the immense shield volcano, Dama Ali, that towers over the western shore of the lake and confirmed the drainage of the Awash into the lake just east of a basalt ridge called Arissa. Thesiger carefully mapped the flow of the Awash, the position of the mountain, and the positions of the major lakes. In his published report to the Royal Geographical Society, he said that his Afar guides showed him the place on the shore of Lake Adobada (Gemeri) where a "Turkish force" had been annihilated years before; this Thesiger believed to have been Munzinger's expedition.[28] In his later memoirs, however, he identified the lake as Abhé and quoted the Afar as saying that cannons belonging to the "Turks" had been thrown into the lake. Beside whichever lake it was that Munzinger perished, we can assume that during the final hours of his life, he learned that the Awash does not evaporate into the desert.*

Maurice and I pulled into Tazieff's camp on the outskirts of Aysaita in the late afternoon of November 18, 1971. The dozen or so French and Italian geologists there

---

*During his travels through Awsa in 1934, Thesiger reported one source as saying that in recent times the Awash had completely dried up before reaching Gallefage (west of Aysaita).[29] This may have occurred during the terrible drought and famine of the early 1890s in Ethiopia.[30] Such severe dry periods could explain some of the accounts of the Awash evaporating in the desert before reaching Lake Abhé.

were not in the most cordial mood, because Tazieff and several others were stuck on the top of a volcano somewhere to the east. The helicopter that had dropped them off in the morning had had a mechanical problem, and it looked like Tazieff and crew were going to spend the night with the elements. The timing of our visit could not have been worse, since Maurice was apprehensive anyway about being around the other French geologists. From a few remarks, I guessed it had something to do with his background.

Maurice was born in Tunisia in 1935 to Sephardic Jewish parents in the port of Bizerta on the Mediterranean coast.[31] His interest in antiquity is not surprising; Tunisia's own rich heritage at the crossroads of the Mediterranean and the Sahara dates to its settlement by the Phoenicians in the twelfth century B.C. Among the settlements was Carthage and nearby Bizerta, one of the first Punic ports on the North African coast. When Tunisia became a French protectorate in 1881, Bizerta became an important naval base, and it was a center of fierce fighting during World War II, at the time when Maurice was in grade school. Following the war, Tunisia went through a violent period of nationalism until the country received its independence in 1956, the same year Maurice got his undergraduate degree from the University of Nice. Over the next few years he received additional degrees in science and geology from the University of Tunis, while also working for the Army Ministry, for the French consul, and for his uncles exploring for petroleum in the Sahara. In 1962, the year before the French navy evacuated its base in Bizerta, Maurice emigrated to France and served in the French army until 1964. That same year he married a German girl, and he soon became the father of two handsome sons, whom I later had occasion to meet. Then, in 1965, he entered a doctoral program at the University of Paris, which led him to Ethiopia a year later.

I can only guess at the social pressures Maurice felt when he emigrated to France. It was not something he discussed, and I only recently learned where he was born. Even in his autobiographical book, *Sur la Terre des Premiers Hommes,*[32] published in 1985, Maurice says nothing about his early life, except for a statement about how being a scientist makes one equal in society.

Whatever the reasons for his discomfort among his adopted countrymen, that night Maurice chose not to sleep in Tazieff's camp; instead we drove into the bush, threw down a couple of cots, and called it a day. Maurice's apparent concerns about his origins are somewhat ironic. As I would soon learn, he had already made one of the greatest single discoveries bearing on human origins on the African continent, or anywhere else.

The next morning we returned to the Tazieff camp, where Maurice was to rendezvous with a fellow Ph.D. student, Françoise Gasse. She was a comely geologist with long blond hair whose field outfit consisted of hiking boots and a well-crafted bikini that perfectly matched her sky-blue eyes. In fact, most of the French and Italian geologists wore bikinis, evidently to have a smashing tan when they returned to Europe. By chance, we arrived just as the helicopter returned from the volcano with Tazieff and crew. The mechanic had worked on the chopper through the night, and as it landed, a cameraman ran up to capture Tazieff's triumphant rescue for French TV. Fair enough. Had the chopper not been repaired, he might still be there, and people would still be arguing about which volcano he had been stranded on.

We sat around and had coffee while the 57-year-old Tazieff—world-famous volcanologist, mountain climber, and still a rugby player[33]—wolfed down a large breakfast and told his story of howling winds during the night followed by howling mosquitoes. As the session ended, the chopper returned from another mission, this time with the corpse of an enormous warthog swinging from its undercarriage by a chain wrapped around its hind feet. Tazieff had spotted the animal when returning to camp and had sent someone to shoot it for a feast that night, to which we were invited. With that, Maurice and I took off into the desert with the bikinied Françoise.

Gasse, a limnologist (a specialist in the study of lakes), uses diatoms to reconstruct paleolake environments. She identifies the abundant species of these single-celled algae and can relate their type and habitat to the rise and fall of lake levels over many thousands of years.[34] The Holocene lake sediments in the central Afar were the subject of her dissertation. Throughout the day we drove around looking at thick diatomite deposits and stromatolites that indicate high lake levels attained during wet periods, and impoverished diatom deposits, gravels, and erosional surfaces that indicate low lake levels and river deposition during dry periods. The previous year Maurice had obtained a carbon-14 date of nearly 9000 years for some mollusks from lake deposits in the northern Tendaho graben, a date close to the one he obtained from similar deposits near Gewani, which suggested an extensive, and probably an interconnected, lake system in the western and central Afar during the early Holocene.[2]

Over the next few years, Gasse carefully mapped the aerial extent of the Holocene lake sediments in the Tendaho graben and of those in the adjoining Lake Abhé basin. She determined that between 4000 and 10,000 years ago, a single paleolake nearly 200 kilometers long and 50 kilometers wide connected the two areas.[35] She also mapped deposits of similarly aged paleolakes in other large grabens farther to the east. During Holocene wet periods, the innumerable grabens of all sizes throughout the Afar must have filled up like a sponge, creating hundreds of lakes. If the abundant artifacts mixed with some of the lake sediments are any indication,[39] the triple junction could have been a vacation wonderland in times past. One can just imagine bands of proto-vacationers tripping from lake shore to lake shore, crossing over one basalt plateau after the next. But because the larger grabens contain numerous deep fissures caused by sea-floor spreading, lake levels could not have been very stable.[36] In addition, climatic changes across eastern Africa during the last 35,000 years have led to a regional reduction in lake levels.[40] Thus there would have been some risk in choosing the right spot for your prehistoric lakeside hut. The other problem, of course, was that bubbling beneath these grabens were emerging midoceanic ridges. While lake shores were being stretched apart by spreading, the lake basins also sat precariously atop the ascending upper mantle.

On November 19, we finally loaded up the Land Rover and drove south to Lake Abhé. We stopped by the plantation clinic and picked up Ali, who was looking reasonably revived. We promised the doctor that he would get plenty of rest. With that we took off on a bone-crushing roadtrip to the southern end of the Tendaho graben. By midafternoon the next day, the three of us were standing on a basalt ridge at Arissa overlooking the marshy western shore of the lake. The Awash entered the lake just to the north and east of us; the border of the French Territory passed north–south through the center of the lake, and we were standing at the intersection of the East African, Red Sea, and Gulf of Aden rifts (see Maps III and IV). The astute explorer Charles Johnson, crossing through the area 129 years earlier, was probably the first to describe the triple junction: "I concluded that [Lake Abhé] occupied one of those numerous depressions, which, like large fissures, here intersect the otherwise level country. . . . These fissures are in variable magnetic directions, but they never cross each other, and I saw evidence sufficient to satisfy myself, that at least in this neighborhood, they form rays from one center."[41]

His words recall those of Arthur Holmes, who described the Afar rifts more than a century later: "Tension clefts in at least three directions, which box the compass."[42] The deep fissures or "depressions" that Johnson noted may not cross each other as such, but there is ample cross-faulting caused by the rendering of the Afro-Arabian superplate into the three smaller plates. Lake Abhé itself sits at the western end of a series of elongated sediment-filled grabens that pass almost without interruption to the Gulf of Tadjoura at the western end of the Gulf of Aden.[35] It is not difficult to understand why some of the early theories on the final course of the Awash suggested that it drained directly into the Gulf of Tadjoura or that it passed through a series of basins until it reached the gulf.[43]

The basalt ridge that we commanded was formed by a lava flow from Dama Ali volcano just to the north of us.[42] The basalts are of a type called tholeiites and are characteristic of ocean floors; in places they cover lake deposits said to be less than 2500 years old.[44] Apparently the ridge has seen multiple occupations by Afar and pre-Afar people. Its surface was covered with stonework, artifacts, pieces of bone, and cowrie shells (the small, glossy marine mollusks commonly used in pre-colonial Africa for trading). Although the Afar regularly use stonework today—for reinforcement of dwellings, thornbush stockades, corrals, water storage vaults, and graves—I had not seen in the Afar before, nor did I see later, stonework that formed a patchwork of squares like I saw here. Very mysterious.

The stone artifacts were made of obsidian (volcanic glass) and were of a Late Stone Age (LSA) type: small blades, scrapers, and rock cores. Obsidian is present on the flank of Dama Ali volcano and may have been the source of the artifacts. LSA technologies in Africa typically date to about 10,000 to 40,000 years ago, although some of the Arissa artifacts may date to the time when trade was established with Arabs or Europeans—or even to a later time. An archeologist friend studying LSA sites around a lake south of Addis Ababa long pondered the age of some of the stone tools he was finding. One morning he was walking along a path and found human hair wrapped around an obsidian blade, as though it had been used and discarded sometime before breakfast. The inhabitants of a nearby village confirmed that, yes indeed, they still made stone tools. And they proceeded to make for him bladelets for shaving and nail trimming, flakes for cutting meat and fabricating weapons, and scrapers used for cleaning animal hides. The archeologist, Jim Gallagher, then with Southern Methodist University, was later able to document present-day stone tool manufacturing in six localities in the rift produced by three separate ethnic groups.[45] In each case an obsidian quarry was found within a day's walk of a "stone age" village.

It would be nice to think that the intriguing patchworks of stones at Arissa represent some kind of an ancient sign to space aliens, perhaps the Afar Triangle's version of the Bermuda Triangle. However, the stones may be nothing more than partitioned-off squares used for trading purposes.

That night Maurice, Ali, and I camped in a small depression on the flank of Asmara volcano, one of the submarine volcanoes described by Enrico Bonatti at the American Geophysical Union meeting I had attended in Washington, D.C., two years before. We intended to camp high enough to catch the night breeze and escape the mosquitoes from the nearby marsh. *Breeze* is not the word. We were practically blown off the volcano by gale-force winds blasting in from the Gulf of Tadjoura. During dinner, as we ate cold chunks of an unknown animal from a box of U.S. military rations Maurice had picked up somewhere, the wind was so strong that it actually blew the slime off the meat. Worse, Maurice was not able to light one of his tar-soaked Gauloises cigarettes, which probably added another three or four minutes to his life.

The next morning, after I had climbed halfway up the 350-meter volcano while Maurice stayed below, we were buzzed by a helicopter. I wanted to see the marine fossils at the summit that Bonatti had described,[46] but without satisfying my curiosity, I clambered back down the volcano and went over to the helicopter. Of course, it was Enrico Bonatti. He was accompanied by two geophysicists, like him from the University of Miami. In collaboration with Tazieff's group, the three were investigating the gaping fissures—large enough to swallow a Land Rover—formed by the westward migration of the Gulf of Aden rift into the central Afar.[47]

Several days later Maurice, Ali, and I once again passed through Tendaho Pass (see Map V, page 30.) This time we headed back to Millé to pick up a second guide, a friend of Ali's, who knew of fossil areas west of the village. Over the next few days we surveyed sedimentary exposures created by stream drainages at places called Weranso, Leadu, and Ahdaitole. Fossils to be sure. Partially buried elephant skulls, one with tusks sticking out of the ground; hippo skulls; jaws or isolated teeth of horse, pig, antelope, carnivore, and monkey; and postcranial bones. As we moved from site to site, the fossils seemed to cover wide areas, eroding from sandstones over a low, hummocky terrain. We collected some specimens, as well as a few stone implements lying on the surface that looked to be "old." I was ecstatic, feeling that we had discovered an abundance of prehistoric riches. Fossils and more fossils. Maurice believed they were about 3 million years old, on the basis of a radiometric date he had obtained from a basalt in the area.[49]

After dropping off Ali and his friend back in Millé, we returned to Gewani. By prior arrangement we rendezvoused with Schönfeld, our German acquaintance, and his colleague. We had agreed to join forces in a survey of a large graben just north of Ayelu volcano that Maurice suspected might contain evidence of another Holocene lake (see Map IV). A problem was that the area was in the territory of the Issa, a Somali tribe that was the traditional enemy of the Afar. The Issa reportedly had been raiding Afar settlements in the Gewani area recently, and the previous year they had fired on some UN geologists in the area we were about to enter. Schönfeld was particularly concerned about safety, having twice been robbed at gunpoint by *shiftas* (bandits) on the Hararghe escarpment, and for that reason he was hauling around several soldiers as an escort.

We met with the lieutenant in charge of the local military garrison, who stressed the danger of going into the area. He insisted that we have an armed escort, which he himself would lead. We were instructed to assemble at the garrison the next morning at daybreak.

That we did.

The lieutenant was a no-nonsense, caffeinated sort of man, probably in his early thirties. He had gathered his entire garrison—40 or 50 soldiers, all highlanders—to serve as our escort, along with a number of armed Afar. All told, about 150 men filled four large Mercedes trucks that would accompany our Land Rover and Schönfeld's. The lieutenant headed the column of six vehicles in a Jeep with a 50-caliber machine gun mounted on the hood. Two trucks loaded with soldiers were positioned ahead of our Land Rovers and two behind us. Schönfeld was decidedly unhappy about all this, convinced that our numbers would provoke a fight with the Issa. He told us that the Ethiopian Air Force had recently napalmed an Issa village. He agreed to accompany us only to the edge of the graben; then he was turning back.

We proceeded east to the graben from Gewani on a track leading around Ayelu volcano. The routine was that when we gave a signal for a geology stop, the lieutenant called the column to a halt using his radio, at which time the troops would fan out around us, taking up defensive positions with their Browning automatics and M-1 rifles. We were forbidden to take photographs. When it was considered safe, we jumped out of our Land Rovers and took notes on the local geology. Then everyone jumped back into the vehicles, and we moved on. Eventually we traveled the 25 or so kilometers to the edge of the graben. At that point the Germans turned back, and Maurice and I were left with our army.

The graben was a spectacular sight, with sheer rock walls of basalt rising 500 meters above its floor, bounded by classic, steep-walled parallel faults like those Hans Cloos replicated in his laboratory using clay models. The surface was spotted with vegetation and appeared to be only lightly mantled with sediment. The entire structure was 100 kilometers long and 20 kilometers wide and pointed toward Aysaita, which lay 150 kilometers to the northeast. Volcanoes stood at either end of the graben and in the middle. The UN team had taken aerial infrared photos of the area, revealing hot ground, hot lava domes, and hot springs.[1]

After Schönfeld turned back, the lieutenant's Jeep zoomed ahead of the column with the rest of the vehicles in pursuit. Shortly we occupied a small volcanic promontory in the center of the graben that had been previously fortified with chunks of volcanic cinder. The soldiers took their assigned positions, while Maurice and I sat in our Land Rover with our feet propped up on the dashboard. For some minutes, nothing broke the silence but the static from the walkie-talkies and the sound of Maurice sucking cigarette smoke into his lungs. After a while several men leading camels appeared on the east side of the graben at the head of a herd of cattle. The signal was given.

The attack was on.

But it was our side doing the attacking! Three truckloads of soldiers and Afar roared off in pursuit. I leaped out of the Land Rover and charged through the clouds of dust over to the lieutenant, who was jumping into his Jeep. In broken Amharic I yelled, "Lieutenant, what the bloody hell are you doing? We are here to collect *dingai bicha* (rocks only), not cattle!"

I was unable to understand his reply, but he shoved a fist into the air as if to say, "Piss off, I'm in charge here!"

What could I do? The man had a plan.

With isolated gunfire echoing across the graben, Maurice and I used the diversion to drive around taking photos and scooping up sediment and rock samples. That done, we decided it was time to get the hell out of Dodge. Back in the Land Rover, Maurice floored it and we shot out of the southern end of the graben onto the track leading back to Gewani. We left our troops to mop up.

All in all, that was not our proudest moment. Maurice devoted about a hundred words to the "Issa Graben" in his dissertation.[49] He thought the sediments along its southwestern margin looked sort of like the younger lake sediments near Gewani. I gathered some volcanic rock samples for Schönfeld, as I had promised him I would. After he returned to Germany, he wrote me a letter describing thin sections of the rocks illuminated under a polarizing microscope, something about "the spherules and matrix often show planar elements with horizontal small-scale dislocations . . .," none of which I understood.

We arrived back in Addis Ababa at the end of November. I later heard that our military escort returned to Gewani late that afternoon. Accounts of their success varied. They captured somewhere between 50 and 200 head of cattle—war booty that reportedly was transported to the ranch of the governor of the province. Apparently there were no casualties. During the calamitous political events that later took place during my years in Ethiopia, I often wondered what had happened to that ambitious lieutenant, but he was soon transferred out of Gewani and I never saw him again, which is just as well.

I was never again party to a military escort.

CHAPTER 4

Louis and Mary Leakey

E thiopia is split in half by its rift valleys (see Map II). The doming effect of the rising upper mantle has uplifted the rift margins and created the adjacent plateaus, which constitute almost half of Africa's highlands above 1500 meters.[1] The country's many rivers either follow the tilt of the plateaus away from the rifts or flow down escarpments into the rifts, forming lakes. Because of its high elevation, Ethiopia has been called "the African Tibet." Because of its many rivers, particularly the Blue Nile, which drains from Lake Tana in the central highlands, Ethiopia has also been called "the water tower of Africa." Rift rivers in Ethiopia include the Omo, which flows into Lake Turkana (a rift lake), and the Awash and its major tributaries. A number of smaller rivers flow into the Lake District of the Main Ethiopian Rift, which hosts nine modest-sized lakes spaced out like giant, water-filled animal tracks from the Kenyan border to just south of Addis Ababa. Dozens of smaller rivers flow off the Ethiopian escarpment and the Danakil Alps into the northern Afar Depression, where they evaporate in the desert, many at below sea level, or flow into small saline lakes.

Most of the major drainages on the Ethiopian plateau originate at elevations above 3000 meters before forming powerful rivers that flow northwest into the White Nile. The river may have gotten its name from the large amount of sediment carried in suspension that appears "white" when reflected in the sunlight. The largest of its tributaries, the Blue Nile, called Abbai in Amharic, originates from a spring above Lake Tana and is clearer and bluer in color, particularly in the dry season. After leaving Lake Tana, the Blue Nile flows briefly southeast, then turns 180 degrees and travels northwest in a nearly straight line through the Sudanese lowlands to Khartoum, where it merges with the White Nile. The Tekeze River (named the Atbara in Sudan), another major tributary of the White Nile, is called the Red Nile by some because of its rusty lateritic hue. It forms part of the northern border with Eritrea and originates in a rugged, mountainous area that hosts the fifth highest mountain in Africa, the 4621-meter Ras Dejen. The rivers on the Hararghe plateau to the southwest originate at elevations above 2500 meters and flow through the Ogaden desert and Somalia before spilling into the Indian Ocean.

Much of the northern and southern plateaus are covered by two kilometers or more of flood basalts that began spilling out of deep fissures some 40 to 50 million years ago.[2] Most erupted over a period of a million years or less about 30 million years ago, just prior to the earliest rifting in the Afar.[3] The basalts overlie Mesozoic rocks formed by ancient seas, the source of Ethiopia's limited known petroleum resources, and the Precambrian basement, the source of the country's gold and other precious metals.[4] Most of the remaining strata in Ethiopia are volcanic rocks less than 15 million years old, produced by rifting, or are sedimentary deposits such as those in the Awash and Omo Valleys.

Addis Ababa lies on the margin of the Ethiopian plateau at an elevation of 2300 meters, midway between the Main Ethiopian Rift and the deeply eroded gorges of the Blue Nile basin. Daytime temperatures are seldom higher than 30°C (85°F) and the nights seldom lower than 7°C (45°F). However, the rainy seasons, typically May through September, and the "small rains," February through March, can be times of penetrating cold, especially if one is desert-acclimated by nature. With a population of about 1.5 million people in 1971 (now triple that), Addis Ababa is ribboned with southward-flowing streams that join the main course of the Awash just south of the city. The streams in the capital can reach flood stage quickly, sweeping away everything and everyone in their paths, from piles of refuse to beggars and small children. I heard of one case in which a small child was swept into an open gutter and under the street at the top of a hill. The mother and several bystanders raced screaming down the street, trying to grab the child at each successive open gutter. After four or five blocks, miraculously, some-one succeeded.

Just north of Addis Ababa are high volcanic hills that offer a breathtaking vista of the margins of the Blue Nile basin.[5] The highest crags of these deeply weathered hills, collectively called Entoto, are covered with mountain heather, of various hues of green and blue, and fragrant wild thyme. For much of the year Entoto is bathed in brilliant sunshine beneath an unblemished blue sky that seems to stretch into eternity. To the north, small medieval-like hamlets can be seen tucked beside rock ridges or on open plains, the thatched roofs and mud-and-wattle walls of the homes blending with the countryside. Much of Entoto is covered with aromatic forests of eucalyptus trees. Originally imported from Australia, they are regularly harvested for firewood, a major source of fuel in the highlands.

The stream waters on the northern flanks of Entoto flow into the Blue Nile and then travel 3500 kilometers to the Mediterranean; to the south, streams flow into the Awash and travel the 700 kilometers to Lake Abhé. It took European explorers several hundred years to figure this out, something the Ethiopians had known for a millennium.

The small apartment that Judy, Justine, and I shared was across the street from the soccer stadium on Ras Desta Damtew Avenue in the central part of Addis Ababa. *Ras* is an honorary title equivalent to *duke;* Desta Damtew was a war hero who fought the Italians in the 1930s and was married to the eldest daughter of Haile Sellassie. Street names, however, were misleading guides, because people seldom knew or used these names when describing locations in Addis Ababa. Rather, they would say, "We live across the street from the soccer stadium," or "next to the Princess Zauditu Hospital," or "out the Asmara road next to the Bulgarian embassy," and so forth.

There was a small, grassy backyard for Justine to play in behind our apartment, which was on the ground floor. By December 1971 she was nearly two years old and very adventuresome. One day she tripped and rolled onto her back in the flower garden. Almost immediately she began screaming. I raced outside and saw to my horror that her hair and clothes were infested with large red ants. I swept her up and dashed inside the apartment and into the shower, where we both flailed away at the ants until they were safely washed down the drain. A stream ran behind a fence beyond the flower garden. On the other side of the stream was Jubilee Palace, the Emperor's home. At night and in the early morning, we could hear the roar of caged lions on the palace grounds. People liked to say that Haile Sellassie threw his enemies into the cages, but there is no evidence of that. On another day, Justine came racing into our apartment to tell us a gray monkey was running around the backyard. It was in fact a royal monkey with a silver collar that had escaped our neighbor and jumped the stream. At night we could hear hyenas prowling the banks of the stream looking for garbage. These fearsome carnivores were one of the most efficient means of maintaining sanitation in the city.

On holidays cannons would boom at sunrise from the palace grounds. Our windows would rattle, and things would fall to the floor. Often on such holidays, dignitaries or heads of state from other countries were invited to Ethiopia. Haile Sellassie, who was devoutly religious and something of an ascetic, would rise well before dawn and escort his somnolent guests to Saint Georgis Church to pray with him for hours. It was said that if there were negotiations to be held later that day, this attendance was part of a ploy to pray his opposition into submission. Prior to church services, we could hear the royal fleet of vehicles warming up their engines; the palace garage was just beyond the stream behind our backyard.

On Sunday afternoons I would sometimes go to a soccer game. Soccer is popular throughout Ethiopia, and the endurance required of athletes by the high altitude of the country produces great soccer players, just as it does long-distance runners. Soccer at the stadium, however, was only part of an afternoon's entertainment. Throughout a game there were running skirmishes between the stadium police and hordes of resourceful street-boys trying to sneak into the stadium. Those who made it inside in turn helped their friends in a variety of ways, by trying to divert the attention of the police or even throwing ropes over the stadium walls for their buddies to climb. When the crowd got excited during a match, the custom was to light newspapers or programs on fire. During half-time the men, or those so inclined, would walk out onto the soccer field and push down divots with their shoes while peeing on the grass.

Shortly after Maurice and I arrived back in Addis Ababa from the Afar in early December of 1971, the Seventh Pan-African Congress of Prehistory and Quaternary Studies[6] was held in Africa Hall on the grounds of the headquarters of the UN Economic Commission for Africa. Before the Congress we had discussed forming a larger team to explore further the fossil deposits of the central Afar. Because Maurice was still involved with his Ph.D. work, he was less enthusiastic about this than I, but he nevertheless saw the Congress as an opportunity to recruit others to join us. Certainly any plans I had had for my project of mapping the middle Awash were postponed; areas to the north had plenty to offer in terms of research interests. Also, I knew that forming a larger team could help me obtain funds for my geological pursuits in the Afar.

In order to get things moving, several days after the Congress began I was able to meet with Louis and Mary Leakey, who were among the honored participants. With a green light from Maurice, I sought their assessment of the age and significance of the fossils and artifacts we had collected west of Millé village. I called them late one afternoon after the day's lectures were over, when they were resting in their room at the Ghion Hotel, which was a few minutes' walk from my apartment. We agreed to meet in the lobby of the hotel early that evening. Maurice was unable to be there because the French were putting on an extravaganza for the Congress participants at the nearby Melka Kontouré archeological site. It was to be a wild night of drinking, feasting, and entertainment by bare-breasted Gambella dancing girls. Louis Leakey, then 68 years old, and Mary, 10 years younger, chose instead to have a restful evening at the hotel.

I was excited. To me Louis was an icon, not only for his great fossil discoveries but also for the breadth of his expertise and interests: ethnology, zoology, linguistics, vertebrate paleontology, archeology, and, of course, human evolution. Moreover, he was one of a kind: a white African born in Kenya near the turn of the century in the heart of British colonialism; raised by missionary parents and Kikuyu tribesmen; a life spent in the bush; a survivor of African independence; and a renowned scholar. Most of all, he represented a way of life—exploration in Africa—that I had dreamed of ever since I was a teenager.

The Ghion was one of the nicest hotels in Ethiopia, with jacaranda and bougainvillea spilling over spacious gardens fluttering with birdlife and butterflies in the daytime. I met Louis and Mary in the lobby, and we quickly moved to the breezy, open bar. We ordered drinks. Mary seemed a bit out of sorts, tense. Perhaps she was concerned about Louis's health, known to be fragile. After I explained who I was, I started telling them about the Afar and the fossils we had found. I described those near Gewani, dated at 4 or more million years old, and those near Millé, which Maurice thought to be 3 million years old. I also told them about the fossils found by Schönfeld near Chorora on the Hararghe escarpment, which he thought were 9 million years old.

Louis was keenly interested. The Afar was one of the few parts of the African rift system he had not visited. He was ebullient and extremely inquisitive, asking question after question. He had a way of sort of giggling when he talked, or chuckling, a bit like Father Christmas with his snow-white hair and his heavy physique. He had an infectious, good-natured enthusiasm that made you think you were talking with someone much younger. With his second drink, however, his complexion reddened, and beads of sweat formed across his forehead. Mary was not happy and wanted the meeting to come to an end. But she too asked questions and thought that Maurice may have known more about fossil areas in the Awash than he had said at his lecture earlier that day. She also said—at least twice—that whatever it was we had found, we should not tell another *single* person until we had obtained a permit for the area, a concession, from the Ethiopian government, lest someone move into the area ahead of us. On this point she was deadly serious, almost foreboding. I had heard this kind of talk before, in my early twenties, when I worked in the Caribbean looking for Spanish treasure ships. To Mary and Louis, we were obviously talking about gold, just gold of a different type. I was not convinced that Maurice and I had yet found the real thing, but the Leakeys were very encouraging in their different ways.

I decided that this was the time to play my trump card. I had brought a few fossils and artifacts with me, which I pulled out of my knapsack. They were from the site of Leadu, west of Millé. I was convinced that the fossils were our first hominid discovery and that the stone tools had been made by "Leadu Man." I laid the fossils and artifacts out on a coffee table between us.

"Aha!" said Louis.

My pulse quickened.

He picked up the fossils one by one, carefully turning each over and over.

My excitement was at fever pitch. I was sure that we would have a press conference the next morning.

Finally, Louis said, "Fossil pig. Yep, fossil pig. Yep, I'm sure they're pig."

Crestfallen, I picked up one of the artifacts from the table and handed it to him, and said, "Well, is there any chance the *pig* could have made the stone tools?"

Louis got a good laugh out of that, but Mary, an archeologist, was not amused—not in the spirit of things so to speak. She said the tools were Middle Stone Age, much younger than the fossils, and probably came from surface deposits. She was right. I had picked them up from gravels overlying the fossil layers. They were coated with "desert varnish," a dark patina that comes from long exposure and weathering. Louis said he was not able to estimate the age of the site without seeing more fossils. I suggested that if Maurice agreed, perhaps they would be willing to examine the rest of the modest collection we had stored at the university. Both agreed.

Terrific.

With that, Mary announced that the meeting was over.

Early the next morning I spoke with a hungover Maurice, and we arranged to take the Leakeys to the university that day during the Congress lunch break. We all met at Africa Hall at noon and piled into a well-kept, vintage Buick sedan the Leakeys had acquired. With an Ethiopian driver at the wheel, we cruised off to the university. Once again, Mary began reciting in a voice of doom that we should keep whatever we found "absolutely secret." We should not trust *anyone*, not even our best friends. With that, she turned around from the front seat and looked at me and then at Maurice. Was she talking about *me?* About Maurice?

A short time later, the old Buick rolled through the gates of the university. We parked in front of the Department of Geology and walked over to a small storage shed where Maurice's field equipment nestled among gardening tools, a wheelbarrow, and piles of rock samples in small cloth bags left by geology students. We all squeezed into the shed, and Maurice pulled a cardboard box off a shelf and began rummaging about, removing fossils wrapped in toilet paper to protect them from breakage. Louis's eyes lit up as he sat down on a wooden packing crate. He looked at the fossils one by one, half muttering to himself while giving us a running commentary. This one looks like a species from the lower levels of Olduvai . . . This one from middle Olduvai. . . . Ah yes, I first found this one at such and such site in 1947. On he went through the collection, handing each specimen to Mary, who offered her comments as well. Finally, they both agreed that the fossils appeared to represent a time span of about 1 to 2 million years old, maybe a little older or a little younger. Most of the specimens came from the "Leadu Man" site near Millé. Evidently they were not 3 million years old, as Maurice had thought, but still old to my reckoning.

On the way back to Africa Hall, where the Leakeys dropped us off before going back to the Ghion Hotel for a rest, Maurice and I told them of our plan to form a larger team, and I told them I was ready to return to the field as soon as possible. We asked if they would be willing to act as consultants for the project. Consultants? Louis asked. Hell, he was ready to go to the field! Let's see, he said, I have this to do and that to do, and I have to go there, and there, but I could join you at this time or that time. As he spoke, he kept glancing at Mary for approval, while she increasingly looked at him as though he were crazy. No doubt still concerned about his health, she told Louis to let us go to the field without him, maybe giving us some of their technicians to help collect fossils which we could bring back to Addis Ababa. Later, after they saw what we came up with, they might join us.

Then we started talking about other experts we would need to bring into the project. Mary was saying okay to this person, no no, not that one, rather this one, and so forth. Louis, however, persisted in discussing field plans. Finally, as we pulled up in front of Africa Hall, I asked him if he would be willing to write us an open letter of recommendation endorsing future field work, which we could then submit to foundations for funding. Sure, sure, he said, of course, of course.

I thought Louis Leakey was a marvel. I still do. Over the next few days at the Congress we met a few more times, briefly, but Maurice and I were going around in high hobnobbing gear, talking to people and listening to presentations. Maurice seemed to be everywhere at once. He gave lectures, led excursions, and bustled about talking with Britishers, Germans, Americans, and the many French archeologists and some geologists. He was not at all the tense Maurice I had seen at the Tazieff camp. The Congress was his opportunity to show that he was an equal among other scholars. However, there was little doubt that his claims about the Awash Valley had been viewed with curiosity by the Congress delegates ever since an interview with him had appeared on the front page of the *Ethiopian Herald,* under the headline:

Scientist Says Man Originated in Ethiopia

So much for keeping things low-key. There was a photo I had taken of Maurice and our guide Ali Axinum at Leadu, lugging a leg bone of a fossil elephant. Maurice was quoted as calling the Awash Valley "the spring of life." I mean we are talking about *the* spring of Life. The *Herald* elaborated:

> Dr. Taieb says the Ethiopian Rift Valley, especially the Awash region . . . gave rise to the emergence of life at one stage followed by the appearance of more and more sophisticated organic structures. The gradual development of primitive cells gave rise to the emergence of primitive animals and from here gradually on to higher forms of life. . . . There is geographical, geological, climatic, and life pattern resemblance throughout the various regions of the Great Rift Valley. Taieb implies that because of these factors and the feasibility of going from one region of the rift to another, the people of Palestine and the other regions adjoining the Ethiopian Rift Valley might have originated in the Awash Valley.[7]

Whoa! If this was what Maurice actually said, then clearly we had our work cut out for us—*the people of Palestine?*

What *about* the modest claim that humanity may have originated in the Awash Valley? Given that French archeologists were already excavating Early Stone Age tools in the upper Awash at Melka Kontouré, the real question was why our early ancestors would not have also lived in the Afar. This seems like a naive question, given what we know about the area today, but it was not so naive at the time. After all, the Afar's best-known resource for thousands of years was *salt,* formed in conditions that hardly suggested the cradle of humankind. Even the work of Tazieff and others described the Afar in "oceanic" terms. Overall, the Afar was best known geologically and historically for its evaporites, marine geology, volcanoes, and sterile desert plains. Certainly such a landscape was in stark contrast to past habitats that were optimal for early humans, such as riverbanks and lakesides supporting wildlife and other food resources.

By late 1971, however, the perception of the Afar as a hostile, uninhabitable, lifeless place through the ages had changed, at least to a very few of us. We now had tantalizing evidence, in isolated patches from Chorora to Gewani to Leadu, that animal populations had lived in the Afar for millions of years. If Maurice knew of other "secret" areas, as Mary Leakey suspected, so much the better.

So which hominids should we be looking for, and how might they reflect the unique history of the Depression?

Let's start from the beginning. Remains of the earliest mammal-like animals appear in the fossil record between 225 and 250 million years ago. Their recovery worldwide supports Alfred Wegener's belief that the continents were then interconnected as a single supercontinent, which he named Pangaea. In a paper published in 1970, Robert Dietz and John Holden proposed that by 190 million years ago Pangaea had split into northern and southern supercontinents, Laurasia and Gondwana.[8] They also proposed that by the end of the Cretaceous 65 million years ago, Gondwana had fragmented into the southern continents we have today, including Africa. At this time the dinosaurs became extinct and our earliest recognized ancestor appears in the fossil record, a primate-like creature the size of a mouse, called *Purgatorius*.[9]

The earliest true primates, lemurs, appeared about 50 million years ago and spread throughout much of Laurasia and parts of Gondwana, including Madagascar before it split from the African mainland and became an island. The oldest known anthropoids—the group comprising monkeys, apes, and humans—are known from fossils 32 to 35 million years old from the once forested Fayum Depression of Egypt. Of these animals, *Aegyptopithecus*, an arboreal, cat-sized animal, is the earliest ape-like anthropoid. Most anthropologists agree that true apes, similar to those we know today, did not appear until the beginning of the Miocene, about 23 million years ago. The subsequent diversification and abundance of apes in Africa in the early Miocene, and their scarcity in Eurasia at that time, suggest that the group to which apes and humans belong, the Hominoidea, originated in Africa. The African great apes, the gorilla and chimpanzee, are considered our nearest relatives. The Miocene itself, which lasted until 5.2 million years ago, is known as the Epoch of the Apes because of their great geographical expansion during this period, which coincided with a global increase in rainfall and forests.

In the 1960s and early 1970s fragmentary fossils, named "*Ramapithecus*," were reported from Pakistan and were believed by many to be our earliest known ancestor. This claim pushed human evolution into the middle Miocene; the fossils were dated at 8 to 15 million years old. Because similarly aged remains were also reported in Africa, this claim suggested that humankind originated in both Asia and Africa and indicated that animal migration routes were open between the two continents during this period.

The great biogeographer and zoologist Alfred Russel Wallace wrote in 1876 that during the Miocene the Eurasian ancestors of present-day African faunas poured across "the shores of the Red Sea" using a "connection . . . by way of Abyssinia and Arabia."[10] If we assume the existence of a Miocene landbridge at the southern end of the Red Sea at that time—connecting the Danakil Alps with Yemen—then Eurasian apes may very well have crossed into Africa by way of the developing Afar Depression (see Map I). In 1965 petroleum geologists discovered that invertebrate fossils of a Mediterranean type were found only in Miocene-age deposits in the Red Sea and that those of a Gulf of Aden type were found only in Pliocene-age deposits, which indicated that the southern end of the Red Sea was blocked off from the Gulf of Aden in the Miocene and that its northern end was blocked off from the Mediterranean in the Pliocene.[11] Hence an Afar–Yemen landbridge connecting the Dark Continent with downtown Bombay had to have existed during the Epoch of the Apes.

The late Miocene, 5.2 to 10 million years ago, experienced a major decline in global temperatures and sea levels, which brought about a reduction of forests and a sharp decrease in the diversity of apes that lived in forests. Many anthropologists not willing to accept the status of "*Ramapithecus*" as a middle-Miocene hominid championed the late Miocene or earliest Pliocene (4 to 6 million years) as the time of the first appearance of upright human ancestors, which had been forced to become bipedal as the forests withdrew. However, the hominid record during this period is very fragmentary. The most notable fossil known in the early 1970s was an australopithecine jaw fragment found at Lothagam in northern Kenya in 1967. The fossil, found by a team led by Bryan Patterson of Harvard University, was believed to be between 5 to 6 million years old.[12]* As far as the Afar–Yemen landbridge is concerned, there was no apparent reason why the immediate ancestors of this hominid could not have crossed into Africa from Asia, or vice versa, using this route.

A global reduction of sea levels at the end of the Miocene, accompanying tectonic movement in the Sinai area, created the landbridge at the northern end of the Red Sea. With the subsequent rise of sea levels that accompanied sea-floor spreading, the Afar–Yemen landbridge was finally severed in the early Pliocene (c. 5.0 million years ago), allowing waters from the Gulf of Aden to pour through the straits of Bab el Mandeb.[13] Thus, at about the time the Afar–Yemen connection was

---

*The jaw has recently been redated between 4.2 and 5.0 million years old, earliest Pliocene in age.[12]

destroyed, the Sinai landbridge was created. Hence there is no apparent reason why early hominids could not have used this route to cross into Africa from Asia, or vice versa, during the Pliocene as well as the Miocene.

The discovery of Baby Taung, the first known australopithecine, in 1924 is now the stuff of legend. How Raymond Dart, a 31-year-old anatomist in South Africa, was given two boxes of fossils recovered from a limestone quarry at Taung, and how, after he found a curious fossil in the second box and spent 73 days chipping away rock matrix, a skull belonging to a juvenile was revealed to Dart two days before Christmas. He named the humanoid *Australopithecus africanus,* the "southern ape from Africa."[14] I had the honor of meeting Dart, briefly ("Hello, I'm Jon." "Hello."), at the 1971 Pan-African Congress in Addis Ababa. The occasion was a feast the Emperor put on for the delegates at the Addis Ababa Restaurant, a large, round traditional building, a *tukul,* with a thatched roof. Dart was then 78 years old and looked to me as though he were tottering on the brink. But later in the evening, after most of the delegates were soused on Ethiopian *tej* (a honey mead) or other spirits, and Ethiopian music had given way to American rock, I saw Dart doing the watusi with a woman half his age. Flushed cheeks, jovial, having a wonderful time. Evidently he knew what he was doing, because he lived to be 95 years old and enjoyed a productive, distinguished life to the end.

Dart's great discovery was, of course, the first of many "near-man" fossils found in South Africa. By the late 1960s the consensus was that there were two types of these hominids: a "gracile" form from the middle Pliocene (c. 2.5 to 3.0 million years old), and younger, "robust" forms from the late Pliocene and early Pleistocene (c. 1.0 to 2.0 million years old) (Appendix I). The fossil evidence shows that both types were bipedal and small-brained. Also, both were found preserved in cave deposits in South Africa. The juvenile skull of *A. africanus* from Taung, and adults of the same species from elsewhere, represent the more slightly built, gracile type. Fossils of *A. robustus* represent the more strongly built type, with larger molars, a thicker mandible, and a broader face. A third type, called hyper-robust, exaggerates these same features and is best known from the famous skull of *"Zinjanthropus" boisei,* found in Olduvai Gorge by Mary Leakey in 1959. Called "Nutcracker Man" by the press because of its large molars, the fossil was originally described as a primitive but *true* man by Louis Leakey and was given a new genus and species name. It was the first early hominid to be dated by radiometric means (in 1963), revealing an age of 1.75 million years, and it was considered remarkable for its antiquity. Following much spirited debate and more fossil discoveries, however, "Zinj," as anthropologists still affectionately call it today, was included with the australopithecines. By 1971 more gracile and robust australopithecines were found in eastern Africa, extending the time ranges of these hominids. These included discoveries by Richard

Leakey at Lake Turkana and finds by French and American teams in the Omo Valley of southern Ethiopia.

Anthropologists believed then, as they still do, that the heavy jaws and large teeth of robust australopithecines were used for crushing food—an adaptation to a more resistant diet brought on by a drier, cooler climate. They also believed that the genus *Homo* diverged from an as yet identified australopithecine in the late Pliocene (c. 2.5 million years old), during a period when global cooling and aridity exerted selective pressure on our early ancestors. One of the behavioral adaptations that resulted was the manufacturing of stone tools. The discovery of *H. habilis* ("Handy Man") at Olduvai Gorge in 1959–1960, at levels similar to those where they found crudely made stone tools dated at 1.5 to 1.8 million years old, convinced Louis Leakey and others that the earliest toolmaker had been discovered.

Homo erectus, a successor of *H. habilis,* was first found in Java in 1890 by a Dutchman, Eugene Dubois. He assigned his find to a new genus, *Pithecanthropus* (later referred to as *Homo*) meaning "ape-man," because he believed the fossil to be the missing link between apes and humans. The first *H. erectus* fossils found in Africa were also found at Olduvai Gorge in the 1960s. The adaptiveness and territorial expansion of *erectus* in Africa probably contributed to the extinction of the australopithecines, which occurred in the early Pleistocene (c. 1.0 million years ago). Certainly, with the expansion of *erectus* into Eurasia, the interest of this gregarious ancestor in real estate was well established. Exactly when *erectus* crossed into Eurasia is still open to debate; however, there were marked reductions in sea levels in the early Pleistocene and late Pliocene (1 to 2 million years), which increased the size of the Sinai landbridge and may once again have left a crossing at the southern end of the Red Sea. It is thus possible that the earliest African *erectus* crossed into Arabia from the Afar and therefore was an "Afar" *erectus.*

In 1971 there was much disagreement, just as there is today, over which fossils constituted the first bona fide *H. sapiens.* Arguments have revolved around whether one skull or another was an "advanced" *erectus,* an "archaic" *sapiens,* or a species separate from, or intermediate between, the two. Nevertheless, differentiation of these late hominids is to be expected; among populations dispersed over 18,000 kilometers between Africa and Eurasia. Judging from a skull found by Richard Leakey in 1968, when he briefly worked with the Omo Research Expedition, and dated (arguably) at about 130,000 years ago, anatomically modern *sapiens* appeared to be present in Africa well before their presence in Europe. More recent humans, such as Cro-Magnons, did not appear in Europe until about 40,000 years ago, well after the appearance of the Neanderthals. The disappearance of the latter may have been due to inbreeding, competition, or both.

When the great faults that formed the Red Sea and the Gulf of Aden sliced their way into the Afar, they exposed the Mesozoic basement along the margins of the Depression.[4] The oldest of these rocks date to the breakup of Pangaea; the youngest are Cretaceous and date to the split of Gondwana into the southern continents. Whether dinosaurs remains or those of other early reptiles, or early mammals, are present in these strata was not known in 1971 (nor is it known today).

Overlying the basement rocks and also cropping out along the Afar margins are Miocene and Oligocene (15 to 28 million years old) flood basalts,[2] which elsewhere in Ethiopia are sometimes found interbedded with ancient lake sediments. Although these deposits commonly contain beautifully preserved plant remains and fossil wood, they have yet to reveal early lemuroids or archaic anthropoids. Sediments described as of a similar age—called the Red Series for their rusty color— date to the earliest rifting of the Afar and lie on either side of the active volcanoes and salt plains of the northern Afar.[15] Fossils from the Red Series could date to the migration of Asian apes crossing into Africa in the Miocene, as envisioned by Wallace, or to African apes crossing into Asia. Otherwise, the base of the Red Series effectively documents the beginning of the Afar's history as a terrestrial basin,[16] which has been a nearly land-locked area of sedimentary deposition for the entire 25-million-year history of hominoid evolution. Thus, in 1971 it was apparent that the Afar could potentially contain ancestors of the African great apes, those of the Lothagam hominid, its successors, more Taung children, "*Zinjanthropus,*" the earliest *Homo,* Leadu Man, "*Afar*" *erectus,* and of course "Palestinian Man."

The problem for the fossil hunter is that the burial and fossilization of our quadrupedal and bipedal ancestors in the Afar, and their ultimate resurfacing through erosion, are determined by several major geological factors. First, given the subsidence that the Afar region has undergone via plate separation, older strata have fallen lower and lower into the Afar Depression. Hence, any pre-Miocene strata unaccounted for are now likely to be deeply buried. Second, even if we were to drill holes around the Afar, there is no guarantee how much of these strata we would find because they were subject to erosion prior to subsidence. Third, though the Depression has been a sedimentary trap for millions of years, much of the sediments are entombed by lavas that have spilled out across the Afar floor from fissures and volcanic centers, such that today a good 60 percent of the Afar is blanketed with volcanic rocks. Finally, much of the remainder of the region is covered with late Pleistocene or Holocene (125,000 years to the present) alluvium or lake beds that mantle older sediments.

Thus, at the time Maurice and I were forming our expedition in late 1971, the question was where to begin looking for fossils. Obviously a good start was in the Awash Valley, where we had left off in December. On a grander scale, older sediments should have been exposed anywhere along the margins of the Afar where erosion from rivers was present and where volcanic rocks were absent. That would encompass the Ethiopian escarpment from Awash Station to Massawa, the Hararghe escarpment from Awash Station to the Somali border, and the margins of the Danakil Alps. Seemingly, anywhere in the interior Afar was fair game for fossils where tributaries of the Awash or other rivers exposed older sediments.

CHAPTER 5

Looking for Leadu Man

B efore the Pan-African Congress adjourned in mid-December, Maurice recruited several participants interested in joining our team: Yves Coppens, whom I had met in Paris, an anthropologist and paleontologist with the Musée de l'Homme; Karl Butzer, a geographer with the University of Chicago; Gudrun Corvinus, an archeologist from the University of Tübingen; and Raymonde Bonnefille, a paleobotanist, who was a fellow doctoral student with Maurice at the University of Paris. Maurice would serve as geologist and leader of the team, and I would serve as another geologist and as a liaison in Addis Ababa. The Leakeys agreed to act as "patrons" and consultants for the expedition. It was left hanging whether they would actually join us in the field, but in view of Louis's health, that seemed unlikely. Otherwise, future field work by any of us would depend on what funds we could raise and how soon.

The day before Maurice returned to France in late December, we met with the Ethiopian minister in charge of the Antiquities Administration, a distinguished, elderly man who was a French-educated historian. Reportedly, he was also one of "the chosen" who regularly had breakfast with Haile Sellassie. The Emperor was noted for keeping in close touch with his government, although with age his control was weakening. We informed the minister of our interest in resuming field work in the Afar with a larger team and assured him that we were certain to make great discoveries. After asking some questions, he responded enthusiastically and offered his support.

Following the meeting, I outlined a three-stage program, with Maurice's agreement, calling for nine months of field work over the next year, using successively larger field teams. The first stage had me returning to the field almost immediately for continued surveys, specifically in areas Maurice had not yet investigated. I sent the outline to Louis Leakey and reminded him of his generous offer to write us a letter of endorsement, which he did indeed write at the end of January 1972.[1] Written on behalf of himself and Mary, the letter was somewhat overstated, given what the Leakeys actually knew about fossils in the Afar, but it was direct:

We were both emphatically of the opinion that everything must be done to pursue the study of these fossiliferous and artifact-bearing horizons. The fossil fauna and the artifacts from this area of Afer [sic] are undoubtedly very comparable to those being found in the Omo Valley of Ethiopia and North East Rudolf in Kenya, and to some extent, those of Olduvai in Tanzania. We will consider it highly probable that, in due course, important fossil hominid remains will be found in association with other mammalian fossils and artifacts.

Louis emphasized that funds should be made available for our project "at the earliest possible moment."

I sent his letter to a small, private foundation in Connecticut, the Fund for Overseas Research Grants and Education (FORGE), that I had learned of before going to Ethiopia; it specialized in supporting small projects in developing countries. I had already sent its director a proposal asking for a modest $2600 to carry me through several months of field work. Within a few days after he received the copy of Louis's letter, I got a telegram from him saying that the funds were approved.

Off and running.

Over the next few weeks, I rounded up field equipment; had repairs made on Maurice's Land Rover, which I would be using; and hired an Eritrean field assistant. Kelati Abraham was from Asmara, spoke English, and had obtained field experience while working with the U.S. Army Mapping Mission in the 1960s. The Mission had made a superb series of topographic maps of Ethiopia, at a scale of 1:250,000, that were just being published.[2] The maps in fact represented the single greatest exploration effort made of the country, before or since. Prior to leaving for the field, I purchased a set of these invaluable maps from the Ethiopian Mapping Institute that covered the central Afar.

Throughout March and early April 1972, Kelati and I surveyed 2500 square kilometers of the lower Awash (see Map V on page 30). We were aided by our guide, Ali Axinum whom we picked up in Millé and who seemed fully recovered from his illness. If he was not, I was soon equipped to care for him, having acquired a lavishly stocked medical kit from Schönfeld on his last day in the Afar before he and his colleagues returned to Germany. By chance our paths had crossed on the Trapp road as they were breezing along in their red Land Rover. They had just picked up a case of Meta beer in Millé to celebrate the completion of their work, and Schönfeld was in an expansive mood. He sold me, at a cut-rate price, not only their remaining medical supplies but also several boxes of canned foods, a large tarp to use as an open work tent, and a large, round desert tent of excellent quality "designed by General Rommel himself."

I also again met the hospitable German engineers at Camp 270, the bridge construction camp next to the Awash River near Millé. While doing so, I befriended the pilot of the Trapp airplane, Urs Carol, a daring Swiss-American who flew a twin-engine Cessna ferrying men and supplies from one camp to the other. Urs flew me on a reconnaissance flight north of the camp as far as the Bati-Assab road. The survey revealed only modest sedimentary exposures but nevertheless promised more fossil localities. Just before landing on the dirt airstrip at Camp 270 on our return, however, I saw what I thought must have been extensive eroded badlands far to the west along the Awash. Maurice had confided to me, before returning to France in December, that an area somewhere west of Camp 270 on the north bank of the Awash was prolific with fossils. He called the area something like "Ahda."

My ground surveys started out at Leadu, a site that I would soon learn was very small by Awash standards—less than a kilometer square—but I wanted to see more of the geology and the fossils that Louis Leakey thought were 1 to 2 million years old. Over several days I surveyed a wide area around Leadu, finding more vertebrate fossils where streams cut through a thick blanket of surface gravels.

Even though I was still certain that the Leadu beds* would yield our first hominid, I was fascinated by the overlying gravels. Composed of quartzite and volcanic rocks originating from the distant Ethiopian escarpment, they formed thick plateau surfaces between streams. The gravels were coated with desert varnish, and their surface revealed more Middle Stone Age artifacts like those identified by Mary Leakey. Acheulean handaxes and cleavers similar to those at Melka Kontouré could also sometimes be found on top of the gravels. Invariably they were heavily worn and abraded, indicating that they had been rolled and tumbled some distance by streams. A basalt handaxe might be found side by side with a Late Stone Age obsidian microlith (a very small tool made from a flake), or a Middle Stone Age flint scraper, or both. The combination of artifacts of different traditions and ages found in the same layer or level is what archeologists call a "mixed assemblage," and the phenomenon has led to much confusion and controversy in archeology literature.

In eastern Africa, such widespread gravels as those at Leadu were carried down from higher elevations by "sheet flooding"—intermittent, fast-flowing, and coalescing streams active in the past during prolonged periods of generally dry, cool weather. At higher latitudes, such as northern Europe, these arid episodes marked the "glacial" periods. The "interglacial" periods were warmer and wetter and in the tropics were similar to conditions in eastern Africa today. During their transport by streams, these gravels abraded everything in their path like enormous sheets of sandpaper, except with grit as large as cobblestones, that ripped back and forth over the surface over thousands of years.

Maurice believed the plateau gravels in the central Afar to be roughly equivalent in age to those at Melka Kontouré, which he and archeologist Jean Chavaillon estimated, on the basis of the presumed ages and types of artifacts below and above the gravels, to be 50,000 to 70,000 years old.[3] They estimated the minimum age of

---

*In geological parlance, the term *bed* means a stratified sequence of sediments that are distinguishable by their composition from layers above and below.

the Acheulean tools underlying the gravels to be 70,000 years and the maximum age of the Middle Stone Age tools overlying the gravels to be 50,000 years.*

From Leadu we moved south, crossing the Ledi River, where we found a small, fossil-rich basin that was especially abundant in elephant remains, including whole skulls. In my field notebook I called the area "Elephant Park." From there we pushed on to the north bank of the Awash at Howoona, then back tracked and crossed the temporary bridge at Camp 270, and traveled west on the south bank as far as Meshellu stream. In both areas, erosion had cut some 100 meters to the level of the Awash. These were the Howoona and Meshellu areas that I had seen on my flight with Urs Carol. I guessed that farther west on the north bank was Maurice's "Ahda" site. But he had said nothing about fossil areas on the south bank, which made it likely that I was the first into the area. Yet, with much regret I recall that day.

I had spent most of the morning and afternoon reconnoitering the Meshellu stream. It flows in nearly a straight line for 30 kilometers along a fault, or a series of faults, aligned with the East African Rift. Near the Awash, the trend intersects another major fault, which is aligned with the Red Sea rift. I spent more time looking at all this than I should have—while also looking out for a pair of leopards I had spotted that morning—because in the late afternoon I faced a dilemma. I climbed to the top of the adjacent plateau to see what lay to the west and saw what looked like great topographical relief in the far distance. Nearby was the top of a hill Ali called Dikika, which looked as though it were dropping off into a deep basin.

Damn, I could just imagine the fossils scattered at the base of that hill! But the sun was getting low on the horizon, and I had to return to Camp 270. I had arranged for Urs to take me on another reconnaissance flight early the next morning. This one was to be east over a 1000-square-kilometer basin called Kariyu, an area I knew Maurice had not surveyed (see Map V on page 30). From my topographical map, it was apparent that the eastern half of the basin was part of a large, diamond-shaped graben formed at the confluence of the East African and Red Sea rifts, walled in by the bow-shaped Magenta Mountains. To the west the basin is blocked off by basalt hills, a large marsh, and the Awash. I was sure I could find a route into Kariyu by going south through what Ali called "lion country," but the way was remote and I was looking for a shorter route that would save me valuable fuel for the Land Rover once I got into the basin. A shorter route would also buy me time, and I soon had to return to Addis Ababa. Over the previous months, I had received a stream of letters from Maurice planning our mid-April expedition, now only a few weeks away.

All of this was going through my mind as I climbed down from the plateau at Meshellu. Back in the Land Rover, I reluctantly retraced our route out of Meshellu and arrived at Camp 270 after dark. Had we gone west a few more kilometers, we

---

*Current radiometric dates of Acheulean industries in Africa now suggest that the main Acheulean level at Melka Kontouré may be between 500,000 and 700,000 years old;[4] estimates for the time period of the Middle Stone Age range between 30,000 and 200,000 years.[5]

would have landed in one of the most extraordinary fossil areas on the African continent—in fact, in the world. The greater Meshellu area on the south bank of the Awash lies directly across from Maurice's fabled "Ahda" on the north bank, the same "secret site" that Mary Leakey suspected he had found.

Early the next morning, March 26, 1972, after the Cessna was fueled and checked out, Urs and I took off. I asked him first to fly west along the Awash to the Meshellu and beyond, which he did. As we soared over the area, I saw the river meandering through a fabric of dark riverine forest between two immense and magnificently eroded badlands, one on the south bank and one on the north. We flew over the area until I had seen enough. Urs banked the plane and we headed toward Kariyu.

The Kariyu basin is named after a semi-nomadic Oromo tribe that once occupied the central Afar but now inhabits the foothills of the Hararghe escarpment 300 kilometers to the southwest. Like the Afar, they are Muslim and mostly pastoralists. You cross Kariyu territory when passing through the Afar funnel and the area near Awash Station. At a distance, Kariyu are hardly distinguishable from Afar, except that the men have their own distinctive facial scarring and the women wear their hair in springlets and are fully clothed.

Flying east toward Kariyu, I saw more badlands, but these were mostly sculpted by lava flows and faults, forming a series of small north–south grabens filled with a hummocky terrain of sediments. The display looked like a large basalt flow carved up with a rock saw, and hardly passable. But with another flyby, I spotted the break in the hills that I had seen on my map. Turning north, we flew to the Awash, then shot east along the margins of a large marsh and grassland onto the treeless Kariyu plain. In our path we scattered flocks of waterbirds, waterbuck, kudu, ostrich, gazelle, oryx, and wild ass. Hippo dropped off the bank into the Awash. I thought that if the Kariyu Oromo had lived here in the past, they damn sure didn't leave willingly.

On turning back to Camp 270 in a wide arc, we were hit with a gust of wind blowing through a gap in the Magenta Mountains. No doubt these were the same winds from the Gulf of Aden that had nearly blown Maurice, Ali, and me off the base of Asmara volcano four months earlier. As the Cessna was buffeted up and down, I could see Dama Ali, the large shield volcano next to Lake Abhé. The gap in the mountains strategically lies at the northern terminus of the East African Rift and the southern terminus of the Red Sea rift, forming the "hyphenated zone" that unites the 8000-kilometer Afro-Arabian rift system. The gap also results from localized uplift created by magmas rising into the center of the triple junction. While Urs piloted the plane through the fierce jerks of wind currents, I felt as though I were looking down at the traces of dark, powerful forces emanating from a subterranean world.

Back in Camp 270, Kalati, Ali, and I loaded our gear and supplies into the Land Rover pickup. We climbed up the basalt ridge just east of camp and then descended into the first of the four north–south grabens. It was filled with former lake sediments and wind-blown silts at least 10 meters deep, forming dune-like crests up to 4 meters high. Crossing them was like trying to drive over great piles of Jell-O. All we needed was to slip off just one of these crests and Maurice would have eventually received a postcard from me saying, "Your Land Rover at southwest margin of Kariyu basin. Sorry."

After slowly working our way east through the gap in the hills that I had seen from the air, we plunged down into the next graben, drained by the Abaco stream. This graben is bounded to the east by another basalt ridge, followed by another graben, and then another, and another.

After two days of this, we camped at the northern end of the Abaco stream on the edge of the white Kariyu plain reflecting in the late afternoon sun. There we found a series of small, northward-dipping hills created by fissural basalt flows that had erupted under water, producing what were once islands in a shallow Kariyu lake. The bases of the hills were made of fault rubble overlain by dark-green glass pebbles formed when molten lava quenched under water.[6] Adjacent hills were made of clay, diatomite, and tuff capped by a limestone representing the floor of the lake. Also present were glassy gravels in stream deposits that had eroded the lake floor after it was drained. Vertebrate fossils were scattered about—elephant, hippo, crocodile, fish, monkey—indicating that terrestrial, aquatic, and arboreal faunas all enjoyed the pleasures of either Lake Kariyu or the streams that later flowed across a dried-up lake bed. Middle Stone Age tools were present, demonstrating that human habitation was also part of story.

I always liked to stop overnight at an interesting spot so that after camp was set up—the cooking area, bedding, mosquito nets, and a pot of water on the fire for tea—I could wander around during the final moments of the day and see what there was to see. As I was strolling around the Abaco hills, enjoying the solitude and trying to sort out which fossils came from which deposits, a dozen or so armed men appeared. Ali had warned me that Kariyu might be crawling with Issa, so imagine my surprise to discover that they were Afar, and friends of Ali's at that.

After the lengthy shaking of hands and kissing of cheeks, the men told Ali that they were part of a larger group camped in a thicket nearby next to the Awash, where they had bedded down a herd of livestock. They also reported having had a shoot-out with Issa the night before, just east of our camp. The upshot of this was that they convinced Ali that we should move from the beautiful spot where we were then camped to a malarial, bug-infested mud-wallow near where his friends were camped. We managed to have our second pot of water for tea boiling just as darkness fell and clouds of ravenous mosquitoes rose into the night.

I took a couple of hours the next morning again to look over the hills along the Abaco stream while drawing a sketch map of the area, taking photographs, and collecting sediment samples and a few fossils. We had discovered another major paleolake basin in the central Afar, to be sure. I guessed it was late Pleistocene, and I estimated from elevations on my topographic map that its highest level was about 465 meters, about the same as the highest level of the Tendaho paleolake, which Gasse believed was also late Pleistocene.*

Were the two paleolakes connected? What caused the Kariyu lake to dry up? Climatic changes, tectonics, or both? Gasse determined that for *climatic* reasons the Tendaho lake had contracted over thousands of years and that present-day Lake Abhé is the end result of this shrinkage. For *tectonic* reasons it is situated at the southeast end of the Tendaho graben, because the floor of the graben slopes in that direction. The floor of the Kariyu graben is also tilted, to the northeast, as the flow of the Awash indicates, which may have caused a northeast migration of the Kariyu lake.

I drove south into Kariyu as far I could before heeding Ali's fears about Issa and turning back. This was unfortunate because my map showed a broad area at the same elevation as that of the Abaco hills, which promised more sedimentary exposures and human occupation sites. In southern Kariyu we did see immense wedges of black basalt dipping northward like half-buried tombs. Passing through the desolate area gave me the eerie feeling that I was trespassing on sacred ground, the sense of intrusion compounded by the staccato bursts from our broken muffler reverberating off the rock walls. Just as we were about to turn around to head back north, we ran into a herd of wild asses. The eastern side of the graben was steeply bounded by rock, so the herd came thundering across our path to pass to the more open, western side. I guessed they were in the area because of grasses in southern Kariyu where small streams and perhaps springs abound.

The African wild ass is a beautiful, lightning-swift animal that is as fast on an open plain as it is agile and fleet on rocky terrain. It is a blend of gray and tan on its upper body with a black mane, white underneath, and black stripes around its legs. It is the size of a small mule, with a head similar to that of a mule or a zebra. That is where any similarity ends, however, because this stock of *Equus africanus* in the central Afar may be the last true breed of its kind in Africa.[7] Experts guess that other populations of wild asses that once existed in the northern Afar, Somalia, and Sudan have been hunted out, caught in the crossfire of warfare, or crossbred with domestic equids, including feral donkeys. We saw some 30 to 40 wild asses in southern Kariyu that morning and, later, easily double that number throughout the entire basin. I was to see more of these splendid animals on occasion elsewhere in the central Afar but never again in such numbers, usually in groups of three or four. In 1972 it was estimated that fewer than 3000 were present in the lower Awash Valley[7]—I would not be surprised if there are a quarter of that many now. Another reason for the reduction in numbers was made apparent later that day.

---

*All elevations are in meters above sea level, unless otherwise noted, and are only estimates based on the U.S. Army topographic maps of Ethiopia.[2]

It was early afternoon and we had been moving along the southeastern foothills of the basin when we stopped for lunch. To escape the sun, Kelati and I suspended a tarp from the back of the pickup to two tent poles. After a meal of canned tuna, shallots, hot peppers, and crackers, I crawled under the Land Rover for more shade to rest my eyes, swollen from the blinding glare of the Kariyu plain.

Visitors arrived.

They were an advance guard of Afar moving a large herd of animals—probably a thousand camels, cattle, goats, and sheep—to the southern end of Kariyu, where we had seen the wild asses. The Afar were on foot, as almost always. Except for small children and the ill and infirm, they seldom ride camels. After exchanging greetings, our friends told us of a shoot-out they had had with Issa two evenings ago. Where was this shoot-out? I asked. They pointed to the area due west of us— the same area where Ali's friends of the previous evening had had a shoot-out with "Issa."

After sorting out "which Afar group was shooting at whom," we packed up and continued north along the basin margin, where we shortly ran into their livestock, which were strung out for a kilometer or more. They were tended mainly by children and old people, followed by young women, some carrying babies on their backs, who were leading camels laden with their portable households. The last in the procession was the rear guard of 15 or more armed men. The next day they would all be in southern Kariyu, where they would camp until the grass was nearly gone, and then move on. Until then, they would raise their children there, tell stories at night, and bury some of the old folks, while keeping a sharp eye out for their traditional enemies over the next hill.

By late afternoon, we had crossed a dozen stream beds coming off the 700- to 1000-meter Magenta Mountains. Most were dry, but some had enough water to be turned into mud bogs by domestic and wild animals. Just south of Magenta Gap, we stopped to look at a hillside covered with volcanic debris. There was ash, formed when magma mixed with volatile gases explodes into the air, and tuff, formed when ash consolidates into layers, such as the water-laid tuff I had seen at Abaco. Along the lower reaches of the hillside were sandstone conglomerates, containing more basaltic glass pebbles, like those at Abaco. The conglomerates were overlain by obsidian flows. Not surprisingly, Late Stone Age obsidian tools were present along the margins of Kariyu, sometimes in great numbers. Finally, scattered about the surface of the Magenta foothills were tiny obsidian pellets, formed when droplets of magma are sprayed into the air like fountains of fire and then chill into glistening glass spheres before hitting the ground, or perhaps the surface of a lake.[8] I found these spheres over much of the northern half of the basin.

But how old were these volcanic products relative to the Abaco hills, and what sequence of events produced them? First, it is likely that the Magenta ash and tuff erupted over a period of time, and that one of the volcanic episodes resulted in the deposition of the tuff present in the paleolake sediments at Abaco. Second, the basalt flows and underlying glass pebbles at Abaco represent a later volcanic event.

The presence of this same glass as a constituent of the stream conglomerate at Magenta means that the formation of the conglomerate has to post-date the eruption of the basalt. It was apparent from vertebrate fossils and stone tools that animals and humans followed the streams into the basin at the time the lake sediments were being eroded. The obsidian must post-date the conglomerate over which it lies, and we can speculate that the obsidian spheres scattered around Kariyu are closely related in time to the obsidian flows. Finally, we can assume that at least many of the Late Stone Age obsidian artifacts in Kariyu post-date the eruption of the obsidian flows, and it is likely that the obsidian was used for tool making, and trade, up to the near present by a multitude of visitors. Hence artifacts found in the proximity of stonework around the margins of the basin may have been left by the pre-Afar inhabitants of Kariyu, by the Afar themselves, or by the Kariyu Oromo when they occupied the area. It is likely that the most recent inhabitants of the area simply built their stonework at sites that had been occupied by earlier stone tool-makers, resulting in a mixture of artifacts made at different times.

A working chronology of the geology and archeology of the Kariyu basin, with intervening tectonic events, appears to be, from older to younger, as follows:

Tectonic episodes that produced the Kariyu basin
Deposition of "early" Kariyu lake sediments
Tectonic event that produced the fissures at Abaco
Eruption of fissural basalts into lake
Tectonic event that tilted the Abaco hills northward
Migration of the Kariyu paleolake to the north
Stream deposition of glassy conglomerates that eroded Abaco hills
Animal remains and stone tools left along streamsides or reduced lake margin
Obsidian flow and eruption of glass spheres
Later Stone Age obsidian artifacts left along streamsides
Afar and Kariyu stonework and artifacts
Tuna fish cans left by geologists

The history of Kariyu is far more complex than this, of course, but the interrelatedness of tectonics, deposition, volcanism, and human occupation in the central Afar should be clear enough.

With the sun nearing the horizon, we moved another 10 kilometers north over an obsidian gravel pavement and ended the day just west of Magenta Gap. We had made camp, put on a pot of tea water, and started to relax, when about 30 Afar men, women, and children appeared from behind a bush. They were friendly, the children running around excitedly, pushing each other and cutting up. One of the elders, watching me write in my field notebook, asked if I could spare some paper to use for rolling *dumbaco,* their word for tobacco. I obliged by giving him Maurice's highway map of Ethiopia, which I assume he later smoked. He told me that in his youth he had seen "smoke rising from inside the Magenta Mountains," not too far from where we were camped. Before I could ask for more details about this, though, he and the rest of his group eased off into the foothills just as darkness closed in.

After dinner, as I lay on my cot reflecting on the events of the day and gazing into the night sky, a breeze picked up, sending gentle eddies of wind dancing around me. Eventually the winds grew stronger, and I could hear an empty canvas water bag flapping against the door of the Land Rover. As my eyes got heavier and I pulled a light blanket around me, I felt as though I were being lifted up by a tempest and carried swirling over Magenta Gap. I heard a rumbling far beneath me, deep within the buried vaults of the volcanic mountains. I know you're in there, Beast, surging, rolling, and waiting, and I know who you are.

———

It was midmorning. We had moved north another 10 kilometers, and I was on my knees gathering up tiny glass spheres in a small cloth bag. Suddenly, I heard a deep and urgent sound behind me. I jumped up and swung around. It was Urs Carol, staring me in the face, from an *airplane.*

I hit the dirt.

"Urs, you bastard!" I yelled, as I jumped up shaking my fist at the tail of the Cessna. Before leaving Camp 270, I had asked him to check on me if he had the time. He just had. The plane rose up into the sky, circled around, and landed nearby. As it taxied toward me, I recalled some of Urs's exploits. He had once buzzed a Trapp Company Land Rover, snapping off the radio antenna with one of the plane's wing tips. One of the passengers was the Trapp general manager making an unannounced visit from Germany. On another occasion, when the large bridge spanning the Awash at Camp 270 was nearly completed, Urs flew under it.

After the Cessna came to a halt, Urs climbed out of the airplane laughing as I was still scraping away the dirt plastered to my sweaty forehead. More laughter and backslapping. Urs gave us a few warm beers as a present. We talked and exchanged news. There had been another student demonstration in Addis Ababa to protest the policies of the government. The cook at Camp 270 was found taking liberties with one of his assistants in the food pantry. I told Urs about the large herd of wild asses in southern Kariyu. This interested him because he had helped a German zoologist survey the animals elsewhere in the central Afar.[9] Urs later flew him up and down Kariyu counting them. The zoologist published several papers on the herd and submitted proposals to the Ethiopian government to turn Kariyu into a national game park. The proposal was approved years later but—knowing how these things go—probably served little purpose.

When in Addis Ababa, Urs would drop by to visit Judy, Justine, and me. The following year he flew Judy to Mombasa, where he was taking his airplane for its annual maintenance. Judy still talks about flying low over the rift valley and seeing the animals. She liked the giraffes especially. Later, on his return from another of these trips, Urs's airplane was caught in a storm and forced off course in southern Ethiopia. He crashed in the Bale Mountains and was killed. People said that he was transporting guns that he intended to smuggle into Djibouti. I have never been impressed with what "people said."

Along the northeastern margin of Kariyu were more relic shorelines, vertebrate fossils, and abundant obsidian artifacts. There were at least two former lake terraces, the lowest indicating a shallower lake concentrated at the northern end of the basin. Nearby basalts looked very fresh, and some Afar told us about "hot ground" nearby. I knew that the UN infrared survey had revealed scores of hot springs and steam vents just to the north of us. The Afar also told us that there were lots of *lafofe,* or bones (meaning fossils) in the area, but time and gas were short. It was time to head back.

———

After returning to Camp 270, and before returning to Addis Ababa, we made a quick trip to see the geology of Tendaho Pass, which is the only outlet through the volcanic plateau for drainages flowing off the central highlands into the central Afar. It is also the only break in the plateaus for traffic between the coast and the highlands.

Immediately east of the pass there are prominent reddish-brown hills 15 to 20 meters high. They contain thick layers of sandstone with massive cross-bedding of sands and gravels deposited during torrential stream flooding. Some of the sediments are silicified by hot springs and contain masses of fossil reed-like plants. On top of the most prominent hill at Tendaho are the ruins of a fort strategically located nearest the pass, protected on one side by a high bluff overlooking the Awash. According to Thesiger, it was built by Tigrayans who conducted a punitive campaign against the Afar early in the century.[10] It is apparent that the fort was also used by the Italians during the Occupation; I found sun-bleached pieces of wine bottles scattered about stamped with "Napoli 1928."

On their published geological maps of the Tendaho graben, both Taieb and Gasse depicted the Tendaho hills as lake beds between 4000 and 10,000 years old.[11] However, below the fort I found a molar of an *Hipparion,* a horse that became extinct at least 125,000 years ago.[12] Although the fossil could well have been transported by streams from older deposits south of the pass, an earlier age for the hills is consistent with the more detailed work of Gasse. She determined that the oldest lake terraces in the graben are *at least* 10,000 years old and occupy elevations of 500 to 400 meters (the fort is at about 450 meters), whereas younger Holocene terraces are present at successively lower elevations of 400 to 300 meters.[13]

Climbing down the hill below the fort, I found the stromatolites just above the Awash that Maurice had pointed out in December. The location of these algal limestones at a lower elevation exactly between the Kariyu and Tendaho lake basins pointed to their connection during the Holocene, maintained by waters from the highlands flowing first into Kariyu, then into the Tendaho graben. The essential questions concerned the timing and duration of such a connection and the overall pattern of lake migrations throughout the central and western Afar. As we headed to the plantation to spend the night, I began putting together a working scenario for this pathway, which I would return to over the coming months and years.

In the later Pleistocene, following major graben formation in the central Afar, an "older" lake filled much or all of the Kariyu basin, including the area of the Abaco hills, to levels between 500 and 450 meters (Appendix II). After more tectonic activity and fissural volcanism, the lake migrated to the north, spilling over at Tendaho Pass and eventually filling Tendaho graben to its maximum levels. Accompanying regional aridity in eastern Africa, lake Kariyu dried up and the Tendaho lake began retreating to the southern end of the graben. Concurrently, the proto-Awash eroded the abandoned Kariyu lake floors and began downcutting Tendaho Pass, dumping sediments in the Tendaho hills to levels of 450 meters or more. Following a return to wet conditions at the end of the Pleistocene or in the early Holocene, lake Tendaho briefly formed a connection with a much-reduced lake Kariyu, depositing the stromatolites below the fort. Kariyu soon drained completely. As lake Tendaho fell below 400 meters, then below 300 meters, to its present level at Lake Abhé (243 meters, in 1970), the Awash River followed the retreat of the lake until it became entrenched in its present position on the border of Djibouti.

Although the scenario was provisional, there was an apparent pattern: The paleolakes became younger and younger *down* the Awash Valley. Lake Kariyu migrated from south to north, and lake Tendaho migrated from north to south. The migration of lakes to lower elevations fit with what we knew elsewhere. According to Schönfeld, the Chorora fossils and paleolake sediments high on the Hararghe escarpment were late Miocene, and those near Gewani in the lowlands were at least 4 million years old. According to Louis Leakey, the fossils farther down the Awash Valley at Leadu were 1 to 2 million years old. Thus, as the Afar floor subsided from the late Miocene to the present, from south to north, the lakes migrated to lower and lower elevations. Eventually the paleolakes migrated through all three rifts of the triple junction.

How do we explain this lake history in terms of plate movements? First, in any rift valley, lakes and rivers occupy lower positions as the rift floor subsides. Likewise, we can conclude that in the western and central Afar, lake waters traveled down its multiple rifts until they reached the lowest point in the structure, ultimately the Lake Abhé basin. In 1970, British oceanographer D. G. Roberts proposed that as the Arabian plate drifted to the northeast, it remained partially attached to the African mainland at the southern end of the Red Sea, creating the grabens in the central Afar[14] (see figures on page 66). In a 1972 paper, Tazieff and colleagues came to

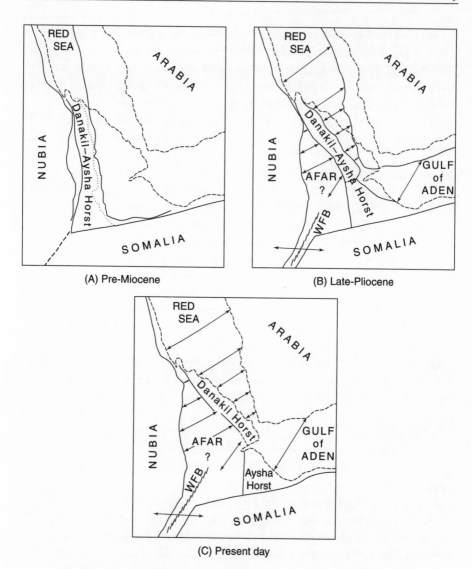

(A) Pre-Miocene

(B) Late-Pliocene

(C) Present day

FIGURE 3. D. G. ROBERTS'S MODEL (1970) of the evolution of the Afar Depression accompanying the relative separation of the Nubian (African), Arabian, and Somalian (East African) plates, and the counterclockwise movement of the Danakil block (micro plate or horst). See Map III in inside backcover showing the numerous mini-grabens filled with sediments created as a result of these plate movements.

a similar conclusion, proposing that the Afar interior "splintered" as the landmass underlying the Danakil Alps remained partially attached to Yemen and rotated counterclockwise.[15] Roberts also concluded that grabens in the southern Afar resulted from the southeast movement of the East African plate. The combined plate movements caused the earliest grabens to form in the southwestern Afar (from Chorora to Leadu), then in the central Afar (Kariyu), and finally in the eastern-central Afar (from Tendaho to Lake Abhé).

What is the significance of all this tectonic history in the real world of grant dollars and fossil hominids? It means that progressively older fossils should be found *up* the Awash Valley.

———

After arriving at the Tendaho Plantation, we checked into the guest house and took a shower—the first ever for Ali, other than rain. Kelati then joined some friends he had met; Ali walked into the adjacent village, Dubti, to mingle with his friends; and I had dinner with the plantation manager and several others next to the swimming pool. Over generous glasses of gin and tonic, they told me of a feasibility study in progress to turn Kariyu into an aquatic park. The idea was to build a dam across Tendaho Pass to use for irrigation and hydroelectric purposes. How much of Kariyu would be flooded they could not yet say, but I had visions of drowned wildlife and submerged archeological sites. As I sipped my second drink, I wondered how well a wild ass could swim. *Schools* of wild asses? Because of tectonic stresses operating across the Pass, however, I was dubious about the prospects of such a dam, unless it were made of rubber. Nevertheless, my hosts told me that "sociological studies" were under way to determine how well the Afar would adapt if forced out of traditional grazing areas flooded or irrigated as a result of the dam. That reminded me of what I had heard about agricultural development in areas south of Gewani. It was apparent that most or all open-range areas along the valley would soon be swallowed by plantations. I could not envision the Afar in Kariyu becoming cotton pickers any more than I could see them transplanted anywhere else.

I hated progress.

After dinner I met up with Kelati, and we strolled into Dubti to retrieve Ali. The only lights there came from lanterns and candles in souks and bars, frequented mainly by plantation workers. As we walked along the bare street with desert silt sifting in and out of my sandals, the rich smells of incense, cooking, and wood smoke drifted across our path. Dark forms passed by, speaking Afar, Amharic, Oromo, and Tigray, men often holding hands or wrapped arm in arm as Ethiopian friends and relatives commonly do. Some were drunk, some were flirting with bar girls or haggling with prostitutes, but most were just relaxing at the end of a long day and enjoying the cool night air. After poking our heads into a few bars and having a beer, we eventually found Ali and walked back to the plantation.

Before heading back to Addis Ababa, I took the Land Rover to the plantation garage for a few repairs. While this was being done, the plantation manager arranged for one of his engineers to drive me over the low-water bridge across the Awash to see some locally famous hot springs west of the plantation. The engineer, Giovanni Oreste, was half Italian and had grown up in Asmara. There were many such Eritreans of mixed marriages from the former Italian colony, the offspring of soldiers, engineers, and tradesmen who had remained in Ethiopia following the defeat of the Italian forces in 1936. When Haile Sellassie regained power, he invited all skilled Italians to stay in the country to live and work. This of course was a shrewd move, because the Italians worked furiously in Ethiopia during the Occupation, building roads and bridges, factories, municipal buildings, and sanitary, irrigation, and communication systems, which enormously modernized the empire. Those remaining in the country helped maintain the product of their labors and passed on their skills to several generations of Ethiopians. Giovanni, a soft-spoken and dignified man, was responsible for running the power plant of the plantation and was keenly interested in the use of steam from the nearby hot springs for creating geothermal power.

Named Allalobad, "the waters of Allah," the springs were identified by UN geologists as part of the most extensive hydrothermal field in the country, covering 120 square kilometers along a fault line beginning just southeast of Tendaho Pass.[16] Regionally, of course, the springs are another manifestation of the recent volcanic activity in the central Afar. The area around the springs is dotted with active steam vents, and small craters of hot mud and hissing gases are known to pop up here and there in the nearby cotton fields.

When we arrived at Allalobad in midmorning, we found several dozen Afar men, women, and children washing clothes and laying them out on the heated ground. The main springs are made striking by pulsating, boiling-hot geysers that shoot 6 meters in the air. Surface water temperatures reach 97°C (206°F) but become tolerable for washing and bathing in side pools and small channels dug by the Afar. These channels, although very small, may account for some of the reports by early visitors to Awsa of "canals" in the area. Everything around the springs had been turned into silica, or microcrystalline quartz: algae, grasses, reeds, nearby Holocene sediments, even insects caught in the mist from the geysers. I found perfectly fossilized dragonflies as delicate as soap bubbles that had landed by the springs hours or minutes before. I walked around the banks looking at the plants by the waters' edge, comparing them with the fossil flora I had seen beneath the Tendaho fort, while being careful not to walk on any algal crusts that might give way. The leader of the UN geothermal team, a New Zealander, had been severely burned earlier in his career by such an accident.

At midday, Giovanni offered to take me to lunch. I assumed we were returning to the plantation. We drove back across the low-water bridge, but instead of turning east into the plantation he turned west, driving us to the small village of Loggia just north of Tendaho Pass on the Bati-Assab road. Loggia consisted of several dozen small buildings, mostly of crumbling masonry left over from the Occupation. One of the buildings next to the road was a cafe and rest stop that catered primarily to truckers. We sat down at a battered wooden table under the shade of an arbor that adjoined the building. Beds made of wooden frames connected by interwoven strips of leather were laid out nearby.

A faint breeze blew across the desert.

After ordering lunch, a bottle of Ambo mineral water, and a couple of beers, we waited for our meal while I listened to my host talk enthusiastically about the prospects of hydroelectric power at Tendaho and geothermal power at Allalobad. The conversation soon shifted to Eritrea and the extensive geothermal deposits in the northern Afar in the below-sea-level salt pan of Dallol.[17] I told Giovanni about my consulting work to try to identify mineral deposits there or elsewhere as a counterpart to those in the Red Sea, which greatly interested him. In turn, he told me of a longtime fanciful scheme by Italian engineers to blast a canal through the volcanic rocks in the extreme northern Afar to allow waters from the Red Sea to flow into the depression. At the point where the marine waters entered the salt pan, a dam could be built to generate electric power, while the waters that reached the Depression would evaporate and the salt be harvested.

Soon our lunch arrived. To my surprise, we were served baby lamb cooked in olive oil and seasoned with rosemary with a side dish of ground *metmeta,* a fiery Ethiopian pepper. The best of both worlds. We also had warm bread and a fresh salad of desert-grown vegetables seasoned with basil. As we were leaving, I complimented the cook. He confided that early that morning, Giovanni had brought him the lamb and vegetables from the plantation. It was one of the great meals of my life—and a fitting end to my second field trip in the Afar.

CHAPTER 6

Hadar

D iscovery is relative. If the dinosaurs that once lived in Mongolia could be brought back to life, imagine their surprise to learn that they were "discovered" in the 1920s by a human that called itself Roy Chapman Andrews. "But that's ridiculous," they would say, "we preceded humans by 100 million years." Consider what the hominids from Bed II at Olduvai would say if told that the stone tools from Bed I were first discovered by Louis Leakey: "But he's a *Homo sapiens!*" And what about the geologist who claims to have found an ancient lake basin whose shorelines are heaped with artifact litter left by hundreds of generations of beach-goers?

The matter is more complicated. There is the claim by James Bruce that in 1772 he was the first to discover Lake Tana, the source of the Blue Nile, when he certainly knew that Portuguese Jesuits had preceded him there by 150 years.[1] And Ethiopians were on the shores of Lake Tana at least 500 years before the Jesuits. Richard Burton claimed to have been the first to discover Lake Tanganyika in 1858, yet when he arrived there, the place was alive with Arabs carrying on a brisk trade in slaves and ivory.[2] In fact, had Burton not attached himself to an Arab caravan traveling from the coast to the interior, he might never have reached the lake. When Henry Stanley first explored the upper reaches of the Congo in 1871, he was also aided greatly by Arab slavers.[3] What Burton really should have said was that he and his companion, John Speke, were the first *Europeans* to discover Lake Tanganyika. What Bruce really meant was that he was the first *Scotsman* to visit Lake Tana. And Stanley, a Welshman turned American, and the Arabs explored the headwaters of the Congo together.

Discovery then, not only is relative to the party doing the discovering, it is also a matter of acknowledging the prior work of others. Despite Bruce's tremendous contributions to the geography and ethnography of Ethiopia, he was skewered by the British press for falsely claiming to be the first European to set eyes on Lake Tana. Even the revered David Livingstone failed to give credit to those, particularly the Portuguese,[4] whose explorations in southern Africa had preceded his own.

Then there is the matter of the collective discovery. It would have been fatuous in 1967 for W. Jason Morgan to claim all the credit for the unifying concept of plate tectonics when his contributions never would have been made without the work of legions of modern explorers. Likewise, recognition of the nature and extent of the East African Rift Valley in 1891 by the Austrian geologist Eduard Suess was surely made possible by the work that numerous explorers carried out during the preceding decades, as he was the first to acknowledge.[5] The mapping of the string of lakes along the western part of the rift by Burton and Speke (Lake Tanganyika, 1858), Samuel Baker (Lake Albert, 1864), Joseph Thomson (Lake Rukwa, 1880), and Henry Stanley (Lake Edward, 1889) delineated what became known as the Western Rift of the Great African Rift Valley (see Map I).[6] The charting of the lakes to the east by Livingstone (Lake Nyasa, 1859), Thomson (Lakes Eyasi, Manyawa, Natron, and Baringo, 1880s), and Samuel Teleki and Ludwig von Höhnel (Lakes Rudolf and Stefanie, 1888) helped define the Eastern Rift. British geologists often refer to this segment of the rift system as the Gregory Rift in honor of J. W. Gregory, a Scottish geologist who greatly expanded the work of Suess and in 1896 coined the term *rift valley*.[7]

In his book *The Rift Valleys and Geology of East Africa* (1921), Gregory was able to draw on the work of numerous expeditions that had trekked across southern Ethiopia during the preceding 25 years to delineate that portion of the rift valley.[8] In doing so, he showed that the Lake District and the Main Ethiopian Rift are one and the same and that the rift reached the doorstep of the newly established capital, Addis Ababa. On his map of the "Abyssinian Section of the Great Rift Valley," Gregory then depicted the rift following the Awash River into the "Plains of the Afar," where it split into two segments, one leading northeast to join the Gulf of Aden, the other north to join the Red Sea.[9]

As Gregory acknowledged, his conclusions about the Afar relied heavily on the pioneering explorations of Italian geologists Giotta Dainelli and Olinto Marinelli, who published extensively on the geology of Eritrea and northern Ethiopia. While engaged in field work, the two were fortunate enough to escape the fate of their fellow countrymen, Giuseppe Giulietti and Gustavo Bianchi, who led separate scientific expeditions across the northern Afar in the 1880s, hoping to open a trade route to the coast. Both parties were slaughtered en route by Afar tribesmen.[10]

This raises the not-too-subtle point that discovery also requires survival skills and attentiveness to changing political winds. The malice shown to the Italians, like that to Werner Munzinger in 1875, was once again mixed up in intrigues between Emperor Yohannes and King Menelik of Shoa. This time the issue concerned the opening of trade routes and, ultimately, the shipment of firearms to the highlands. If the Afar had any suspicions of ulterior motives behind "trade" or "scientific missions," they were justified: An alleged scientific expedition to the port of Assab on the Red Sea coast in 1869 was followed by its annexation by Italy in 1882, then by the occupation of the port of Massawa in 1885.[11] The whole of Eritrea was colonized in 1889, including much Afar territory in the north and along the coast.

During Nesbitt's exploration of the Afar in 1928, he claimed to have found the spot, located 130 kilometers north of Tendaho, where Bianchi's expedition was massacred in 1884.[12] In 1929 an Italian aristocrat, Baron Raimondo Franchetti, led another expedition through the area that included a small army of soldiers. He also claimed to have found the spot where Bianchi perished, as well as the place where Giulietti's mission was massacred in 1880.[13] Franchetti concluded that the Bianchi and Giulietti expeditions met their fate on opposite sides of the same mountain, later mapped by geologists as a mountain of granite called Affara Dara.[14] It may be that when Bianchi was killed, he was seeking information on Giulietti's death, an enterprise the local Afar may have regarded as threatening. At the spot where Giulietti reportedly had died, Franchetti's men built a tomb of boulders with a large stone that, translated, read[15]

HERE PERISHED THE GIULIETTI EXPEDITION
14 ITALIANS
THEY WERE BARBAROUSLY SLAUGHTERED

From there Franchetti and his men marched 40 kilometers north to the shores of a small saline lake, Lake Afdera, which the baron solemnly renamed Lake Giulietti.* The trade route that Bianchi and Giulietti were hoping to open was later established well south of where the two explorers died. It is the present-day Bati-Assab road, which was completed in 1939 by the Italian forces that had invaded Ethiopia four years earlier.[17]

The occupation of Ethiopia by the Italians from early 1935 until the end of 1941 signaled a new phase in the exploration of the country, the Afar Depression, and the Awash Valley. This work followed the efforts and sacrifices of others, some of whom lived to see their contributions acknowledged and rewarded, and some of whom did not. But as each new hill and stream was "discovered" and mapped by newcomers, it was always under the watchful eyes of the Afar nomads, who had named these features hundreds of years before and who rightly feared that each new intruder was in some way a threat to their livelihood and future.

After leaving the Tendaho Plantation following our survey of the Kariyu basin, and after dropping off Ali in Millé, Kelati and I arrived back in Addis Ababa on April 4, 1972. Kelati disappeared into the Mercato area of the city—which included a vast open-air market called the largest in Africa—where he lived with an uncle. I rejoined Judy and Justine.

Both were doing well after a year in Ethiopia. Judy was working part-time as a volunteer in the gift shop of the leprosarium. She later got a job with an Armenian

---

*The lake, discovered in 1920 at the foot of Afdera volcano by an Italian geologist, lies 103 meters below sea level and is 160 meters deep.[16]

merchant who owned a fabric shop in the Piazza (a name left over from the Occupation), the main shopping area in the central city. Judy's university degree in political science was not much in demand in a country run as an autocracy, although I was certain that our backdoor neighbor, the Emperor, would have benefited from her outspoken liberal views. Eventually she began teaching English in a Muslim school established for Adere children, a people from the Hararghe plateau living in the area of the ancient city of Harar. Like most Ethiopians, the Adere are multilingual, and Judy's job was to teach 6- to-10-year-olds English while they were also learning Amharic, Arabic, and French.

Justine, now just over two years old, was also working on her job description, which was to learn English while playing much of the day with children who spoke only Amharic. She was becoming Ethiopianized in other ways too. Instead of carrying her doll in her arms, she carried it on her back wrapped in a *shamma,* a white shawl-like cloth that is worn by Ethiopian men and women and serves many purposes. She also ate the spicy hot native food with gusto, and she ate it Ethiopian-style, scooping it up with pieces of *injera,* a thin, light, pancake-like bread made of *teff,* a native cereal.

I had more letters waiting for me from Maurice when we returned from the field, with more details about preparations for our next field trip, due to begin in two weeks. His letters also concerned the participants on our team, in particular whether the Leakeys would join us in the field. Much of this seemed to revolve around a doctoral student in paleoanthropology at the University of Chicago, Don Johanson, who would be joining us. He came recommended by geographer Karl Butzer, who had known him as a student at Chicago, and by Maurice's colleague, paleobotanist Raymonde Bonnefille, who turned out to be Johanson's girlfriend. The two had met in the Omo Valley, where she was working with the French team, directed by Yves Coppens, and he was working with the American team, directed by his graduate supervisor F. Clark Howell, formerly of Chicago but now at the University of California at Berkeley. Johanson's interest was paleoanthropology, the study of human origins. He opposed the participation of the Leakeys in the Afar work, and, as it turned out, that of Coppens as well, because he feared their participation would detract from his own role. All of this was moot, however, because Louis's health precluded his going to the field, and Maurice had no intention of uninviting Coppens.

In mid-April 1972 Maurice and Johanson arrived in Addis Ababa. Coppens would join us later. While completing final preparations for the field over the next few days, Johanson and I became acquainted, and over the next few weeks friends. He was born in Chicago in 1943, an only child.[18] Don was bright, ambitious, and full of interesting stories about the work in the Omo. He relished gossiping about this researcher or that, often with humor that was irreverent and engaging. More ominous, however, was his persistent criticism of the French scientists working in the Omo, whom he viewed as incompetent and lazy. As I would observe more and more over the coming weeks, Johanson's opinions and complaints about people were oddly bitter and came in spurts, as though he were periodically seized with vitriol or some dire affliction. The only time he seemed really free from this was when he was stoned. Then he was often funny and entertaining. In the evenings, Johanson and a few friends of mine and I would sit around my apartment, have a smoke, and either eat whatever Judy had prepared or order takeout from the China Bar Restaurant around the corner. We would laugh at Don's stories and his obscene ZAP comic books.

On April 22 Maurice, Don, and I took off for the field accompanied by Kelati and Getachew Ayele, a representative of the Antiquities Administration, the government organization that regulated archeological and prehistory research in the country. All scientific teams conducting prehistory field work were assigned a representative to deal with local officials and to see that regulations were followed. In Millé we picked up Ali and recruited another field assistant, Meles Kassa, a Tigrayan who was part Afar. Shortly we were joined in the field by Coppens, who had been held up in Nairobi studying fossils at the National Museum. There he had met Louis Leakey, who confirmed that he was not up to field work but said he would still gladly help evaluate anything we found.

Over the next month, the six of us revisited many of the localities Maurice or I had surveyed. Maurice drove a Land Rover station wagon on loan from the French embassy, with Coppens, Kelati, Getachew, and Meles, while Johanson and Ali rode with me in the Land Rover pickup. Either Maurice led the way or I did, depending on which site we were visiting. Overall, we visited 63 fossil localities and collected nearly 900 specimens and a sampling of artifacts.

Once again, we started out at Leadu and within minutes Kelati found a sandstone cast of the brain of a monkey (see Map V on page 30). The skull lay amid much of the rest of the skeleton. Johanson was elated and identified the fossil as a *Cercopithecus,* similar to the modern vervet monkey. He believed the skeleton to be the most nearly complete of its kind known. On the basis of other fossils we recovered from Leadu, we later estimated the find to be about 2 million years old. In retrospect, this discovery was an omen of extraordinary things to come for Johanson.

Following dinner, Johanson and I stayed up late celebrating the discovery by sitting around drinking wine and sharing a smoke. After darkness settled in, the central Afar night winds that I had come to know arrived. Sparks from our resurgent dinner fire blew across the dark sky like drops of fluorescent paint smeared across a black canvas. The sun tarp that I had put up that afternoon began fluttering. After a couple of hours of conversation, we turned in. Later that night, however, I was awakened by a persistent flapping sound. I was not sure whether I was dreaming or pieces of the moon were dropping from the sky. Soon I recognized the sound of my sun tarp beating in the winds that had picked up greatly in the night. Afraid that the tarp would be torn, I got up and unsteadily made my way over to pull it down. Just as I reached the tarp, however, a steel stake attached to one of the nylon cords ripped from the ground and began to thrash violently about. Before I had the presence of mind to get out of the way, the stake slammed into my forehead, knocking me off my feet. I staggered up and groped along to the other side of the wildly beating tarp and with some effort hauled it down. I removed the tent poles, wadded up the tarp, and threw a spare Land Rover tire on top of it. I made my way back to my mattress and flopped down. Just as I slipped into unconsciousness, I felt a trickle of blood running down my face, cooled in the night winds.

———

As we traveled from Leadu to Ledi, Howoona, and every place in between, we began to get a better idea of the ages of the sites, which seemed to cluster between about 1.5 and 2.5 million years. Coppens was a well-rounded vertebrate paleontologist as well as a paleoanthropologist, and was able to estimate the ages of the sites from the similarity of key fossils to those elsewhere in eastern Africa that had been dated by radiometric means. Such *relative* dating is understood to be approximate but is often very accurate. Coppens was especially familiar with the ages of fossils similar to those recovered from carefully dated stratigraphic levels in the Omo.

Elephant fossils are particularly useful for relative dating, because their large teeth are commonly well preserved and possess morphological features that change incrementally over time. The three main elephant lineages—*Elephas, Loxodonta, Mammuthus*—diversified rapidly over the last 4 or more million years in response to changing climates and environments, and they are well represented in the African fossil record. Nearly every half-million-year increment during this period is associated with a different combination of elephant taxa.[19] Thus radiometric dates may reveal that in the Omo Valley, Species A lived between 2.0 and 4.0 million years ago and Species B lived between 1.0 and 2.5 million years ago. Because these two species overlapped between 2.0 and 2.5 million years ago in the Omo, we could guess that if we found them together in the Afar, they reflected a similar time period. Such relative dating is also called *faunal* dating, and when used for correlation purposes, it is called *biostratigraphy*. Before radiometric and other laboratory dating techniques were developed, biostratigraphy was used exclusively in Africa and elsewhere to estimate the age of geological strata. It is still widely used when modern laboratory dating techniques cannot be applied, such as when volcanic materials for radiometric dating are unavailable, as is the case in southern Africa, or when field workers simply want a preliminary "fix" on the age of a site.

Certain features of the molars of the most commonly found fossil elephant in Africa, *Elephas recki*, have gradually changed over its history of nearly 4 million years, again as an adaptation to climatic and environmental changes, and that makes this species particularly useful for dating. Molars of *E. recki*, like those of all elephants, are made of packages of vertical enamel "plates" stacked much like a loaf of sliced bread and held together with cement. The plates vary over geological time in number, height, width, thickness, and other features. In 1971 paleontologist Vincent Maglio of Princeton University was able to categorize these variations into four broad "stages" of development from the most primitive and oldest (Stage 1) to the most progressive and youngest (Stage 4).[20] When he assigned time periods to these stages using radiometric dates from East African sites, researchers such as Coppens immediately began using the *E. recki* stages for dating. Thus fossils of *recki* Stage 1 suggested a geological age of approximately 3 to 4 million years, Stage 2 an age of 2 to 3 million years, and so on.

I learned a lot from Coppens about recognizing key fossils for dating purposes. In 1972 he was 38 years old, and I found him to be urbane, congenial, and diplomatic, qualities that over the years helped him rise steadily in French scientific circles. Although he had published widely in diverse fields of vertebrate paleontology, he considered himself first and foremost a paleoanthropologist. In 1965 he described the first early hominid from central Africa, a partial skull from Chad, which he named *Tchadanthropus uxoris*, now regarded by scholars as *Homo erectus*.[21] His work there over six years gave him the credentials in 1967 to become co-leader, and eventually leader, of the French team working in the Omo Valley.

Johanson thought Coppens "didn't know shit about anything" and regarded him as opportunistic, devious, overambitious, and publicity-hungry—qualities that I now know Johanson himself admired. The only purpose Coppens served in the Omo, according to Johanson, was to suck up to the Americans, who were doing the real scientific work and outspending the French in research dollars three to one. Day after day, as we collected fossils at one locality or another, Johanson ranted about Coppens, usually behind his back. After a week in the field, Johanson served up the same plate about Taieb, whose work he called sloppy and simplistic—nothing compared to the "exceptionally excellent" work (Johanson loved superlatives) of the Americans working in the Omo.

Johanson saved his most rabid criticisms, however, for all things Leakey, especially Richard, who by then was making major discoveries at Lake Turkana.[22] Richard had previously worked one field season with the Omo Research Expedition, as leader of a Nairobi-based team, a position he had inherited from his father. Originally Louis was supposed to lead the Kenyan team, but his health made this impossible. According to Johanson, the notion that Richard was qualified to lead anything scientific was ludicrous. When Richard took over leadership of the team in 1967, he was 23 years old (one year younger than Johanson), lacked a college degree, and had not published a single scientific paper (but then neither had Johanson). His only claim to fame, according to Johanson, was his daddy's name.

The Omo Valley and the Lake Turkana area were among the last areas of eastern Africa to be explored by Europeans.[23] In 1888 two Austrians, Count Samuel Teleki and Ludwig von Höhnel, reached Turkana, the last of the major African lakes to be discovered by Europeans. They named it Lake Rudolf after the Crown Prince of Austria.* Over the next 20 years, a dozen expeditions marched up and down the shores of the lake, extending the work of the Austrians. One, led by the Italian military officer Vittorio Bottego from 1895 to 1897, was the first to confirm that the lower Omo River drained into the lake. The expedition ended in tragedy, however, when Bottego was later killed by Ethiopians while exploring areas to the north. It was a bad time for Italians to be traveling around the country, because in 1895 a large Italian-led army invaded northern Ethiopia from Eritrea with notions of expanding the newly established colony. In the famous battle of Adwa, the Italians were slaughtered, with 9000 casualties and prisoners taken, including 4000 Italians killed, in the greatest single defeat ever of a European army in Africa.[25]

In 1902 a French aristocrat, Robert Bourg de Bozas, was the first to discover rich fossiliferous deposits in the lower Omo, but he died of fever en route to explore the Congo. It was not until 1933 that systematic investigations of the Omo fossil deposits were undertaken, again by a French team, led by paleontologist Camille Arambourg. No fossil hominids were found, but the French carted off several tons of vertebrate fossils to Paris, where they were duly described and illustrated in lengthy scientific publications.[26]

Years passed.

In 1959 F. Clark Howell, then an assistant professor of anthropology at the University of Chicago, followed up on the work of Arambourg and again surveyed the Omo fossil beds. At the end of his field work, however, he was unable to leave Ethiopia with the fossils he had collected, nor was he successful in obtaining from Ethiopian authorities a permit to return to the Omo.[26] After turning his attention to Spain, where he worked for several years, he met in 1965 with Louis Leakey, then visiting the United States, and discussed his interest in returning to the Omo. There are several versions of what happened next, but over the next year Louis twice met with Haile Sellassie, once in Nairobi and again in Addis Ababa, and obtained permission to lead an international expedition to the Omo made up of teams from Kenya, the United States, and France. It is likely that the French got into the act because French archeologists were then serving as advisors on antiquities matters to the Ethiopian government.[27] By 1967 Howell's Omo ambitions had evolved from an intimate date with his girlfriend to a triple date that included her grandparents.

Scheduled to lead the French and Kenyan teams, respectively, were the 82-year-old Arambourg, still very active and eager to return to the Omo, and Louis, then 64, whose health was rapidly declining. Earlier in the year Louis had collapsed

*Within months after the lake was named after him, Prince Rudolf met a tragic and scandalous end.[24] At the age of 31, he died of a gunshot wound in a secluded hunting lodge while in the company of his beautiful 17-year-old mistress. She died similarly, apparently in a double suicide. In 1974 the Kenyan government changed the name of Lake Rudolf to Lake Turkana, after the predominant tribe in the area. The Omo River drains into Lake Turkana 50 kilometers due west of a small lake in Ethiopia also discovered by Teleki and von Höhnel in 1888, which they named after Princess Stephanie, Rudolf's wife and soon his widow.[23] The Ethiopian name for this lake is Chew Bahir.

while on a speaking tour in the United States, possibly from a heart attack. Nevertheless, he was to be the nominal director of the overall effort, while Howell was to be leader of the American contingent. Given Arambourg's age and Louis's health, Coppens was named co-leader of the French team and Richard Leakey co-leader of the Kenyan team.

What was billed as the Omo Research Expedition in the summer of 1967 quickly revealed itself to be three expeditions, with the Kenyans, Americans, and French all setting up camps in separate concession areas to search for fossils. The French and the Americans worked on one side of the Omo River, the Kenyan team on the other (see Map I). Louis Leakey only occasionally visited the Omo. Arambourg was overcome by the heat and returned to France, after bravely sticking it out in the field for two months. He died two years later.

From all accounts, the relationships among the teams during that first field season were not a model of collegiality. Competition for fossils and territory led each to increasingly criticize and snipe at the others. Richard Leakey thought the French were effete and lazy, as did the Americans; the French thought the Americans were arrogant and impetuous; the Americans thought the French treated them like second-class citizens and interlopers; and the Americans and the French thought Richard Leakey was untrained and uneducated. The Americans and French also chafed when the Kenyan team found the first hominid, a largely complete skull and skeleton of a remarkably early *Homo sapiens* (believed to be about 130,000 years old, although the date is still disputed). Finally, the Kenyans and the Americans chafed when the French found the first australopithecine, a robust jaw over 2 million years old.[28]

After nearly three months of feeling he was being treated like a "tent boy," Richard had had enough.[29] By then he had also realized that the deposits his team had been assigned were the youngest and would not produce older hominids. In mid-August, he found a way to leave the Omo project—and as a consequence garnered lasting international fame and the smoldering envy of his former associates. In a story that is now part of the Richard Leakey legend, he fortuitously discovered rich fossil deposits along the eastern shore of Lake Turkana while returning to the Omo in a chartered airplane after a quick trip to Nairobi. A thunderstorm forced the pilot off a more westerly course, and Richard saw that vast areas to the east, previously mapped as volcanic rocks, were in fact extensive sedimentary exposures. What followed has been told and retold in the *New York Times,* in the polychromatic pages of the *National Geographic,* and in scores of books.

In the summer of 1968, Richard renamed his team the "East Rudolf Expedition,"* which over the next four field seasons recovered 49 fossil hominids, 13 more than Louis and Mary had found in 36 years at Olduvai. Aided immeasurably

---

*Later the name was changed to the East Rudolf Research Project, and later still, with the name change of the lake, to the East Turkana Research Project. In the early 1980s the name was changed yet again to the Koobi Fora Research Project, for the picturesque area on the east side of Lake Turkana where the project's base camp was established.

by Mary's highly skilled Kamba fossil hunters, who had been trained at Olduvai, Richard found several spectacular crania at Turkana, as well as mandibles and postcranial bones. Most were robust australopithecines, but some tantalizing fragmentary specimens appeared to be very early *Homo.* Most of the hominids were between 1.0 and 2.0 million years old (though they were initially thought to be older), as old as the lowest levels at Olduvai. Early artifacts that were also found proved to be almost 2 million years old—among the oldest known.[30]

As director of the Turkana project, there was nothing "ludicrous" about Richard. His leadership skills, as well as his ability to make unabashed use of his father's name to raise money for his team, served the project admirably. It scarcely mattered whether Richard had any kind of academic degree, because by 1972 he had gathered around him a small army of 70 professionals—some of the best experts, graduate students, and technicians in the world—to help him at Turkana.[31] It is probably true that he was piss-poor at algebra, but he put together an enterprise that took the hominid business to new heights. And very soon he would make a discovery that would rock the world of anthropology and vindicate one of his father's most controversial claims about human evolution.

Meanwhile, back in the recesses of the Omo, the American and French teams were struggling to find any hominid fossils and arguing about those few they did find.[32] At the end of the 1967 field season, Arambourg and Coppens gave a preemptive press conference announcing discovery of their australopithecine jaw, even though they had previously agreed to share any such announcements with the Americans. If Howell and his colleagues were livid over this, they would have exploded had they known of the treachery of their former cohort, young Richard.

As revealed in Virginia Morell's tell-all biography of the Leakey family, *Ancestral Passions,* in 1968 Richard secretly wrote to an Ethiopian Antiquities Administration official telling him that Howell's graduate students were incompetent and that the American team lacked a single competent paleontologist.[32] Therefore, Richard advised, the Ethiopians should prevent the Americans from expanding their explorations into new areas east of the Omo River. In a separate letter, he requested that the Ethiopians give *his own team* the concession to this area, one that he thought would be rich in fossils and artifacts. So impressed was the official with Richard's concern for scientific excellence that he gave him permission to extend his operations east of the Omo, while sending him a list of Howell's team members for his scrutiny and approval.

Morell presents no evidence that Howell was ever aware of this perfidious intrigue, but it is highly likely that the French archeologists in Addis Ababa advising the government on antiquities matters knew about each and every communication that Richard had with the Ethiopians. Furthermore, we can speculate that this information was passed on to Arambourg and Coppens. If so, both were probably delighted with the dalliance of Richard. Whether Louis was involved in his son's scheme we can only guess, but judging from Morell's account of Louis's career, it would have been entirely in character.

There is nothing in Howell's illustrious career to indicate that the man was slow-witted, and we can surmise that he could smell the fetid fumes blowing from Nairobi to Addis Ababa, and perhaps from Paris. As it turned out, however, Richard had all the bones and artifacts he needed at East Turkana, and he never took advantage of the unexplored area east of the Omo that the Ethiopians handed him. After Arambourg died, in late 1969, Coppens magnanimously invited Howell's team to expand his operations onto a small piece of the French concession during the 1970 field season, evidently as a means of placating Howell's territorial frustrations. Nevertheless, suspicions remained between the Americans and the Kenyans, as between the Americans and the French who were then left to work side by side in the Omo. The strained relations in the Omo were symbolized by the nickname reportedly given to the boundary separating the French area from the American area: the Demilitarized Zone.[33]

That was the setting in the Omo when Johanson joined the American team in the summer of 1970 as part of Howell's stable of devoted graduate assistants. To hear Johanson tell it, the Americans spent much of their time screaming at the French, as each team became increasingly contemptuous of the other. I soon realized, however, that Johanson had his own agenda and was quick to exaggerate or distort anything if it served his own ends. He was intent on perpetuating the antagonisms that developed in the Omo, and he never lost an opportunity to needle Coppens or Taieb about some perceived outrage the French nation had committed over the last 300 years. But because Taieb, Coppens, and I had Johanson outnumbered, I wasn't worried about his creating any serious divisions on our incipient expedition. Besides, Coppens did not seem overly concerned about Johanson's increasingly vocal insults. Rather, somewhat to my surprise, he simply endured whatever abuse Johanson heaped on him, while Taieb treated Johanson as though he were a worrisome insect. I was rather amused by it all. I thought, well, this is the way hotdogs from the University of Chicago do science.

I do not recall anyone excavating a single fossil during the month we were in the field; there were plenty to recover on the surface exposed by erosion, finds that usually represent only a fraction of those present at any one locality. Also, at this stage we were simply sampling what fossils were present. As we moved from locality to locality, specimens were put in plastic bags and labeled on the spot with locality numbers. In camp the more delicate specimens were wrapped in toilet paper, and the larger ones in newspaper, before being packed in wooden boxes. Maurice, or the two of us, would sketch in our field notebooks measured geological sections of each site, depicting the stratigraphy and geological levels of fossils. My principal task was to document precisely the geographical position of each fossil locality, which I did by using a combination of sketch maps, aerial photographs, and topographic maps.

Late every afternoon in camp, whatever fossils were collected that day were carefully labeled with India ink by locality and specimen number, then recorded in the fossil catalog by Coppens or Johanson. Entries included the locality number, a specimen description (molar, right mandible, femur, etc.), and assignment of the specimen to the nearest taxonomic level possible. If Johanson knew that a particular fossil was a *Cercopithecus* monkey, that identification would be recorded in the faunal catalog; if he knew it was a primate but did not know which one, he would simply write down "primate."

All of this was essential to the integrity of the fossil documentation process. These procedures would allow us, and others who might follow us, to return, if necessary, to the spot where a fossil was collected to confirm its stratigraphic position, or to re-examine its paleoenvironmental context, to investigate the association of other taxa, or to look for more fossils in that locality. Such information could then be used to reconstruct the history of the site. At central Ledi ("Elephant Park"), for instance, the expedition catalog shows that over a few days, fossils of the following fauna were collected:

elephant	rhino	crocodile
bovid	equid	turtle
hippo	giraffe	fish
suid	monkey	

It can readily be seen that we have animals here that were partly or fully aquatic (hippo, crocodile, fish, turtle) and those that frequented a waterside (all the rest). A more refined identification of the Ledi fossils would show that their modern counterparts in eastern Africa lived in a combination of open habitats (elephants, horses, rhinos, giraffes, large bovids, some pigs), such as a savanna or a lightly wooded area, and more closed woodlands (forest pigs, monkeys, some species of antelopes), such as a forest fringe along a river or a lakeside. Rigorous collecting and a comprehensive analysis of the types and completeness of bones and teeth preserved could give us added information about the environment of deposition, and a similar analysis of proportional numbers of animals could give us information about species diversity and habitat.

The geologist can provide the paleontologist with additional information about past habitats, such as whether the sediments reflect a stream channel, a riverbank, a floodplain, or a combination of these environments. Archeologists may then come along and find artifacts and associated animal bones scattered throughout the riverbank sediments, which launches an entirely new and more complex documentation process. If a fossil hominid is discovered, this too begins a more complicated process, although in practice all fossils should be collected and documented with the same care, as we shall see.

Thus it is important and often essential that data be documented in such a manner by the paleontologist, geologist, archeologist, and paleoanthropologist to allow information to be interpreted and used interchangeably by all concerned, so that the final analysis provides a coherent picture of the past. This is called team-work or, in the mumbo jumbo of the grant application, "multidisciplined, inte-grated, and interactive research."

From central Ledi we moved south to Howoona, where we were hit by a massive rainstorm just as we were setting up camp. After a soggy night, we surveyed the area and collected more fossils. When the ground had dried sufficiently, we pushed south to the Awash. The upper Howoona sequence is probably slightly older than central Ledi, close to 2.5 million years old, and like Ledi may prove fruitful for finding primitive artifacts and the earliest toolmakers, but we found neither. At Howoona, Coppens was always picking up rocks that he thought were stone tools, and Johanson always tried to convince him that he was "full of shit." Closer to the Awash, the lower Howoona beds are deeply exposed by erosion and are likely to be at least several hundred thousand years older than the upper Howoona beds, and nearer in stratigraphic position to the lowest strata at Meshellu, on the other side of the Awash.

After Howoona, Maurice and I felt we needed a change of scenery, so we all headed for the Tendaho Plantation and parts south. It was May 4, 1972, and we had spent two weeks enduring indescribable hardships: listening to Johanson's conspiracy theories about the Omo and his views on the fall of the French empire. Also, by then Maurice needed to make repairs on his Land Rover, so a visit to the plantation garage would be timely.

That evening at the plantation we were treated to a hot shower, a square meal, and a dry bed. The next morning, after retrieving the now repaired Land Rover, we headed to the southern end of Tendaho graben. Maurice and I wanted Cop-pens's opinion of the age of fossils at several sites just east of Magenta Gap that we had looked at briefly the previous year. They were all within the central part of the triple junction, an area of the greatest tectonic instability.[34] In the brief time we were there, we saw basalts that had erupted under water, forming droopy *pil-low lavas,* and we saw sediments and fossils—which Coppens thought were mid-dle Pleistocene—that looked as though they had been tumbled and dumped by turbulent waters. It would have been interesting to explore this further, but time was short.

By midafternoon the next day we had crossed back through Tendaho Pass and had continued west past Millé on the Bati-Assab road (see Map IV). Near Weranso village, we turned southeast into the bush and traveled 20 kilometers across the gravel plateau until we came to an abrupt halt. There the bottom dropped out of the earth.

"Holy Moses!" I said, as we all piled out of the vehicles.

Johanson started saying over and over again, "Far out. Far out! I mean far *frig-ging* out!"

Coppens, always the cool one, was whistling under his breath, while Maurice was strutting around throwing his arms up in the air saying, "See, what I tell you? It is good, no? It is good, no? This is *Hadar!*" Otherwise known as Ahda or the "mystery area."

All was badlands as far as we could see to the south, west, and east: an immense, seemingly endless landscape eroded and chiseled into an exquisite sedimentary complex of stratified sand castles and cliff dwellings. Deep ravines and precipitous slopes descended from the plateau toward the Awash, creating steep ridges of clay and sandstone layered into multiple shades of brown, red, gray, yellow, and white. Nearer the river the hills fell away, then gradually rose again on the south bank in the greater Meshellu basin until they reached the top of the opposite plateau nearly 10 kilometers away. A large dry wadi crossed obliquely in front of us, intersecting one of the sweeping meanders of the Awash, which course back and forth through the thick strip of dark green riverine forest that I had seen from Urs Carol's airplane. Although Maurice called the area Hadar, the name is actually *Ahdi d'ar,* meaning "treaty *(ahdi)* stream *(d'ar)*." Perhaps this refers to the settlement of some previous dispute in the area among the Afar, or it may be that the Ahdi stream is a long-standing boundary between two Afar clans.[35] When saying the name, the Afar run the four syllables of *Ahdi d'ar* together into a contraction that sounds roughly like "Ahda" or "Hadar," which is how Maurice came up with the name.*

The breadth of the combined exposures at Hadar and Meshellu is as wide as the main gorge at Olduvai is long, and the vertical depth is three times greater, over 100 meters. The total area of Hadar is 20 square kilometers, twice the size of the main gorge at Olduvai. As we looked over the area, it was immediately apparent that there were decades of work to be done there, not counting the other areas we had surveyed or those that might lie to the west and south around the bend of the river.

From all indications Maurice was not only the first non-African to discover fossils at Hadar; he may have been the first to set foot in the place. Some day someone may come up with an obscure report from an early traveler who actually descended into the Hadar badlands, but for my money Maurice gets the credit. On early maps, caravan routes are depicted as passing within 10 or 15 kilometers of Hadar, but not through it[37] (see Map IV). Nesbitt passed 15 kilometers west of Hadar in 1928, and Thesiger came as close as eastern Meshellu in 1934, as I did in March 1972.[38] Nesbitt traveled north from Gewani for 17 kilometers on the east bank of the Awash, crossed over to the west bank, and then made a wide arc west and north around Hadar before turning toward Tendaho Pass. Thesiger traveled on the east bank of the Awash from Gewani for 55 kilometers and then bypassed the eastward bend of the river by cutting across to Meshellu. When he again reached the Awash,

---

*Although "Hadar" is incorrect, the name will forever remain in the scientific literature. In Maurice's first maps depicting Hadar, he also placed the name on the wrong stream, putting it too far to the west by 5 to 10 kilometers.[36]

across from Howoona, he continued east on the south bank, passing the area of
Camp 270.

By avoiding the eastward bend of the Awash, Nesbitt and Thesiger missed a 50-
kilometer stretch of the river that bisects the most rugged area in the entire Awash
Valley. It covers some 1000 square kilometers and is deeply eroded by rivers and
scores of streams flowing off the adjacent plateaus.[39] It is understandable why
these pioneer explorers, with their burdensome caravans, bypassed the area and
why the site that is now called Hadar was spared exploration until the 1970s. In
fact the immediate "Hadar" area is only a fraction of the total badlands in the
region, much of which remains unexplored today.

Maurice apparently found Hadar in December 1970 by following one of the
tributaries of the Ledi River, which originates in the highlands north of Bati and
intersects the Awash just north of Camp 270.[40] At the point where they come
closest together, the Ledi and Hadar drainages are only a few kilometers apart.
Maurice had followed one of the Ledi branches south to the gravel plateau separating the Ledi basin from the Hadar basin, which he crossed, and he was there. I
can imagine the exultant feeling he experienced that day, much like John Speke's
first glimpse of Lake Victoria.

We made camp on the rim of the basin. The next morning, with our vehicles
stripped of all weight except essential gear, we descended into Hadar along a small,
winding streambed. Almost immediately we found fossils at the base of ravines and
cascading down hillsides from thick sand layers. We found whole or partial skulls
of elephants, horses, rhinos, hippos, bovids, and pigs and occasionally the remains
of smaller animals such as monkeys and rodents. Soon Coppens was telling us that
two species of elephants, *Loxodonta adaurora* and *Elephas recki,* looked distinctly
older than those we had found at other sites. He guessed that the *E. recki* was a
Stage 1 or 2 and that the lowest Hadar beds might be 4 million years old, or as old
as the lowest levels in the Omo.

As we raced around looking at this or that, one person would call out to another
to come see some new discovery. Soon we declared that the first one to find an
entire elephant skeleton standing on four legs would win a case of beer.

Hadar was elaborately laid out like an immense, tiered amphitheater nearly
devoid of vegetation. The geology was textbook-perfect through and through.
There were stratified lake beds and stromatolites, stream channels choked with
volcanic tuffs, laminated lagoonal deposits with carbonaceous clays and lignites,
mudstones with desiccation cracks and gypsum, fluvial sandstones with broken
bones and fossil wood, beach deposits with bioclastic debris and detritus, and limestones packed with gastropods. The site was like an enormous, flat-lying encyclopedia of natural history with part of one page exposed on this hill, another in that
ravine, another on the crest of a ridge. The formidable task ahead of us was to put
together the pieces and see how much of any one page we could read.

On occasion, when near the Awash over the next few days, I waded into the
river to cool down and wash off the layers of dirt and salt accumulated from clam-

bering over the outcrops to see what lay beyond the next rise and then the next. The swims gave me the chance to take a closer look at the shoreline and the surrounding stands of gallery forest along the river with their branches drooping over the water's edge. Blue-green tamarisk trees lined the banks between isolated patches of knee-high, pale green grass and sweeping loops of brown waters. Approaching the river, I could hear the occasional plop of turtles or crocodiles tumbling off the banks, sending small ripples into the main currents. The crocodiles were seldom longer than a few meters, nothing like the monsters I had heard about in the Omo River. Abundant birdlife soared overhead and darted between trees, accompanied by the chatter of monkeys and the barks of retreating troops of baboons. Along the shore, the canopy of trees was so thick in places that most direct sunlight was blocked out, save the occasional rays breaking through the branches to the dappled earth. It was also so high and spread out that the Afar could drive their camel herds through the naturally made tunnels. Animal tracks left over from the night before, or from the early morning, were abundant where the floor of the forest was bare of foliage or where a trail led down to the river. The sweet, clayey smell of the Awash was mixed with the fungal and fermenting smells of decaying vegetation, the droppings of animals, and the musky odor of wild pigs. When standing still in the forest or along its edge, I could hear the faint swishing sounds of the winds blowing through the tamarisk in rhythm with the subtle lap of water along the shores.

Surrounded by this extraordinary abundance of fossil riches, Johanson was more paranoid than ever about "the frog menace," particularly the prospect that Coppens would find a fossil hominid before he did. Johanson's eyes would frantically follow his movements as he walked from locality to locality, seemingly ready to pounce on any specimen Coppens picked up that might be The One. As we sat around during our last evening at Hadar, planning a larger expedition the next year, Johanson made it clear that he was worried that Taieb and Coppens would bring in a horde of French researchers—disregarding the fact that Hadar was a *French* discovery. In turn, Coppens and Taieb were worried that Johanson would unilaterally invite in a horde of Americans, particularly his Omo pals. To avoid such possibilities and to form a basis of understanding over the coming field seasons, Coppens and Taieb proposed that we draw up an agreement or protocol describing our respective responsibilities on the expedition. The document we drafted outlined procedures for the recruitment of more researchers, the sharing of credit for discoveries, the coordination of field work, and the publication of results. We would all sign the protocol and give the original to the Antiquities Administration. Copies would go to our respective financial sponsors as a description of our organization and a statement of our purpose—and as a demonstration of our efforts to avoid another Omo-like schism.

The next morning we loaded up and headed for Camp 270, where Johanson sat down with his portable typewriter and tapped out the hand-written agreement. The accord named Maurice head of the "International Geological and Anthropological Expedition to the South and Central Afar" and spelled out the duties of the four of us.[41] Coppens would be in charge of "vertebrates other than primates"; Johanson, "primates other than hominids"; and Coppens and Johanson together, "hominids." In addition to working with Maurice on regional mapping and stratigraphy, I was in charge of the "geographical location of sites," which suited me just fine. As it turned out, I was also in charge of mapping the location of fossil localities. For now, the Leakeys were named simply as patrons of the project. Archeologists would be named at a later date.

Johanson was decidedly not happy with the agreement. Coppens's name was not only listed next to hominids in the protocol alongside his own, but it was also listed first. Johanson wanted exclusive control of all hominids found, but he had no choice but to share this responsibility with Coppens. We all signed the protocol.

Coppens returned to Addis Ababa that same day on a Trapp truck and from there returned to France. The rest of us spent a few more days in the Camp 270 area that included another trip to Meshellu. I led the way, but once again we failed to go farther west. The terrain was difficult, and neither Johanson nor Maurice was willing to invest the time and effort to search out a route. Instead, we decided to end our field trip by going to Camp Arba and proceed from there to the Hararghe escarpment to look briefly at the late Miocene sediments described by Schönfeld.

We did not get away from Camp 270 until after dark, however, and en route to Arba, Maurice sideswiped a dump truck. The truck had only one working headlight and was apparently hugging the center of the newly completed Trapp road. Apparently thinking it was a motorcycle, Maurice narrowly escaped a head-on collision. Johanson, who was traveling with him at the time, immediately jumped out of the Land Rover and began screaming at the Ethiopian driver of the truck, saying, "We will report you to the *American embassy* and have your driver's license revoked *forever!*" The Ethiopian, responding in perfect Oxford English, told him that perhaps he was mistaken because he must be thinking he was in the Congo or South Africa, and that maybe he did not realize that Ethiopia was not a colony beholden to the United States or anyone else. Realizing that this might be someone with authority, Johanson immediately became obsequious and apologetic, finally asking the Ethiopian to *please* be more careful next time.

Enough damage was done to the Land Rover to keep us stuck at Camp Arba until it was repaired. Our time was well spent, however. It was the weekend and many of our German friends from the various camps were there, so there were plenty of festivities, food, and drink.

Three days later, May 17, 1972, with only two days left on our field trip, we headed south and then east along the foothills of the southern escarpment. The road ran parallel to the French-built railway connecting the port of Djibouti with Addis Ababa. Commissioned by Emperor Menelik (crowned in 1889) in 1898, the railroad took 20 years to build.[42] One reason for the lengthy construction period was periodic attacks on the work crews by the Afar and Issa, who apparently realized that the railroad would severely undermine their control of trade to and from the coast. Eighteen years after its completion, the railroad served as Haile Sellassie's means of escape from the advancing Italian armies. He and members of the royal family boarded the train on the outskirts of Addis Ababa on May 2, 1936, at 4:00 A.M. and managed to cross through the Afar funnel into the Hararghe foothills just hours before the rail line was cut by an advance patrol of Italian soldiers. Three days later the First and Second Eritrean Brigades of the Italian army seized the capital.[43] From Djibouti, the Emperor and his entourage boarded a British war vessel bound for England, while Ethiopian patriots were left behind to fight a bloody five-year guerrilla war against the Italian forces. Although criticized by many for fleeing the country, Haile Sellassie continued fighting, on European diplomatic turf, and remained free to return to Ethiopia another day.

We stopped periodically along the road, looking at exposures of diatomites and tuffs, just as Schönfeld had reported. We found no Miocene fossils, but we did find abundant relic hot spring deposits, fragments of Pleistocene fossils, and stone artifacts, all indicating that the geology and prehistory of the area are more complex than Schönfeld had indicated.

By midafternoon we reached the small town of Afdem, which lay at the foot of a 2125-meter volcano. There we ran into a professional big-game hunter I knew, Thomas Matanovich, a Yugoslavian who had been in Ethiopia for some years. Tom was in town purchasing a donkey to use as bait for a lion hunt planned the next day. We told him that we were looking for white sedimentary strata, and he suggested we take a track to the north around the volcano, where he had seen some "limestone-looking" rocks. As we were driving off, he urged us to be careful because "there are lions over there."

We took the track north but saw no diatomite or limestone. After a short while we entered a gorgeous rolling savanna with waist-high grass amid patches of rock teeming with animals and birdlife. There were oryx, Soemmering's gazelles, greater kudu, gerenuks, anubis baboons, secretary birds, bustards, ground hornbills, sacred ibis, guineafowl, bee-eaters, and the carrion-eating marabou storks. By then it was late afternoon, so we camped on a small knoll that offered a spectacular view of the open plains before us and the volcano to our backs. This was to be our last night before heading back to Addis Ababa, and Johanson and I had already decided that we were going to end it with a cultural experience.

Earlier in the day, each of us had paid the equivalent of $3.00 to buy a leafy bundle of the narcotic plant known as *chat (Catha edulis)* or variously as *khat, qat,* and *cat.*[44] It is sold in the form of slender twigs with fresh, glossy green leaves that are pulled off one by one and chewed.[45] The masticated pulp is spit out or (more commonly) swallowed. A native plant containing amthetamine, chat is used widely throughout the Horn of Africa and Arabia and is a multimillion-dollar cash crop. In the towns of Gewani, Millé, and Aysaita, I frequently saw Afar men chewing chat or, afterwards, sporting traces of green leaf residue around their lips or coating their teeth.

As Richard Burton was traveling around Somalia and the Hararghe plateau in the mid-1850s, he was apparently a regular user of the drug, when he was not pursuing his life's work of investigating sexual customs, or perhaps when he was. Regular users of chat describe it as a potent aphrodisiac and "boast of unsurpassed vigor and endurance," according to one writer. Others speak of it as "enlivening the imagination, clearing the ideas . . . taking the place of food" and producing "great hilarity of spirits" or an "agreeable state of wakefulness."[44] It is a socially acceptable drug, plays an important role in the culture, and is chiefly used among Muslims as a refreshment, or for its stimulating effect while working or praying. In the region enjoyment of chat is not equated with "reefer madness," although, like anything else it can have harmful effects if used in excess.

Johanson and I had tried chat several times over the previous weeks with no noticeable effect, because, we concluded, we had chewed too few leaves. This time we were serious. As soon as we had set up camp in the resplendent late afternoon and Maurice had begun busying himself with dinner, Johanson and I prepared a large kettle of sweetened tea, which we intended to sip as we chewed the bitter chat leaves, as we had observed Afar men do. That done, we got right down to the bovine business of masticating mouthful after mouthful of leaves while sipping our tea. As the sun faded behind the hills, Maurice looked on somewhat disdainfully at the evening's entertainment, while he leisurely ate his dinner and puttered around camp.

Over the course of hours, it seemed, we chewed on. As we did, we carried on an animated debate beside the fire about just how badly Maurice had destroyed his Land Rover over the last month. This included the time he flipped it on its side at Hadar, and then there was the time he broke two axles in one day. Intermittently, we punctuated this account with comments on our current mission:

"Do you feel anything from this shit?—because I sure as hell don't."

"No, hell no, I don't feel a damn thing. Do you feel anything?"

"No, I don't feel a thing, not a damn thing. Do you feel anything?—because I don't feel a thing."

This went on long after Maurice had gone to bed. Seemingly hours later, we decided to investigate the animal sounds around camp. We grabbed flashlights and walked to the perimeter of camp and began shining them into the darkness.

There were luminous eyes all around us, some *very* large.

"Jesus, we're surrounded!"

"Don't make any sudden moves or *we're dead!*"

"Look at those eyes! What do you think that is?"

"Damned if I know, but look at *these* over here. What do you think *that* is?"

How long this went on I have no idea, but we slept late the next morning, until the flies woke us up.

CHAPTER 7

Transition

Maurice deposited the expedition protocol and a map of our proposed study area with the Antiquities Administration. As he explained it to me, the administration gave out area study permits much as oil companies are given concessions for exploration. The area Maurice requested in his concession spanned most of the Awash Valley from Awash Station to Tendaho, including part of the Hararghe escarpment, the entire middle Awash, the Issa graben, the Kariyu graben, the Meshellu basin, Hadar, and areas north, east, and west of Hadar (see Map VI, page 91).[1] In all, the permit requested covered 33,000 square kilometers, an area larger than the neighboring French Territory of Afars and Issas. When Maurice later wrote me to say that the area was approved, I was not surprised, because the chief advisor to the Antiquities Administration was a Frenchman. The concession was absurdly large, but no other teams were engaged in prehistory work in the Afar, and I was delighted that we had the liberty to explore where we wanted. Not only was I pleased, but I had helped Maurice draw the boundaries of the permit area, so later, when the size of the permit became a hotly disputed issue, I had no one to thank but myself.

Maurice and Johanson flew to Nairobi, where they met Louis Leakey to get his opinion again on a sample of the fossils we had collected. In mid-June 1972, I received a letter from Don saying that Louis was "just too old and committed to really have any role in our project." Louis took them to his home in Langata outside of Nairobi for lunch and then on a tour of a nearby game park. Don said he "rattle[d] on about all sorts of things, and to have a solid, coherent conversation with him [was] most difficult." Apparently this was Louis's normal behavior, perhaps more so in his later years—always jumping from one subject to the next, just as he jumped from one project to the next. It was only a few months later that I received a letter from Maurice with a message scribbled in the margin, "P.S. Louis Leakey died in London."

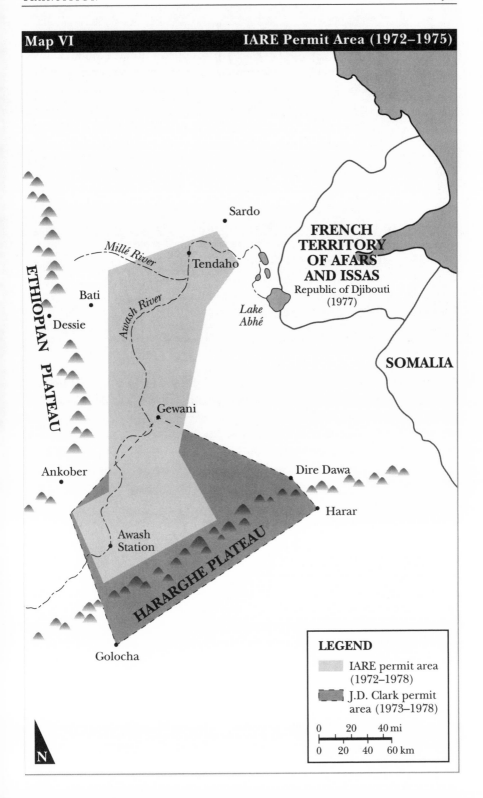

Map VI — IARE Permit Area (1972–1975)

Sardo

Millé River

Tendaho

FRENCH
TERRITORY
OF AFARS
AND ISSAS
Republic of Djibouti
(1977)

Bati

Awash River

Lake
Abhé

Dessie

ETHIOPIAN PLATEAU

SOMALIA

Gewani

Ankober

Dire Dawa

Harar

Awash
Station

HARARGHE PLATEAU

Golocha

LEGEND

IARE permit area
(1972–1978)

J.D. Clark permit
area (1973–1978)

0   20   40 mi

0   20   40   60 km

N

On October 1, 1972, at the age of 69, Louis had collapsed of a heart attack while in the apartment of Vanne Goodall, Jane Goodall's mother.[2] Vanne, a divor-cée, had been Louis's companion out-of-Africa for years and was long the subject of rumors. One was that Jane was Louis's illegitimate daughter, a story well trav-eled but untrue.[3] By all accounts Louis *was* a notorious womanizer. The most famous of his liaisons was with an artist and budding archeologist, Mary Nicol, who was his mistress while he was at Cambridge University in the early 1930s. Mary later became his second wife, following a scandalous courtship that spanned Louis's years at Cambridge and a field season together in East Africa.[4]

Louis Leakey's personal life, however, is hardly the legacy for which he will be remembered. He was "notorious" in just about everything he did, whether for carving up animals with stone tools to demonstrate Stone Age technology, running guns to Ethiopian guerrilla fighters to help them defeat the Italians, spying for the British during the Mau Mau rebellion, or sensationalizing his many discoveries and controversial theories about human evolution. The legacy that I and legions of others will remember him for, however, was his great willingness to inspire and help young people undertake research in Africa.

The rest of 1972 and early 1973 I spent in transition. The first order of business was to move my family from our apartment next to the soccer stadium to a small, rented house located off a gravel road in front of the Ministry of Mines. It was in the south-central part of Addis Ababa near Mexico Square, midway between the train station and the headquarters of the Organization of African Unity. Our land-lords, and neighbors on one side, were part of the extended aristocratic Kassa family. On the other side was an Italian family. Mr. Alberti, a road builder, had chosen to remain in Ethiopia following the Occupation.

The house, previously a bakery, was a shambles—broken windows, leaky plumbing, a porous roof—and in desperate need of paint. After several weeks of renovation, however, and with the help of a Norwegian friend who managed a paint company, it fit our modest needs perfectly. After hauling off truckloads of trash in the compound and restoring a garden that contained coffee trees, papyrus, and shoulder-high poinsettias, we even had a large, attractive yard for Justine and her friends to play in. And for security, a high fence surrounded it all.

Like nearly all house-dwellers in the city, we hired a *zabanya,* or guard, to live on the property. Eschetu lived in a thatch-roofed outbuilding with his wife and baby. A rural carpenter by trade, he kept up repairs, built Justine a swing out of eucalyptus poles, and maintained the garden, which we soon expanded to include vegetables that our two families shared. His wife, Mamitu, helped Judy with housework, cooking and keeping Justine out of mischief, while their young daugh-ter, Aselefech, and Justine became playmates. We also retained the housekeeper, Dinkinesh, whom we had employed at our apartment. She invariably brought one or more of her children to stay during the day, so almost immediately the Americans in our new household were a minority. To complete the picture, an Israeli friend gave us a puppy to raise as a watchdog, which I soon realized was a cross between a wolf and a wolf.

The only disappointment Judy had with our home was that just before we moved in, our landlord dug up the pomegranate tree planted by the back door. Also, regrettably, we no longer had the Emperor next door to rely on for the occasional cup of sugar, but I swear one of the royal monkeys followed us across the city. Soon after we moved, Justine came running into the house to tell us breathlessly that there was a gray monkey in the garden. Yes, it was wearing a silver collar. We knew the routine. We called the Jubilee Palace, and a couple of men came to pick it up. They arrived in a vintage black Cadillac and coaxed the monkey down from a tree with food. We could no longer hear the lions at night roaring across the back fence, but for entertainment there was a munitions factory across the road behind a high stone wall, where gunfire periodically erupted on ammunition-testing days. Finally, even though we could no longer hear the royal fleet warming up their engines to take the Emperor to church at ungodly hours, we were periodically treated to streams of limousines traveling on the nearby main road to sessions of the Organization of African Unity.

Haile Sellassie played a prominent role in the establishment of the OAU and was an ardent supporter of the UN. This was despite the failure of the League of Nations to come to Ethiopia's aid in 1936, when the country was overrun by several hundred thousand Italian soldiers.[5] When the Emperor addressed the League on June 30 of that year in Geneva, he told the European delegates in a defining moment, "if Europe considers the matter . . . over, it has to take into account the fate that awaits itself."[6] His words were prescient. Five years later, when he triumphantly returned to Ethiopia through the Sudan, accompanying a British–Ethiopian army,[7] Europe was exploding under the guns of the Third Reich. Following World War II, the Emperor invited the UN to establish its African headquarters in Addis Ababa, which was built on Menelik II Avenue, across the street from Jubilee Palace.

Our team was now called the International Afar Research Expedition (IARE), at my suggestion, to highlight the intended multinational character of our organization. Our plan was to resume field work in October 1973 with an expanded team of specialists, graduate students, and technicians. In the interim, I would return to the Afar in July for several months to complete the mapping of the IARE fossil localities and to explore for new sites. For financial backing, I again applied to the Fund for Overseas Research Grants and Education for a grant more substantial than the one FORGE had given me the previous year. Taieb and Coppens applied to the Centre National de la Recherche Scientifique. Johanson and geographer Karl Butzer put together a $130,000 proposal for the U.S. National Science Foundation, which if awarded was to provide me a modest salary. Regrettably, because of ongoing fighting between the Eritreans and Ethiopia, my consulting work went no further than creating summaries of known mineral occurrences in the northern Afar. The fight for independence had increased to the extent that the Ethiopian government declared a state of emergency in Eritrea and replaced the civilian governor with a general.[8]

Meanwhile, Taieb and Johanson were busy finishing up their doctoral dissertations, and I was busy writing up field notes and reports for the Geological Survey and FORGE. In August 1972, Taieb, Coppens, Johanson, and I had a short paper published by the French Academy of Sciences that summarized the geology and paleontology of the sites we had surveyed from Leadu to Hadar to Meshellu.[9] Most significantly, the paper established the priority of our scientific claim to the rich fossil deposits.

My greatest concern was to create a financially stable research base that would allow me to continue working in the country. It was apparent that the IARE had years—rather, decades—of work to do, and a base of operations was needed in Addis Ababa for our research. We also needed an equipment storage facility, which would also serve as a staging area for the expedition—one better suited than the grounds of the French embassy. The Antiquities Administration was inadequate for our purposes, because it was devoted to just that, administering research, mostly concerning the country's many ancient churches and ruins, and to managing the National Museum. Housed in the former home of the Italian viceroy during the Occupation, the museum had limited storage space for collections and was used almost exclusively for its modest exhibits. Besides our other needs, the IARE required considerable space to curate and temporarily house fossils. Most important, we needed trained field assistants and technicians. Ethiopian officials had made it clear that they would not allow us to use Kenyan technicians and fossil hunters in the Afar, as Mary Leakey had suggested, nor would it make sense to store Afar fossils in the Kenyan National Museum, where Omo fossils, were stored.

Finally, we needed to bring young Ethiopian scholars into our work. For the Geological Survey I had prepared lengthy reports and bibliographies of current, foreign-based geological research in the country—most of it conducted in the Afar and the Omo—with practically no substantive involvement of Ethiopians.[10] Haile Sellassie I University was awarding an increasing number of degrees in geology, biology, and other fields of science to undergraduates who were desperate for jobs and further training. At minimum we could provide them with practical field experience, while also opening up new research opportunities for the university.

All of these factors made it plausible to establish a modest research facility in Addis Ababa to serve as a long-term base of operations for the IARE. Operating funds could initially be pooled from the money we were all raising for our research. Taieb was against the idea, because he wanted the research based in France as much as possible. Johanson was generally supportive, mainly because he was outspokenly critical of Richard Leakey's idea of housing fossils from other African countries in Nairobi for study. The director of FORGE agreed to try to help fund the training of students, because this was more in keeping with the projects the foundation supported. Ethiopian friends and officials, as well as people in the U.S. embassy, thought that just about any proposal that attracted research dollars to the country and involved student training would be supported by the Ethiopian government.

In March 1973, two years after coming to Ethiopia, I left the Geological Survey and rented, for $125 per month, a large old house in the heart of Addis Ababa that was to become the "Afar Research Center." To make the transition, I borrowed $5000 from my bank in Houston. The house was located on Teodros Circle at the intersection of two main roads, one leading to the Mercato and the other, Churchill Avenue, running from the train station to the nearby city hall. The building was set low from the street, only its high-peaked roof visible to passers-by.

Built for an Ethiopian bishop early in the century, I was told, the house had eight rooms, the largest with a fireplace, and comprised 1200 square meters, with an entrance that opened onto a verandah and a garden. The walls were made of *chica,* a mixture of mud and straw, plastered with white lime inside and out, with wooden trim painted green and brown. Two outbuildings added another 1000 square meters of space. My plan was to use the main house for offices, two of which could double as guest quarters, and the outbuildings for equipment storage and sleeping quarters for an attendant. The combined buildings faced onto a large compound that could easily accommodate five or six vehicles. It also had its own water well, and a high fence surrounded the property.

But there were a few problems. The house had been vacant for several years, and trespassers were using the compound as a public toilet. The grounds were buried in human excrement that could be smelled for blocks around. Large portions of the mud walls had dissolved during the rains, and many of the windows in the house were broken. The roof leaked in a dozen places, and the electrical wiring looked like it had been installed by a spider. At one time the house had been used as a school, and the floors were pebbled with fossilized chewing gum. The last occupants had been French teachers from the nearby lycée and had covered an entire wall—in the room I would use as my office—with an immense red and black Picasso-esque drawing of female genitalia. When I tracked down the owner of the house and told him I wanted to rent it, he looked at me like I had been sent by God.

The first thing I did was employ a senior student from the university to help me negotiate a three-year rental contract for the building that allowed me to deduct repair costs from the rent. The student, Dejene Aneme, a Combatan from south of the capital, was studying public administration. He would serve as the administrator for the "Center" at $25 per month for a year, in fulfillment of a university requirement that all seniors work for one year performing national service at a subsistence salary. I convinced his professor that Dejene would indeed serve the country. Among Dejene's first accomplishments was to petition the city successfully to build a 2-meter fence along the east side of the property to prevent foot traffic from continuing to use our compound as a rest stop.

Next I leased a typewriter from a Mennonite nun, who ran a nearby bookstore, for 50 cents per month (for the next three years) and put Dejene to work typing reports while four workmen spent the next six weeks cleaning up the mountains of refuse in the compound and renovating the house. We made countless repairs and painted the house and annexes, including the thought-provoking anatomical mural, which required eight coats of watered-down yellow paint to cover. We built tables, shelves, cabinets, and two single beds; hired a gardener to replant the garden; and acquired a guard dog from a Kenyan friend who had raised it in the same pen with a leopard. I employed one of the workmen, Menda, a carpenter, as a full-time caretaker and sent him to a trade school to learn how to rewire the electricity in the house, which he did. Finally, the Trapp Company, whose roadwork was all but completed, sold me file cabinets, a large map table, chairs, lamps, and a big map cabinet for $200. They piled everything onto a truck at Camp Arba and kindly transported it to Addis Ababa.

When the renovation was all but complete, I was fortunate enough to recruit an American, Dennis Peak, to work with me as a cartographer. He was a student in environmental sciences at the University of California at Santa Cruz and was taking a year off to spend with his family in Addis Ababa, where his father served with the U.S. embassy. Dennis was superbly talented and worked with me at no salary on the condition that he could join me in the field. He came to play an invaluable role in the field as an aerial photo specialist for purposes of drawing site maps and mapping the positions of fossil localities.

For transportation, I bought a nearly new, long-wheel-base Land Rover for $2500 from a wealthy American who had driven it across the Sahara. By prearrangement I sold it to Johanson at cost, which he later paid for with grant money. The condition was that I had exclusive use of the vehicle when he was out of the country.

In late April 1973, the "Afar Research Center" was ready for business.

To celebrate this milestone, Judy, Justine, and I, with Dennis and three other friends, took a vacation to the upper Omo River gorge in southwestern Ethiopia near the town of Jimma in Kaffa Province. The river flows south into the Main Ethiopian Rift and then into Lake Turkana, cutting through massive layers of plateau basalts. I had the idea that the gorge would be deep enough to expose sediments underlying the basalts, possibly with fossils, that pre-dated the formation of the rift valleys, the Afar included. Of scenic interest was the prospect of hiking through a bamboo forest depicted on early British military topographic maps; and of historical interest, this was the portion of the Omo Valley that Vittorio Bottego explored before he was killed in 1897. On Italian maps dating from the early part of the twentieth century, the Omo is named the Bottego River.

It took us six hours to hike the 1300 meters to the bottom of the gorge, with a congenial porter carrying Justine on his shoulders much of the way and several others carrying our gear. We trekked through a magnificent, brilliant green bamboo forest streaked through with afternoon sunlight, and then along a trail that passed between steeply terraced cultivation plots. Farmers astounded at having foreign visitors successively shouted out excitedly, to their neighbors down the gorge, that we were coming. One festive group of small children after another followed us, jumping up on the porter carrying Justine, trying to touch her long, blond hair. At the very bottom of the gorge, before darkness descended, I found terrestrial sediments, as I had hoped. They contained enormous sections of opalized fossil trees, testifying to the existence of another forest there some 40 to 50 million years ago.

That night we slept on the rocky banks of the surging river, but sometime after midnight we were awakened by a steady rain, and complaints from our porters, that continued until dawn. After eating breakfast in a light drizzle, we packed up and spent the next eight hours climbing out of the gorge. By the time we reached the top, Justine had spent over ten hours on the shoulders of her sure-footed friend, who made the entire trek barefooted.

We were met by a throng from a nearby village, led by a man who claimed to be the grandson of the former king of the Janjero people. A small Oromo tribe conquered by King Menelik in the 1870s, the Janjero live along the steep slopes and abyssal valleys of the upper Omo and its northern tributary, the Gibe River. In a gesture befitting royalty, he invited us all to his home for a feast, after correctly anticipating that we would be exhausted and famished at the end of our climb. After we readily accepted his offer, he put Judy and Justine atop a mule and in procession we all walked to the village. Our host lived in an immense, smoke-filled *tukul*, the traditional, round, thatch-roofed home that consisted of one large partitioned room. It was filled with his extended family, servants, and a variety of animals, including goats, chickens, a milk cow, and a donkey. Off to one side was the cooking area, with several wood-burning fires beneath pots and platters of steaming vegetables, spicy meat dishes, and platters of freshly made injera. After meeting the many members of his family and washing our hands in bowls of water offered to us, we fell ravenously to the feast. The occasion was made all the more memorable by glasses of homemade beer and numerous toasts to the hospitable Janjero people.

CHAPTER 8

Famine and Fame

A t the end of June 1973, I received a telegram from the director of FORGE telling me that I had been awarded $4200, which I would receive immediately, and that the additional $2500 I requested would be given later. Over the next three months I returned to the field and surveyed more areas in the Awash and on the Hararghe escarpment. I was joined at various times by Dennis, Kelati and one Antiquities Administration representative or another, and we were always picking up guides here and there. Since our last field work together, Kelati had married a Tigrayan girl he had met at Camp Arba, who was now pregnant. It must have been something we drank in the Awash, because soon Judy was expecting as well.

We began by surveying areas along the southern escarpment that Schönfeld described as late Miocene, but we found very few fossils and none readily indicative of their age. Much of the area was wooded, and outcrops were scarce. On one occasion, as we were crossing a stream, I spotted a modern human skull at the bottom of several meters of recent river gravels, suggesting that some hapless soul had been swept into a torrent. Another time, we were sitting on a riverbank eating lunch when I realized that human teeth were scattered all around us. Of more vital interest were reports of tribal skirmishes in the area. Because of its geography at the interface between lowland and highland tribes, the escarpment lends itself to territorial disputes. In this case the cause of friction was more pervasive: A widespread drought had struck the lowlands, and the Afar, Kariyu, and Issa were pushing their herds into higher and higher elevations, seeking forage.

Moving down into the Awash Valley, we found dried-up waterholes with dead livestock swarming with vultures. Even wildlife in the nearby Awash National Park was dying.[1] Along the river we came upon recently abandoned Afar encampments littered with the bones of domestic animals, apparently killed as a last resort for food. Up and down the valley the story was the same. Freshly dug graves were present around settlements and towns, and there were reports of widespread cholera and some typhus. I later came down with typhus, and I am convinced that our guide Ali Axinum had cholera when Dennis and I picked him up in Millé. He soon started throwing up black vomit and had acute diarrhea, signs of the disease. We took him to the Trapp clinic at Camp Arba and probably once again saved his life.

The "small rains" of February and March had never come. The drought, apparently part of a larger cycle of diminished rains across the Horn of Africa, caused severe crop failure and livestock loss.[2] The highland towns of Bati and Dessie that we drove through in Wallo Province, the area hardest hit, would eventually fill with thousands of refugees. Relief workers made dire predictions of widespread food shortages. But for me and many others at the time, it was hard to grasp just how bad the situation was, because by some definitions the lowlands are always in a state of drought. Furthermore, the Ethiopian government did not seem overly concerned.

Over the months I had received a number of letters from Johanson, who in the fall of 1972 had accepted a teaching position at Case Western Reserve University while still finishing his dissertation. His letters concerned his intrigues with Coppens and Taieb, funding, and plans to visit Addis Ababa in the summer before returning to the Omo. Our expedition into the Afar with an expanded team was still planned for the fall. I was particularly interested in discussing plans for the "Center" with him and in learning the specifics of his NSF grant and what support it would provide me. I knew his proposal had been funded some months previously with a reduced budget, but Johanson had not been forthcoming about the details. First he said he was coming to Ethiopia in June, then it was early July; next he wrote to say the approval of his dissertation was held up pending major revisions, and he would not be in Addis Ababa until later in August. Twice I made the long, disruptive drive from the field to meet him, at considerable cost in time and limited grant money, only to receive word days later that he was delayed. In July I took one of his graduate students, Tom Gray, to the field at more expense, and when Johanson showed up in early September, en route to a conference in Nairobi, I was not in a very understanding mood. Nor was he, because the problems with his dissertation had killed his plans for returning to the Omo.

We had a colossal blowup.

Suddenly we seemed to disagree on just about everything—from my perspective, his secretive handling of his NSF funding, the disturbing news that Karl Butzer had resigned from the project over related issues, the Land Rover I sold him, equipment, and his relationship with Taieb and Coppens. I accused Johanson of breaking past agreements and causing divisiveness on our team, and he accused me of his own litany of abuses. Shortly he left for Nairobi. On his return several weeks later we more or less patched things up, but from then on I was no longer tolerant of his dark moods and intrigues.

In early October 1973, the International Afar Research Expedition left for the field in four Land Rovers loaded with 18 people: eight Frenchmen, Johanson, his two students, myself, Dennis Peak, an Antiquities representative, a cook, a field assistant, and two Afar guides we picked up en route. Among the French, aside from Taieb and Coppens, were Claude Guérin, a paleontologist, three field technicians, a mechanic, and a documentary filmmaker. Johanson's worst fears had come true: The expedition was "crawling with frogs," and they were going to make a film of it.

Our first stop, Geraru, was a new site Dennis and I had found in July, lying 40 kilometers south of Tendaho (see Map IV). It is only 5 kilometers square but consists of tens of meters in thickness of deeply eroded sediments containing abundant fossils, and numerous volcanic tuffs. The site is bounded on two sides by basalts, the Geraru stream crosses its southern end, and a gravel-capped plateau lies at its northern end, where we made camp. The strata are intensely chopped up by crisscrossing faults that made its stratigraphy impossible to decipher during the week we were there. Nevertheless, the paleontologists and anthropologists collected 200 fossils from 28 localities, which Dennis and I diligently mapped, as part of my responsibility to document each locality geographically. Guérin, an expert on fossil horses and rhinos, and Coppens thought that Geraru was between 2 and 3 million years old.

That first week everything went smoothly, with everyone collecting things, measuring things, mapping things, taking photographs and notes, making drawings, cataloging, and at night drinking the bottles of brandy and whiskey that people had brought with them, and smoking things.

In mid-October, we moved camp to another new site, Amado, that an Afar guide told me about. Located 45 kilometers northwest of Geraru, Amado is half the size of Geraru and lies in the upper Millé River basin, just south of a 1600-meter volcano called Gura Ale.[3] Amado is noteworthy for its thick, fluviatile, crystalline tuffs containing beautiful translucent fossil wood and hundreds of monkey and baboon fossils. Coppens and Guérin judged the elephants and rhinos to be between 3 and 4 million years old. Some of the associated sediments looked like hot spring deposits, and the fossil wood looked like palm. Together the geology and fossils suggested a proto-Millé River passing through a forested area, perhaps with adjoining hot springs, at the time when monkeys, baboons, other animals, trees, and volcanic ash were swept into the river. By analogy, there is today, at the foot of Fantale volcano in the Afar funnel, a small palm forest surrounding hot springs that drain into the Awash River. Baboons live around the springs, and monkeys inhabit the gallery forest along the river.

In five days at Amado we collected nearly 400 fossils, a quarter of them monkeys. At this point Coppens and Guérin returned to Addis Ababa and then France, where Coppens had urgent business. The rest of us moved on to Hadar, stopping en route at Camp 270 for repairs on Maurice's Land Rover. That night we ate dinner with our German friends in the camp mess and then retired to the "Amoeba Bar," overlooking the Awash, for a few beers. The bar, a small building with a wooden verandah, was built on the hill above the camp by an enterprising merchant from Millé named Ephrem. He made a killing catering to thirsty Germans and Ethiopians and later opened another bar in Bati, in addition to a bar and cafe he owned in Millé.

After a few beers, Taieb, Johanson, and a few others retired to one of the thatch huts that had sprung up around the camp. It was here that the "girlfriends" lived—a handful of prostitutes brought in from Bati to help the Germans and Ethiopian high-landers overcome the boredom of desert life. In the semidarkness before dawn, the women could be seen in their brightly colored dresses, returning to their lodgings from the camp sleeping quarters. After checking on the status of our expedition cook, Kebede, who was down with malaria, I joined my colleagues and found them in one of the huts having a grand old time drinking beer, dancing, and fooling around with the women. Just then we heard a terrible ruckus on the road. We raced out of the hut to find Joseph, one of our guides, fully drunk and sitting behind the wheel of Mau-rice's Land Rover, with the motor running. Somehow Joseph had pinched the keys, apparently from the repair shop, and had been driving around camp crazily until he was stopped by one of the Germans, who was trying to coax him out of the vehicle. Maurice instantly appraised the situation and flew into a rage, dragging Joseph out of the Land Rover and screaming at him as Joseph screamed back.

Now, Joseph was my prized linguist. I had hired him during the summer because he spoke the languages of the Hararghe escarpment: Itu Oromo, Adere, Somali, and Afar, as well as Amharic, Greek, French, English, and some German. As he explained it to me, his mother was part Itu and part Somali Issa, and his father was part Afar and part Greek from Dire Dawa, the largest town in Hararghe, where Joseph was born and where Adere is spoken. There Joseph received a secondary education and learned English and French. Later he moved from town to town on the escarpment with his father, who worked for the French-run railroad. I met Joseph in Arba, where he had been employed for several years by the Trapp Com-pany, just after they had fired him for drinking, a problem he apparently still had. Another problem: When he got drunk, Joseph could never keep his languages straight, so when he was yelling at Maurice that night he sounded like a meeting at the United Nations.

The next morning I had a heart-to-heart talk with Joseph and fired him. I liked Joseph and later saw him on occasion in Addis Ababa, where he sometimes worked as a tourist guide. From time to time he would drop by the office when he was out of work. I would hire him for the odd chore around town, and he would further my knowledge of Ethiopian ethnography, about which he knew much. The last time I saw him, he greeted me as he was coming out of the Hilton Hotel while escorting some elderly German tourists. He had picked up a sport coat somewhere that was about five sizes too big for him. Joseph may have been imperfect, but the man was unique. And the disruption he caused that night at Camp 270 probably saved some of my colleagues from catching the clap.

We stayed at Hadar for nearly two months; expedition members and visitors came and went. Our sprawling camp with my large open work tent in the center overlooked one of the widely looping meanders of the Awash. The view north was of the Hadar badlands; to the south we overlooked the steady, brown waters of the river, its sentinel forest, and the distant hills in the greater Meshellu basin. I thought Hadar was beauty itself, with all of its mysteries and promises of discovery at arm's length. It was not really a question of "making a discovery" at Hadar; rather it was a matter of which one you chose to record in your notebook, or in the fossil catalog, or on a map. There were also the smells of the Awash to take notice of, the chill left over from nightfall, and the scarlet sunrise creeping its way across the eastern sky at dawn.

I was never happy about our scientific method on this expedition—people running around collecting fossils, carting them off to camp, and then moving on to the next locality to collect more fossils. All the while, Dennis and I were playing catch-up to map the geographical and stratigraphical position of fossil localities. Sloppy stuff, and Geraru was the worst. With the multiple faults there, we scarcely had a clue as to where we were stratigraphically from one locality to the next. Ideally, the stratigraphy should be worked out *before* fossils are collected, making it possible to pay greater attention to geological context. I was determined to make stratigraphy my first priority at Hadar. Maurice and I were supposed to do this together, but he had a problem delegating authority and was always bouncing around attending to vehicles and supplies and keeping his field assistants busy.

Hadar is cut by a modest number of faults, by Afar standards, and is not complicated in this respect. Nor, unlike Amado, does it have the complex time breaks or unconformities; that is, major pieces of the geological section are not eroded away, except in the uppermost levels. Hadar is what geologists call "layer-cake geology." The strata are nearly flat-lying, and distinctive marker beds used for correlation purposes—tuffs, sandstones, shell beds, and colored clays—can be traced over wide distances. Using a 2-meter piece of string with knots tied in it every 10 centimeters, a measuring stick, a compass and level, and an army shovel, it took me two weeks to determine that there were approximately 160 meters in thickness of exposed strata in the central Hadar area.

By the end of the field season, Dennis and I had mapped the location and stratigraphical position of all 79 fossil-collecting localities that had been established. Fossils recovered from Locality 120, for instance, were eroding from a thick sand layer 9 meters below a distinctive green clay in the northern part of the site; fossils from Locality 131 were eroding from sands directly above a gray clay in the southern part of the site; and so on. All told, we measured and described 51 geological sections that we correlated from one place to the other using 17 marker beds.[4] Because Maurice returned to France midway through the field season to complete his dissertation, the description of the Hadar stratigraphy contained in his dissertation was far from complete.[5] When he finally published the stratigraphy of Hadar three years later, it was nearly identical to the version I had given him at the end of the 1973 field season.[6] At the time, such differences in our work product was not an issue: I understood the limitations on Maurice's time and, especially, the continuing distractions caused by Johanson.

Don was an unending source of divisiveness and tension in camp, except at night when he was stoned. Then he would relax by sitting around the work tent spewing out his latest gossip or dirt about researchers far and near. Johanson was unsurpassed at saying something in an offhand manner, with just the precise amount of innuendo. On one occasion, he told an elaborate story about how a Belgian geologist working in the Omo with the American team mysteriously lost his field notebook, "just at the time that *Coppens* was visiting camp. . . ." More than once his stories would gravitate to Richard Leakey. Johanson was beside himself over Richard's latest headline-making find at Lake Turkana, the famous "1470 skull" of an early *Homo,* announced in November of 1972.[7] Named after its catalog number, the skull helped confirm one of Louis Leakey's career doctrines, the great antiquity of the genus *Homo,* originally reported from the lowest levels of Olduvai. Johanson reveled in Richard's subsequent problems in dating 1470, which was first dated radiometrically at 2.6 million years but was later proved to be 700,000 years younger.[8]

Looking back for insights that might explain Johanson's particular disdain for Coppens and Taieb, I recall one early episode that I thought was especially telling. On occasion at night, Johanson would bring out the novel he was reading, which that season was *The Stranger* by Albert Camus. One evening he asked Coppens to explain a particularly complex passage to him.

"Ah, but you must be *French* to understand that book," Coppens said.

Later, when Johanson asked Taieb about the same passage, he replied, "Ah, but you must be *Algerian* to understand that book."

Some days went by, and again and again he would ask Coppens or Taieb questions about the book, and always he would receive the same answer. "Ah, but you must be. . . ."

I could sense Johanson's frustration and anger. Then one morning, just before daybreak, I was crawling out of my mosquito net when I heard two people near the cook's tent engaged in heated discussion. It was too dark to see, so I crept up to listen. I was taken aback to recognize one of the voices as Johanson's, because he was seldom up before breakfast. "You *must* explain that passage to me," he was saying, "or you will be sorry, *very* sorry."

"But Mr. Don," the other voice replied, "I haven't read the book. I don't even *read!*" *What the hell?* I realized he was talking to Kebede, the cook! I was shocked.

"You're lying, I know you can read! I've seen you reading grocery lists," Johanson hissed.

"No, Mr. Don, please, another time, *please.* I'm begging you, I must fix breakfast for the camp."

"I warned you Kebede, I warned you!"

Realizing that something terrible was about to happen, I leaped from behind the tent just in time to see Johanson spit into the pancake batter.

"No, Mr. Don—no, no, you'll poison the entire camp!" Kebede screamed.

At that moment I awoke covered in sweat. God, I had been having a horrible nightmare. But just to make sure, I got up to check on Kebede. Sure enough, there he was, just getting out of his tent to start preparing breakfast.

At the end of October, Maurice and I loaded two vehicles with colleagues and crew and left to survey areas west of camp along the Gona and Busidima Rivers, leaving Johanson alone at Hadar for the first time, with his two students. From the Busidima, Maurice and I made our way north to the Bati-Assab road and then drove up to Bati on the escarpment. There I called Judy to check on the status of two more expedition participants who would be joining us: Maurice's colleague, paleobotanist Raymonde Bonnefille, and Gudrun Corvinus, the German archeologist Maurice had recruited at the 1971 Pan-African Congress. Both had arrived in Addis Ababa on schedule and were staying with Judy. Over the telephone, we made arrangements for Maurice to meet Gudrun and Raymonde at Camp 270 in two days. The next morning, October 30, Maurice headed for the rendezvous, while Dennis and I returned to Hadar.

According to an article he later published in *Nature,* Johanson found the expedition's first hominid fossils that same day, discovering four partial leg bones: two joining at the knee and two upper femurs, or thighbones.[9] He said the finds were made at two adjoining localities, L128 and L129, just west of camp along the drainages of a small stream called Denen Dora, near the base of the Hadar section, thus making the fossils among the oldest at the site. The associated fauna at that level, we then estimated, would make the hominid bones at least 3 million years old, or middle Pliocene. Most significantly, Johanson concluded from their morphology that the hominid had walked upright; given their age, the leg bones thus represented the oldest known evidence of human bipedality. From their similar size and proximity to one another, he guessed that the leg bones came from a single individual. It was an important discovery and cause for great celebration—except that Johanson made no mention of the find to Dennis or me when we returned to camp that afternoon.

Instead of returning to Denen Dora the next morning to renew surveys of L128 and L129, Johanson took his student, Tom Gray, Dennis, and a few others to renew surveys of a different locality, L116. This was the single richest fossil locality yet found at Hadar, one located in the higher (and therefore younger) part of the stratigraphical section at the north end of Hadar. I had found the locality before leaving for Bati and had shown it to Johanson as a potential dissertation project for Gray in taphonomy, the study of processes that affect the preservation of animal remains, then a hot topic in East African research. It was late in the afternoon of October 31, after the group had moved on from L116, that Johanson told Dennis and me in camp that he had discovered hominid fossils at Denen Dora. He showed us the leg bones, which he retrieved from a small box. At the time, we assumed that he had just found the fossils, and we thought his subdued behavior was odd for someone who had just made a great discovery. He would allow no photographs of the specimens until he had "confirmed they were hominid in London or Nairobi" following the field season, and he even raised the possibility that the fossils "might be baboon." Later, when I looked at the expedition catalog, I found that Johanson had entered the fossils simply as "primate"—not "hominid." Our standing procedure was to catalog fossils to the nearest taxonomic level to which they could be identified. Yet in *Lucy*, Johanson's book about the Hadar discoveries, he said that after picking up the specimens and examining them, he immediately recognized that they were hominid.[10]

That evening, because it was Halloween, we celebrated the discovery by making popcorn. I was not sure whether Johanson's find was a "trick" or a "treat."

In *Lucy*, the story continues the next day, when Taieb returned from Camp 270 with Bonnefille and Corvinus. Late that afternoon Johanson led Gray to an Afar stone grave, where they removed a leg bone (a femur) to compare with the fossil leg bones, to be "absolutely sure" that the Denen Dora fossils were hominid.[11] Later, apparently in the secrecy of his tent, Johanson compared the specimens, concluding that this was all the confirmation he needed to ensure the hominid status of the fossils: The Afar leg and the corresponding fossils leg bones were "virtually identical." Whether Taieb was aware of the nocturnal grave robbing I do not know—I seriously doubt it—but until the end of the expedition, I believed that confirmation of the hominid identity of the fossils was to await comparative studies in London or Nairobi.

In retrospect, I gave Johanson the benefit of the doubt in many of his actions, as regards scientific matters, but on this occasion, in the context of his overall behavior, I suspected there was something rotten in Denmark.

The arrival of Corvinus and Bonnefille gave Johanson new opportunities to exercise his leadership skills. Whereas he chose to regard Corvinus as incompetent, he paid special attention to Bonnefille, with whom he still had a personal relationship from their days together in Omo.

I thought Corvinus was very competent, both as a field archeologist and as a geologist, for which she was trained in the rigorous German tradition of prehistory studies. During two field seasons with the expedition, she documented dozens of stone tool localities in the greater Hadar area.[12] Her surveys west of Hadar during the next field season, in the upper Gona River area, resulted in the discovery of artifact-bearing deposits that have since yielded the oldest human artifacts known, 2.5 to 2.6 million years old.[13] In *Lucy*, Johanson credited the initial discoveries at Gona to others, including himself, failing to recognize Corvinus's pioneering work there, or at Hadar, which she described in two papers published in *Nature*.[14]

Johanson's particular complaint about Corvinus was her interest in the Acheulean bifaces found on the slopes of the much older deposits. The tools were characteristic of the middle Pleistocene type, about 400,000 to 700,00 years old. He regarded them as "junk." This was not an altogether unjustified criticism, at first glance, because the stone tools were worn and patinated and were found on the surface. But what Johanson did not appreciate was that the stone tools were not quite *that* worn.

After Corvinus arrived at Hadar, she began carefully to search the higher levels of the site, where she thought the bifaces might have originated. In particular, she looked for their source along the contact of the plateau gravels with the underlying older Hadar strata. Soon she began finding large numbers of handaxes in even better condition than the original finds, although they were still found on the surface. Interestingly, they were not in the higher Hadar levels but at lower elevations closer to the Awash. Although she did not find their source during that field season, the next year (1974) she hit the jackpot.[15] Of all places, it was in the Denen Dora hills, the same general area as Johanson's hominid locality. She found fresh, sharp handaxes and associated flakes and chips raining down from a sand layer within a 10-meter wedge of stream deposits. The sequence unconformably overlay the fossil-rich Pliocene beds and was unconformably overlain by the plateau gravels. The wedge thinned and disappeared toward higher elevations, which explained why Acheulean tools were rare in the upper Hadar levels. Corvinus described the artifacts as probably representing a stone tool factory site situated on the edge of a floodplain. The river had carved into the Pliocene beds during the middle Pleistocene, depositing a fluvial terrace; in the later Pleistocene, rivers fed by intermittent highland storms eroded any overlying strata, leaving the wedge of remnant sediments buried under the gravels.

The contempt that Johanson showed for Corvinus's work on the Acheulean was mild compared with his views on another set of stone tools she found at Denen Dora. These came from *on top of* and from *within* the plateau gravels, which she also described in her *Nature* papers. She identified these "plateau artifacts" as protobifaces, modified pebbles, choppers, and crude flakes, which she characterized as a "degenerate biface-flake industry."[12] The term *degenerate* referred to stone tools that were poorly manufactured compared with the more sophisticated Acheulean tools originating beneath the gravels. The quality of toolmaking thus appeared to decline from the lower to the upper levels: Well-crafted handaxes and refined flakes beneath the gravels became "proto-handaxes" and "crudely made flakes" within and above the gravels. By analogy, Corvinus compared these disparate technologies with those described by Mary Leakey from the middle and upper beds of Olduvai. There, Acheulean bifaces lay at the same level as stone tools of a more primitive tradition called Developed Oldowan (see Appendix I on page 319). But at Denen Dora, Corvinus described an Acheulean industry that was overlain by a "pre-Acheulean industry" made by "a somewhat later group of people."[12]

Could this make sense? Yes and no.

Corvinus was simply saying, although in confusing terms, that the later group of stone toolmakers put less effort into their craft than had the earlier group. It is also likely that at least some of the more worn primitive stone tools in the Denen Dora gravels originated from older (genuinely pre-Acheulean) deposits swept up by streams in the later Pleistocene. Pliocene tools from the higher levels of Gona, for instance, could have been eroded and transported by Pleistocene streams to lower levels at Hadar and elsewhere and redeposited with the younger stream gravels.

Overall, Corvinus's work was far more right than wrong. And, as we shall see, she laid the groundwork for archeological discoveries in the Afar of an epic nature. But these were made in 1975, the year after the forces of evil caused her to abandon her pioneering work at Hadar. The last I heard, she was investigating Acheulean sites in the diamond fields of South Africa, but that was years ago.

During the first few weeks of November, Dennis and I continued work on the stratigraphy of Hadar. At one point I took Bonnefille across the Awash to see the stratigraphy along the Meshellu stream. She collected samples of a thin layer of lignite, a very low grade of coal, that I hoped would correlate with a similar layer on the north bank of the Awash, thus linking the stratigraphy on both sides of the river. Her specialty was the study of fossil pollen, called palynology. She was interested in the lignite for the abundance of pollens these deposits often preserve, which are invaluable for identifying paleofloras and reconstructing past environments.

From laboratory studies, Bonnefille later determined that the lignite and other pollen-bearing sediments from the lower Hadar beds were deposited in a treeless marsh lying next to a shallow lake or delta.[15] By comparing the paleoflora with modern flora, Bonnefille concluded that during earlier Hadar times, some 2.9 to 3.3 million years ago, the marsh environment was more like that of the wetter, montane conditions of the present-day highlands than that of the lowlands. By contrast, the conditions during the middle Hadar period, about 2.5 to 2.8 million years ago, were comparable to the arid grasslands or subdesert steppe present today in the southern Afar. She postulated that the shift toward aridity during the later Hadar times accompanied tectonic lowering of the area by as much as a kilometer. A curious find came from a clay layer just below the Pliocene artifact levels discovered in the adjacent Gona area. There, Bonnefille had found fossil pollens from herbs and small shrubs similar to those that grow at *higher* altitudes, like those found today on the shady floors of bamboo forests in the uppermost Omo Valley. She was unable to explain such plants at Gona. Perhaps the hospitable Janjero people we visited in the upper Omo Valley, who live among bamboo forests, could tell us something about the first toolmakers that the rest of us could not know.

In mid-November, Maurice returned to France with his assistants, leaving a leadership vacuum, which Johanson sought to fill. I would tell the cook one thing;

Johanson would tell the cook something else. I would instruct the camp attendants to do something; Johanson would tell them to do something else. No big deal. Concerning the Afar, however, his idea was to turn them into "hominid hunters," like the Kamba tribesmen trained at Olduvai by Mary Leakey. For incentive, Johanson wanted to institute a reward system like that he said Richard Leakey used at Lake Turkana—so much money for a tooth, more for a mandible, a bonus for a skull. The problem was that Johanson did not speak the Afar language, and he had little knowledge of their culture, customs, or value system. Another problem was that the Afar guards who were supposed to be resting during the day were now spending their time following Johanson around looking for fossils and were sleeping at night.

At the end of November thieves crept into our camp late in the night and made off with clothes, field gear, notebooks, a camera, and a medical kit and supplies belonging to a German doctor then visiting us. One of the field notebooks that was stolen belonged to Corvinus; the other belonged to Johanson. Soon he began making remarks like "Why wasn't *Kalb's* notebook stolen?"

That same morning the doctor, Jürgen Knoblauch, who ran the Trapp clinic at Arba, was summoned to an Afar encampment at Hourda, just east of Hadar, to treat one of our attendants who had returned there gravely ill. Most of Jürgen's patients at Arba were Afar, and he was very familiar with their medical problems. They went to his clinic from miles around for treatment of everything from tropical ulcers to cholera. As a means of ensuring that his Afar patients would return to the clinic for follow-up treatments or medicine when needed, he required that the men leave their rifles or spears at the clinic as a reminder. As a result, every corner of the clinic was stacked with weapons, which added up to one of the more impressive arsenals in the Awash Valley.

I went with Jürgen to Hourda, where we found Mohammed, probably about 19 or 20 years old, lying on a pile of tamarisk boughs in a small clearing in the forest next to the Awash, with the morning sunlight streaking through the trees. Several men were in attendance. Mohammed's gums and fingernails were white. Jürgen's diagnosis was severe anemia, probably caused by chronic malaria and requiring immediate and massive doses of iron, a treatment not available in the camp's medical supplies.

Mohammed died two days later. That same afternoon I went back to Hourda and watched from a distance as four Afar men buried him along the forest fringe among other graves. They were covered with dust from the digging, which stuck to their bodies as they perspired in the afternoon sun. After putting him in the ground, they stacked large stones over the grave in a circular mound, as was their custom. Just as they finished a breeze picked up, causing eddies of wind to carry off some of the loose soil around the grave.

I spent much of that week trying to run down the thieves. Using ten or so Afar from the Hourda camp, we followed their sandal prints—two sets—north to the Ledi River, then east, and then south back toward Hourda, where we lost them on the plateau gravels. Close enough: The thieves had come from the same camp as the trackers, which did not surprise me, because they were the nearest Afar around. I then spent two days locating the Afar chief for the area, Melo Seco, who happened to be in Bati. After he agreed to help, I took him back to Hadar, and from there we walked to Hourda in the late afternoon, accompanied by Dennis and our Antiquities representative, Alemayehu Asfaw. There seemed to be about 75 people in the Afar camp. It was tucked into the gallery forest where the nomads kept their livestock at night, grazing them during the day in isolated meadows up and down the river. It was a beautiful spot with the oval huts of the Afar scattered beneath the trees.

The men were waiting for us when we got there. There were about 20 of them standing in a semicircle in a clearing. Each came up in turn and kissed the hand of Melo Seco. The chief talked to them for about an hour as the sun fell below the trees. While he did so, I focused my attention on five or six Afar draped in cloths lying along the edges of the clearing. Occasionally someone would walk up to one of them and say a few words or offer a sip of water. As far as I could guess, they were ill or dying. Surely they were not starving to death from the famine before our very eyes?

Melo Seco spent the night at Hourda, and Dennis, Alemayehu, and I walked the 5 kilometers back to Hadar in the dark. Early the next morning, the chief walked into our camp with one of the thieves and another Afar carrying a few small items that had been taken. The accused was a strong, handsome young man, probably in his late teens, whom we took to Bati and put in jail. We were told that his accomplice was long gone, and we never recovered the rest of the items, including Jürgen's medical supplies. The police in Bati said they would hold the young Afar in jail for a month, which was a month too long as far as I was concerned. The incident never should have happened.

On December 9, 1973, the day before we packed up camp and left Hadar, Dennis and I tied the last fossil locality, L174, into the Hadar stratigraphical section in the moonlight. We were at the northern end of the basin, racing to finish up our work during the final moments of daylight, when I found what I was looking for: an aqua-colored clay that I had used as a marker bed in the upper part of the section. As we were tracing out the clay to the fossil locality and starting to take a few measurements marking our position in the section, I dropped my prized Brunton compass and watched it tumble down a steep embankment. Like a fool, I immediately charged after it, half falling and rolling down the hill—and in the process burying the compass in the loose sediment that cascaded down with me. Dennis joined me and we searched and searched, digging through the entire hillside until it was fully dark. Finally, we sat down covered in sweat and dirt and laughed. I had waited until the last goddamned *night* of the field season to lose my compass. We rested a while and drained the last water in our canteens, while listening to the ubiquitous *wooo-uph* of hyenas that comes with nightfall. Shortly, we realized that it was getting lighter: The moon was rising in the eastern sky. After a little more searching, we clambered back up the hill and finished our work on the fossil locality in the lunar glow.

A few days after returning to Addis Ababa, I picked up a copy of the Sunday edition of the *Ethiopian Herald* and read,

3-Million-Year-Old Human Fossils Found in Wallo Governorate[16]

So they were hominid after all. Taieb and Coppens read about the discovery in the Paris newspapers. Johanson had given the press conference he had so desperately wanted the previous day at the Antiquities Administration. I had actually believed that he would wait until he compared the Hadar fossils with those in Nairobi or London, as he had said he would. In *Lucy,* Johanson said that he had no choice but to give the press conference, because by the terms of our agreement with the Ethiopian government, "any fossil deemed important enough to be taken out of the country had to be described at a press conference before removal."[17]

Hardly.

There was a requirement that press releases be first submitted to the Ethiopian press through the Antiquities Administration, not through some other organization in the country. The Ethiopians were as aware as anyone that press announcements of scientific findings prior to confirmation of their significance would be nonsensical. They also understood that more often than not such confirmations would have to be made at museums or laboratories outside the country. Announcements to the media could then be made anywhere—Cleveland, Nairobi, or Beirut—as long as they were *first* announced to the Ethiopian press through the Antiquities Administration. In short, the Ethiopians were not imbeciles.

When Johanson's two students conversed with Ethiopian officials, they always referred to him as "Professor Johanson," a title used in the *Ethiopian Herald* article. For someone who had received his Ph.D. only a few months before, Johanson was rapidly moving up in the world. The *Herald* also referred to him as "D. Carl Johanson," which is how he then liked to be referred to, apparently to emulate other notables, such as L. Ron Hubbard, J. Edgar Hoover, and his mentor F. Clark Howell. Evidently, Johanson thought this would further distinguish his rising career. *Professor D. Carl Johanson.* A problem, of course, was that when friends of F. Clark Howell referred to Clark as "Clark," people knew whom they were talking about, but when you referred to Don Johanson as "Carl," people were confused.

But D. Carl had a plan to fix this.

CHAPTER 9
Storm Clouds

On October 18, 1973, 12 million viewers saw a report on British television that showed terrible scenes of starvation from the famine in Wallo Province, where we were then working.[1] The film, called "The Hidden Famine," was produced by English journalist Jonathan Dimbleby, using footage taken only a few weeks before. Over the previous months, I had witnessed nothing like the scenes reported in the film, perhaps because we worked in remote lowland areas. Dimbleby had visited villages and relief camps on the margin of the Ethiopian plateau around the town of Dessie, just west of Bati. Farmers from the western escarpment and the Afar nomads had moved into higher elevations seeking relief, as had the impoverished tribes along the Hararghe escarpment.

Historically, famines have not been infrequent in Ethiopia, occurring on the order of every 10 to 15 years,[2] but what turned this one into an international scandal was the refusal of the Haile Sellassie government to acknowledge its existence.[3] Large numbers of people were dying or in extreme peril. Dimbleby estimated that 100 people a day were dying from the famine and that more than 100,000 people had already perished. Relief workers later estimated that as many as 300,000 people died in 1973 and 1974.[3] One report from Great Britain referred to Ethiopia as "the land of death."

Addis Ababa's denial of the disaster and its initial refusal to allow relief organizations to mobilize brought immediate condemnation to the Emperor and his government. The cover-up proved to be just the catalyst that dissidents needed to show the fading regime's inability to cope with the country's enormous problems, which were exacerbated by a sharp rise in international oil prices brought on by the Middle East War. Ethiopia's problems, however, were not limited to the famine and the global oil market but were deeply rooted in the country's history.

When Haile Sellassie was crowned emperor in 1930, he inherited an empire welded together by force and and governed via feudalism. Beginning in the mid-nineteenth century, successive rulers expanded the empire to the west, south, and east into areas previously ruled by Oromo, Somali, and Afar.[4] Emperors Tewodros (1855–1868), Johannes (1872–1889), and Menelik II (1889–1913)—the founders of modern Ethiopia—all conquered new kingdoms, and each earned the title of *Negus negast,* King of Kings.[5] By far the greatest expansion came under Menelik, renowned for his wily statesmanship, obsessive acquisition of modern weapons, and brutality. During the African colonization period, Ethiopia under Menelik pursued its own scramble for territory by bludgeoning its weaker neighbors into submission. In 1889 Italy entered the picture by expanding its foothold from the ports of Assab and Massawa to include all of Eritrea. The attempt by the Italians to expand the colony further in 1895, which was crushed by Menelik at the battle of Adwa, was a defeat that Italy vowed to avenge. The opportunity came in 1936 with the invasion of Ethiopia, six years after Haile Sellassie assumed the throne. During the five-year Occupation that followed, Italian geologists combed the country for mineral resources, while agriculturists made preparations to convert Ethiopia into the "granary of Italy."[6] Mussolini's forces never fully controlled the country, however, and in 1941 lost it all to Ethiopian insurgents and a British-led liberation army.[7]

Sellassie's triumphant return to Addis Ababa from self-imposed exile made him a hero in the eyes of many. Ethiopia was the first country to battle the Axis powers and the first to emerge victorious. After the war, the Emperor's international reputation grew as he modernized the government, expanded its communications system, built hospitals, established an airline, and enlarged the army—one that contributed to United Nations efforts in the Congo and Korea.[8] He also greatly expanded Ethiopia's educational system, which included establishment of the country's first university. The country's economy, based largely on the export of coffee, also prospered.

Despite this progress, by most standards Ethiopia remained one of the poorest countries in the world, with a feudal system of agriculture comprising millions of peasants. In the late 1950s, the country's postwar honeymoon ended as its fragile economy suffered from lowered coffee prices, closure of the Suez Canal, and severe drought in the eastern provinces.[9] These factors and widespread dissatisfaction with the country's oligarchy led in 1960 to an attempted coup, which failed and was followed by bloody reprisals. The 1960s and early 1970s witnessed continued social unrest, economic problems, and deterioration of the power of the aging Emperor, who turned 80 in July 1973.[10] Increasingly, the Emperor relied on the military and his internal security apparatus to control the growing opposition to his regime. I felt a sence of these conflicts in 1972 when a bright young geologist and friend from Eritrea, Temesgan Hailu, an outspoken critic of the government, was thrown off the top of an office building in Mescal Square by security agents. Official sources called his death a suicide. That same year a group of seven young dissidents attempted to hijack an Ethiopian Airlines jet bound for Europe but were overwhelmed by six security guards in a frantic midair shoot-out.[11] All the hijackers were killed, including a young woman who reportedly was riddled with bullets while throwing herself across a wounded male comrade. The incident was reported to be the fifth attempted hijacking of an Ethiopian Airlines plane in three years. When news of the failed hijacking was reported on the radio, the wailing and screaming from friends and sympathizers of the slain dissidents echoed across the courtyard of the Ministry of Mines. After the traditional three-day mourning period, many reappeared at the Ministry with shaved heads, as is Ethiopian custom.

Many of the country's problems during this time were tied to its bloody history of territorial expansion. These problems were greatly heightened by its most recent acquisition, Eritrea, a land that includes the northern highlands, the Danakil Alps, the northernmost Afar Depression, and the coastal lowlands. Much of Eritrea is potentially mineral-rich, and the Red Sea ports are critical to the highlands for trade. Following World War II, Ethiopia agreed to a UN mandate making Eritrea an autonomous federation, but in 1962 it revoked the agreement, arguing that Eritrea was part of its centuries-old empire. While it was an Italian colony, however, Eritrea's Christian and Muslim populations had increasingly adopted their own national identity, and they were determined to remain independent. The conflicting perspectives of the Ethiopians and the Eritreans soon erupted into open conflict, followed by a full-scale war of independence that ultimately claimed the lives of an estimated 60,000 Eritreans and at least several hundred thousand Ethiopians.[12]

When I went to Ethiopia in 1971, the Eritrean rebellion was in its tenth year, and Ethiopia's resources were fast eroding as its Arab neighbors aided the Eritrean freedom fighters.[13] Insurgencies had developed in other parts of the country as well. Somalis in the Ogaden region, fired by nationalism from across the border, were attempting to reclaim ancestral lands, and Oromo in the southern provinces of Bale and Sidamo were rebelling over increased taxes.[10] By late 1973, Ethiopia was in grave trouble, confronted with widespread rebellion, internal dissension, economic peril, and famine. Because of Haile Sellassie's age, more and more people were asking, "What's going to happen when the Emperor dies?" There was no heir apparent, because the crown prince, Asfa Wossen, had suffered a massive stroke in early 1973.[14]

At a time when Ethiopia desperately needed friends, its most important ally, the United States was turning away despite the Emperor's long-standing popularity there.[9] In 1936 Haile Sellassie had been named *Time* magazine's Man of the Year for standing up to Mussolini.[15] At the end of World War II, he had met with President Roosevelt in Egypt to chart Ethiopia's foreign policy as the dying president was on his way home from Yalta.[16] In 1954 he was warmly received in the United States, where he gave a speech before Congress, and in 1963 he was a solemn and distinguished figure at the funeral of President Kennedy.[17] Haile Sellassie gained added prestige in the United States and elsewhere in 1971 when he mediated a peace agreement in Sudan's civil war.[18]

Throughout these years, the United States had become Ethiopia's greatest ally and had supported its claims to Eritrea and the Somali-populated Ogaden.[19] In exchange, the United States maintained an important military communication base in the Eritrean capital of Asmara throughout the Cold War and used the port of Massawa for its navy. The United States also became Ethiopia's greatest market for coffee, and American oil companies were given generous concessions for petroleum exploration in the Ogaden and Eritrea. During 1950s and 1960s, the United States was Ethiopia's chief source of military assistance and was deeply involved in the country's economic development.

In May 1973, however, Haile Sellassie was unsuccessful when he went to Washington to plead with President Nixon for increased military aid to hold his empire together. By then the military base in Eritrea was no longer needed, and the Emperor's *ancien régime* had become a geopolitical liability.[20] There was some irony to the meeting between the two leaders. The U.S. president was fighting for his political life trying to cover up Watergate, and the Emperor would soon be doing the same for covering up the famine. Both leaders were desperate to find solutions to wars—Vietnam and Eritrea—that their countries would lose, and both would be forced to leave office in disgrace.

By late 1973, Haile Sellassie surely sensed that the end of his reign was near. To add to his troubles, that December, Ethiopia's army suffered a devastating defeat by the Eritreans.[21] It was said that the Emperor had become senile. When I had met him two years earlier at a palace reception for participants of the Pan-African Congress for Prehistory, he had looked every bit as regal and imposing as his reputation. As I approached him in the reception line, he looked straight into my eyes, and there was a message there, however fleeting. It was one of dead seriousness: Be a guest in my country if you wish, but remember, I will expect your allegiance.

Sitting beside the Emperor was Louis Leakey. With his famous shock of white hair and crooked tie, he was red-faced and perspiring heavily. After we were all served champagne, I stood off to one side and watched them converse. They both spoke French. Haile Sellassie had been educated by French Jesuits in Harar. Louis, with a terrible accent, could be heard across the ornate reception hall, while the Emperor spoke in a barely audible voice. I guessed that Louis was having difficulty hearing, as he was leaning forward on his cane. This was an open audience, and we were all encouraged to speak with the Emperor. Few ventured to do so. I wanted to talk with him about the need for a prehistory research facility in Ethiopia but decided that the opportunity would come again. It did not.

Even though the Emperor was ten years older, he was to outlive Louis by three years. They say that when Haile Sellassie died a few years later, it was not a natural death. The tumultuous end had come, and the enemies of his 44-year reign, the longest ever for an Ethiopian monarch, had caught up with him.

At the end of the 1973 field season, and a few days after Johanson's press conference at the Antiquities Administration in late December, I left Ethiopia with my
family for a vacation in the United States, where I planned to devote much of my
time to fund raising. By then Johanson and I were fully estranged. Toward the end
of the field season, we were barely on speaking terms. As a parting shot, he told
officials with the Antiquities Administration that I would not be receiving continued support from FORGE or anyone else. This led the Administration to delay
sponsoring my re-entry visa until I provided proof of FORGE's support, which I
did, but it required a lengthy exchange of telegrams and letters.

En route to the United States, I stopped with my family at the new De Gaulle Airport in Paris, where we checked into an airport hotel overnight. There I met Taieb
and Coppens, who had come to the airport for a late-night meeting. I briefed them on
the progress of the expedition during the month following Maurice's return to
France, concentrating on Johanson's performance. As Maurice chain-smoked, we
reviewed Johanson's handling of the hominid discovery step by step, leading up to the
preemptive press announcement. In absolute terms, I voted to have him kicked off the
expedition. I felt that if Johanson was not dealt with decisively then, we risked seeing
the expedition divided into separate camps, as had occurred in the Omo Valley. The
previous year he had frequently talked about such a division between the French and
Americans on either side of the Awash, and I knew that he had met secretly with
Ethiopian officials toward this end. Taieb and Coppens preferred to adopt a wait-
and-see posture with Johanson, fearing that an ugly break would jeopardize their
future relations with American researchers. They also feared that such a schism could
lead to just the kind of team division that Johanson wanted.

Taieb and Coppens's deliberation over Johanson's future did not last long. After
all, Johanson held the cards—the hominid fossils, which the Antiquities Administration allowed him to take to Cleveland for study. Shortly after our airport meeting, Johanson also visited Paris, which enabled Taieb and Coppens to stage their
own face-saving press conference, with Johanson and the hominid fossils in tow.
No doubt an American anthropologist's capitalizing on an extraordinary French
discovery, Hadar, raised eyebrows in some French circles.

In early February, Johanson held another press conference. This one was given at
the U.S. National Academy of Sciences in Washington, D.C., sponsored by the
National Science Foundation. NSF was eager to display the accomplishments of one
of its home-grown grantees. In a typical account, the *National Observer* reported,
"D. Carl Johanson found what may be the middle link in man's evolutionary chain
. . . the oldest, most complete specimen of fossil man anywhere in the world."[22]
Johanson was quoted as saying the fossils represented two species of *Australopithecus,* the leg bones of *A. africanus* (the gracile australopithecine) and—as I learned
then for the first time—a skull fragment, which he called *A. robustus.*

While in the United States, I gave a series of lectures from New York to Texas
about the work and discoveries of our expedition. I also talked with a number of
foundations, but without Johanson's cooperation my efforts to raise significant

funds were unsuccessful, although FORGE agreed to continue supporting my work. When visiting Washington, D.C., I met with the director of NSF's anthropology program, Iwao Ishino, an oddly cynical man, about the problems of the expedition, but he had helped arrange Johanson's press conference and was convinced that Johanson would add luster to his program.

On the way back to Ethiopia at the end of February, my family and I again stopped in Paris. I visited Taieb for several days at his laboratory at nearby Meudon-Bellevue, where we went over Hadar geology and, with Coppens, discussed future field plans and funding. Both felt they had no choice but to continue working with Johanson, and that was that—I was the odd man out. By then, however, I had more pressing concerns.

On February 28, 1974, French television reported that mutinous Ethiopian troops had taken control of Addis Ababa, and there were fears that the airport would be closed. Since mid-January Ethiopia had experienced more military and civil unrest, and the country seemed about to explode into a full-scale revolt. The last thing Judy and I wanted was to be stuck in Paris. For better or worse, Ethiopia was our home.

I booked the first available flight back to Addis Ababa.

───────

By several accounts, the Ethiopian revolution began with a broken water pump. Soldiers at a small garrison at Neghele in southern Ethiopia mutinied because of a shortage of drinkable water. The soldiers also complained of miserable food, poor working conditions, and low pay. They forced a personal envoy sent by the Emperor to drink the same water and eat the same food as the men. After becoming ill, the envoy, a lieutenant general, was sent back to Addis Ababa. Another revolt followed a month later, at the Debre Zeit air force base south of Addis Ababa. Then two weeks later a third took place, in Asmara, involving the entire Second Division. The succession of rebellions spread quickly to the civilian population. Students and teachers in Addis Ababa demonstrated against education policies, and taxi drivers demanded lower gasoline prices. Buses and cars were stoned and police fired on a mob, killing a student. At the end of February, Haile Sellassie's newly appointed prime minister and cabinet resigned, and concessions were made to placate the protesters.[23]

A few days after our return to Addis Ababa, on March 2, we heard what sounded like a shoot-out at the munitions factory across the street from our house. The factory was part of the Fourth Division headquarters—we later learned that several of its units had mutinied. The U.S. embassy warned all U.S. citizens to stay inside their homes. Plans were made for evacuation. At night we heard sporadic gunfire around the city.

Between March 7 and 11, the first general strike in Ethiopia's history was staged. Rumors of more mutinies spread, and panic grew within the imperial court. Over the next few months more demonstrations and strikes followed, as a newly appointed prime minister and cabinet tried desperately to still the mounting demands for reform and a new order. Plans to overhaul the tax system and land tenure were announced, along with an end to press censorship. Suddenly the newspapers were filled with articles about corruption and demands for the arrest of the enemies of the common man. My office at Teodros Circle was on the corner of two major thoroughfares, and I frequently heard the chants of demonstrators passing by. In mid-April, some 100,000 Muslims marched through this intersection, demanding equal representation in the Christian-dominated government.

The city was gripped by an odd combination of euphoria and fear, but little violence. This period and the coming months would later be called the "creeping revolution," but the pace was to quicken all too soon. Meanwhile, I doggedly pursued discussions with officials about an Afar Research Center, but with all the political uncertainty my efforts went nowhere. Then, in early May, my future with the Afar expedition looked bleak when I learned that Taieb had accepted an invitation to visit Johanson in Cleveland. Shortly thereafter, I received a letter from Taieb, written in uncharacteristically good English, advising me to return to the United States to obtain a Ph.D. I assumed he had thrown in his lot with Johanson.

More problems followed when our landlady dropped by for a friendly chat. She told Judy that, by the way, while we were out of the country she had sold our newly renovated house to the Brothers of Verona Mission. With all of the bad news and with my future in Ethiopia in jeopardy, we nevertheless moved to another house in early May. Our new home was located on a hill in a secluded neighborhood a few blocks from the office. The rent was reasonable, the house was in good repair, and a beautiful, well-kept garden, encircled by a stone wall, surrounded our compound. The former occupant was the outgoing headmaster of the nearby lycée, who, like many foreigners, was nervous about Ethiopia's future. The landlady was a lovely, stylish aristocrat who seemed to drive a different color Mercedes each time she visited us, until her last visit, when she showed up in a Renault. On one side our neighbors were Tigrayans, who owned a large, stupid dog that began barking precisely at sundown. On the other side was a French-speaking family from Cameroon, whose children became Justine's playmates. A few houses down the lane lived an English entomologist who owned an organ and traveled around the country collecting beetles. Periodically, a likable and eccentric German anthropologist, Ivo Strecker, stayed with the Englishman. Ivo had what appeared to be a long dueling scar across one cheek and liked to play classical dirges on the organ late at night with all the lights out. He studied the culture and language of a small, remote tribe in the Omo Valley. Ivo once returned from the bush with one of his Hamer subjects, whom he had outfitted with truck tire sandals and a U.S. Army jacket.

The move to our new home was timely because two weeks later, at Zauditu Hospital, Judy gave birth to our second daughter. Because she was born during the heady days of a revolution, in May, and because she had green eyes, we named her Spring. When Judy and Spring came home from the hospital the next day, they were greeted with a large banner stretched across the front of the house that Justine had colored with flowers. It read: "Welcome Home Spring. Happy Birthday."

Three weeks later I submitted an ambitious 50-page grant proposal to the National Geographic Society, requesting $57,000 for work at Hadar. This included funds for one geophysicist and two geologists with backgrounds in plate tectonics and oceanic studies particularly suited to Afar geology. I was sure that imprinted in the Hadar sedimentary record were the spreading histories of the triple junction and that plate movements hundreds of kilometers away from Hadar had affected the geometry of paleolakes and the course of paleorivers. By combining these investigations with faunal and paleoclimatic studies, and rigorous dating, I thought we could produce a unique, integrated oceanic and continental history of the region. The grant application emphasized that special precautions would be taken in 1974 to ensure that hominid finds were documented and collected according to orthodox procedures.

The proposal was an attempt to turn the work of our expedition into something more than a glorified treasure hunt for hominids. It was also foolhardy and doomed to fail. Of course the National Geographical Society contacted Johanson and Taieb about its merits. Although Taieb had agreed with the concept of the proposal in Paris, that was before his Cleveland visit, and we had had very little communication afterward. It was also before I had sent him my report on Hadar stratigraphy, revealing that his own work failed to account for fully one-third of the Hadar section, including a massive, nearby basalt flow present in the middle of the sequence.[24] I sent copies of the report to FORGE, and to the anthropology program of the National Science Foundation, because I had received partial field support at Hadar from Johanson's grant. Included in my report to NSF was my assessment of their rising star. I received no reply.

———

With an additional $2500 from FORGE, I returned to the field for the month of July. This time I spent the entire time at Geraru attempting to unravel its intensely faulted stratigraphy (see Map V on page 30). I was joined by Dennis, briefly, and by a friend of his, Liz Oswald, a senior undergraduate geology student at the University of California at Santa Cruz, who began working with us in March. We were also joined by a student in physics, Yesahak Wurku, from Addis Ababa University (until lately Haile Sellassie I University), who served as our representative from the Antiquities Administration. In addition, we had with us a cook, two camp assistants, and two Afar guides. With sadness I learned that Ali Axinum had died the previous month from an unknown illness. Twice we had saved his life, and perhaps I could have done so again had I been around.

For transportation we used the long-wheel-base Land Rover that I had sold to Johanson the previous October; our arrangement gave me use of the vehicle while he was out of the country. He was also responsible for paying the duty, which he had not done. The duty was due in one month, or the Land Rover would be impounded by customs authorities. I planned one month of field work.

Our first night at Geraru turned out to be a cram course in central Afar meteorology during the rainy season. We had set up camp on the plateau at the north end of the site, and by sundown we were sitting around drinking Ethiopian gin and enjoying a steady breeze from the east. Suddenly our two guides jumped up animatedly, and within moments we were clobbered by gale force winds that came ripping across the open desert. Our large Aruba tent was the first to go. We frantically held on to the tent poles until the entire tent and everything in it exploded around us. The nylon fabric shredded, and the aluminum poles flailed wildly as the lethal steel stakes popped out of the ground like carrots jerked from a garden plot. The winds then turned into a massive, heaving dust storm that caught the underside of our open-ended work tent and began snapping the lines one by one. I raced over, grabbed a machete, and cut the remaining lines to save the tent from being ripped in half. As darkness fell, we were pummeled by a hail storm, then by rain.

At the end of it all, only the round German tent ("designed by Rommel himself") was standing—and that was only because I had driven the Land Rover to its windward side and secured the center pole to the vehicle with ropes. After a while the rains stopped; dead stillness set in, followed by a mosquito storm that lasted until dawn.

Despite the enormous volume of water that poured into the Geraru basin that night, by mid-morning the ground was dry, and by midday we were assailed by a heat storm. I am not one of those earnest types who carries a rain gauge and a thermometer to the field (to my regret), and I think the explorer Nesbitt surely exaggerated when he claimed to have recorded a temperature of 66°C (151°F) in nearby Awsa in May 1928.[25] But I believed the people at the Tendaho Plantation two weeks later when they told me the daytime high was 54°C (129°F). At the Geraru site itself, which is nearly a closed basin, the temperature could well have been four or five degrees higher, with a heat index closer to Nesbitt's figure.

The weather pattern we experienced that first 24 hours basically repeated itself the rest of the month. Sometimes the rains were lighter, and hail was rare, but violent storms were frequent. Lightning was also part of the pattern, with fiery power surges across the blackened sky. During such times it was unclear whether the flashes came from the heavens or burst from the earth. Sometimes Liz and I would pile into the back of the Land Rover and watch the dazzling show, defying the Furies of the Universe to blow us and the Land Rover into eternity.

It was not surprising, after a few days at Geraru, to see the nomads with their animals and packed-up belongings passing by our camp to higher ground. They enthusiastically waved at us, as though to say, "Have fun, suckers!" Dennis also took off, for Kenya, leaving the rest of us to secure the camp at the end of day for the doom that would come with nightfall.

On one occasion a wind devil—a small tornedo—roared into camp at midnight, despite shrieks from our Afar guides to Allah to make the blast impotent. By this

time we had double nylon lines secured on all sides of the work tent and ropes
lashed across its top secured to thick acacia stakes driven deep into the ground. The
funnel tore into the tent like marrow sucked from a bone. I had left a pile of field
notes and drawings on the work table in the center of the tent with a heavy storage
box sitting on top. The box was dumped aside like Styrofoam and my papers were
sent rocketing into the night sky, to Liz's perverse delight. The next morning we
retrieved the papers from all over the northern end of Geraru.

To contend with the heat we arose before dawn, and after a breakfast of bread,
jam, and dark Ethiopian coffee, we hiked into the Geraru badlands at first light to
renew our work from the previous afternoon. We worked on the geology until
early afternoon and then retreated to the nearest waterhole to soak and have lunch.
We also bathed at the end of the day and sometimes more often, depending on
where we were. Our favorite waterhole was a large depression in the basalt at the
point where the Geraru stream cuts through the rock a few kilometers southwest
of camp. We would leap off the bank into the cool waters or, when the stream was
flowing, sit in the waterfall at the point where the stream rolled across the polished
stone.

Both the Geraru and the nearby Weranso streams flowed after particularly big
downpours, but not necessarily from rains in the immediate area. I recall one
morning taking some notes on a hill near the Geraru stream when I heard a "plop-
plop" sound. Strange. Plop, plop-plop, plop. I was accompanied by one of our
guides, Sheik Mohammed, an affable older man, with whom I communicated in a
mixture of broken Afar and Amharic.

"Sheik, what the hell is that sound?"

"It's water flowing in the Geraru," he said matter-of-factly.

"That's ridiculous, the streambed is as dry as the top of your bald head."

He stood up without saying a word and motioned for me to follow him.

We walked down the hill and came to a high bank of the Geraru. There was no
water in it.

"See, no water."

"Take a another look, *firenji* (foreigner)," he said.

Ooops, oh yes there was!

At some distance up the streambed, I saw the glistening flow slowly snaking its
way down the center of the channel. Below us a narrow tongue of water had
broken away from the main course and was stealthily inching its way along the
bank, undercutting small lumps of dirt that were falling into the water: plop, plop-
plop, plop.

We sat on the bank watching the channel slowly fill. In the far distance, 40 or
50 kilometers to the west, there were heavy rain clouds on the escarpment dumping water into the catchment of the Geraru stream. Above us there was not a cloud
to be seen.

While at Geraru, Liz and I learned a lot from Sheik Mohammed and from our second guide, Hebah, a handsome young devil, who was forever grooming his Afro hairdo in the mirror of the Land Rover. He explained that this was in anticipation of the moment when the gorgeous girl of his dreams would come strolling by in the desert. It was Hebah who had guided Dennis and me to Geraru the previous year.

Sheik Mohammed was probably in his early forties, a tall, erect man with a shiny bald head, which he often adorned with a skullcap. He was devoutly religious and could often be seen praying under the work tent or off by himself, in addition to bowing to Mecca the prescribed five times a day. He was not so devout, however, that he lost his sense of reality. On one occasion when we were low on food I sent him to Millé, 15 kilometers away by foot, to buy a goat, some salt, *birtan* (incense), and a few other critical items. After he had been gone for four days and we had begun to worry about him, he sauntered back into camp one afternoon, drunk, bringing with him nothing but a nearly empty bottle of arake. I was furious. After tossing down the rest of the bottle with Liz and Yesahak, I questioned him intently about the missing goat and other items. He told me, frankly, that he had become overwhelmed by the sins of the city. And that he had had a terrific time and spent all my money. What could I say?

We learned much about the mysterious Afar and its forthright people from Mohammed. He taught us the names of different acacia, which ones have edible fruit or sap; the names of plants used for medicinal purposes; those that can be brewed into tea or have edible roots, leaves, or flowers; those that can be made into a love potion. He showed us how to use an obsidian chip for shaving and how to identify the pattern of stonework around a rock grave that signifies whether an Afar warrior had killed one, two, or more Issa. He also told us the meaning of the names of the surrounding hills and streams. Geraru means "line of hills" or, literally, "chain of hills (*geèra*) one after the other (*ròor*)" like a line of camels, in apparent reference to the north–south series of basalt hills passing through the area. Data Koma means "black (*data*) hill (*kooma*)" and is the name of a small basalt-capped hill at the south end of the site, which is actually two contiguous hills bisected by a fault and the Geraru stream.[26]

Many names identify edible or usable resources that make the Afar landscape sound like a veritable desert supermarket. At Malab Hill near Millé there is *malab,* honey, and in the Badole area south of Hadar there is *ba'do,* white clay used for soap. Along the banks of the Adaytu stream just east of Camp 270 is a type of tree (*qàday*) whose twigs are chewed into a fibrous pulp and used as toothbrushes. Then there are landmarks named after people, such as Agàbu Hill west of Leadu, which I was told is named after three women (*agàbu*) murdered there by Issa; and after animals, such as Denen Dora at Hadar, a stream (*dora*) named for the scattered wild asses (*danan*) that still inhabited the area. Most common are geographical features, such as Amado, meaning "white (*ado*) head (*amo*)," in apparent reference to the striking profile of the tuff-capped hills at the site; Gona stream west of Hadar, a large tributary on the "outside bend" (*gona*) of the Awash; and Galifagi near Aysaita, an Awash river crossing or ford (*faage*) used for camels (*gaàli*).

Not surprisingly, according to linguists Enid Parker and Richard Hayward, there are numerous words describing the type and availability of water (*lee*), whether it is bitter (*badlìyta*) or muddy (*caro*), originates from a spring (*abqa*) or a geyser (*gucum*), or is found in puddles (*galac*), in rock pools (*xaxxaqo*), or in hand-dug wells (*maddar*).[7] There are single words for the sound of falling rain (*baraar-aco*), dripping water (*xoyya*), and pouring water (*koxxoxxooqo*), and words for a steady rain (*tataataco*), a short rain (*sugum*), intermittent rain (*faagise*), a torrential rain (*weèqa*), and the first rain at the end of a drought (*anfaxxuuga*). There are also words signifying animal watering days (*aràki*), the appearance of falling rain at a distance (*xiimàytu*), and drinking more than ones's share of water (*kabqo*).

Seemingly every creature and many plants in the Afar Depression have been given a name by the Afar, including at least 33 species of birds, 9 types of lizards, 7 snakes, and a long list of mammals, including some that are no longer present in the region, such as the elephant (*dakànu*). I include myself in this accounting: The Afar word for "wild animals" (*àla*) is the same word used for "white man." Of the plants with names recorded by Parker and Hayward, 80 have some kind of use, 36 are edible, and 6 are medicinal. According to Afar folklore, the region also has its share of spirits (*kùbur*), demons (*afatgiraytùli*), man-eating monsters (*gàbbay*), werewolves (*nafur*), and evil genii that speak in highland tongues (*amcarigìnni*).

Who are these people? What do we know of their origins in this land, where streams flow when there is no water, hail appears when there is no cold, famines exist amid floods, and storms are born from other storms? Ali Axinum once told me that the word *Afar* comes from "Ja'far," the legendary first Afar man, and that the fossils we so eagerly carried off date from the days before the prophet Mohammad. Was there really a Ja'far, and if so, did his ancestors live in this wild and enigmatic desert in the days before the Prophet?

CHAPTER 10

Desert Origins

W ith no reported archeological record to speak of, discussion of the origins of the Afar people must start with their language, which linguists assign to the Lowland East Cushitic family. This language group is believed to have diverged from a Proto-Cushitic core, along with Agaw, an ancient but living language of the Ethiopian highlands. In most scholarly accounts of Ethiopia and Eritrea, the early history of the Afar people is described peripherally to that of the highlanders and successive migrations of Semitic tribes from South Arabia. The first migration began some 3500 to 4000 years ago when small groups of Semitic colonists and farmers settled on the northern Ethiopian plateau. That the Afar were around then is considered unlikely, but hostile nomadic tribes lurking in the lowlands probably influenced the initial settlement of these immigrants in the northernmost highlands, which was directly accessible from the coast in the general area of the present-day port of Massawa[1] (see Map II).

By 3000 years ago, colonists of probable Sabaean origin from Yemen had begun settling on the northern plateau. The kingdom of Saba was known in the ancient world for its agriculture, incense, and gold—and for its control of Red Sea trade routes. The source of Saba's gold has long been a mystery, because there are no significant gold deposits known in Yemen today. Nevertheless, some scholars suggest that the gold of the land of Ophir frequently referred to in the Old Testament came from there or elsewhere in South Arabia. Others propose that Ophir and Afar are the same place, but there is no gold known historically from the Afar area either. First Kings 9:28 tells us that 3000 years ago, King Solomon sent his ships "to Ophir, and brought from there gold." First Kings 10:1–2 describes the travels of the Queen of Sheba, or Saba, to Jerusalem to seek Solomon's worldly wisdom, at which time she gave him more gold.[2]

It may be that Solomon sent his ships to the *coast* of Ophir (Afar) to pick up gold carried down from the Ethiopian plateau. Apparently the coast was within the trading area controlled by the Sabaeans and was nominally part of the Saba kingdom. A logical entrepôt for this gold would have been the northern embayments on the Eritrean coast used by the Sabaean colonists for commerce. As for the source of Ophir gold, numerous gold deposits of Precambrian origin in the northern highlands have been documented from ancient times to the present.[3] Even today, farmers on the northern plateau supplement their income by panning for gold in the highland rivers. There are also major precious metal deposits in southwestern Ethiopia that could have been a source of trade with the Sabaeans, though evidence for this is lacking.

Although Ophir and its wealth are more legend than documented history, comparisons of languages with historical chronologies suggest that an Afar language had originated at least 1500 years ago.[4] Contact with South Arabian traders at that time, or earlier, may help explain the primitive irrigation system and some of the bridges at Awsa, described by explorers as constructed in ancient times by Arabs. Most of these works were probably built much later, but South Arabians have been noted since antiquity for their engineering and irrigation skills. One possible source of early Arab influence in the Awash lowlands of Awsa was the traders and colonists of the Ausan kingdom, one of the great trading powers of South Arabia in the early first millennium B.C.[5] Ausan is believed to have rivaled Saba by building an extensive trading network along the eastern African coast before it was conquered by the Sabaeans in A.D. 410.

Towards the end of the first millennium B.C., the eastern Afar was occupied by the Adal kingdom. In the third century B.C. it was conquered and cut in half by more colonists from Arabia, the Ablé, who occupied the present-day Djibouti region. The people to the north, living along the coast, came to be known as the Ankala; those to the south retained the name Adal. Some scholars have suggested that the name Dankali—Danakil in its plural form—may be an Arabized version of *Ankala*.[6] Today the Ankala, or "Ankaala," are a Christianized Afar clan that live on the eastern margin of the Ethiopian plateau in Tigray Province.[7]

By 2000 years ago, the Sabaean colony in the northern highlands had been transformed into the Axumite kingdom. Its center was the town of Axum (located 130 kilometers south of present day Asmara), which soon grew into a major trading and military power on the Red Sea. Axum was strong enough to conquer its Sabaean motherland in South Arabia and to launch punitive expeditions against lowland nomads who were interfering with inland and coastal trade.[8] Auximite records refer to these nomads by various names, which suggests that the northern Afar Depession was occupied at the time by at least three tribes: the Rousoi, the Solate, and the Shamhar.[9] Stone inscriptions at Axum identify four additional tribes—the Sarene, Sawante, Gema, and Zahtan—who were punished because one attacked and annihilated an Axumite caravan, apparently while it was traveling from the salt deposits in the northern Depression to Awsa. Scholars assign these tribes to the pagan kingdom of "Afan," which some have suggested was the Afar.[10]

In the fourth century, Axum's ruler embraced Christianity; in the ninth century, the desert kingdoms of northeast Africa began adopting Islam.[11] There ensued a struggle between the Christianized Ethiopian highlanders and the surrounding Islamic lowlanders that would last for more than a thousand years. By the tenth century, an increasingly powerful and hostile Muslim world, probably including the Afar, helped bring about the fall of Axum by choking off its trade to the Red Sea.[12]

From the ruins of Axum eventually grew a new highland empire, which became known as Abyssinia, a name derived from the original settlers from Arabia, the Habtasha.[13] The Abyssinians expanded southward across the mountainous plateau, bringing with them Christianity and, in the fourteenth century, the divine legend that the first Abyssinian king, Menelik I, was the biblical offspring of Solomon and the Queen of Sheba. At the core of the Christian resurgence was the powerful Amhara tribe of the present-day central highlands, which traded with Muslim merchants via the lucrative caravan routes of the Afar Depression. Among the chief exports were gold from southwest Ethiopia and slaves. The weaker tribes from western and southern Ethiopia were routinely enslaved, a practice that continued to some degree into the twentieth century.[14]

As Christianity grew in strength, so did Islam, with converts in the Horn of Africa adopting their own divine myth of Arab ancestry going back to the prophet Mohammad. By the sixteenth century, a series of Muslim states stretched from the coast to the southern Ethiopian plateau, eventually occupying most of the Awash Valley and the southern Afar Depression. A number of smaller principalities or sheikdoms that formed in the desert lowlands were likely to have been Afar.[15] Described as "Ethiopian bedouin," these pastoralists were introduced to Islam in the ninth century and had nominally accepted it by the eleventh century.[16] When not tending to their herds, the nomads preyed upon the rich caravans crossing the desert. From time to time they also joined forces with the larger Muslim states to fight the Christian highlanders. Conflict was inevitable between the Amharas, who continued to move southward, and the Muslims, who were pushing westward. By the early 1500s, the stage was set for an epic collision between the two groups. The Muslims formed a confederation of Somali and Afar tribes led by Ahmed Gran from Adal, a sultanate located on the Hararghe plateau, near Harar (close to present-day Dire Dawa). Its capital had just moved from Awsa in the Afar heartland. Gran was of Somali or Afar ancestry and was said to be a brilliant military strategist.[17]

In 1529 he led a massive army into the central highlands and destroyed the Christian forces.[18] For the next 14 years, the jihad ravaged and looted the Christian state in a conquest that one historian has described as "devastating in its destruction, irresistible in its ferocity, and appalling in its cruelties."[19] One by one, churches and sanctuaries fell victim to the savage harvest and were systematically plundered of their priceless treasures and torched. Finally, in 1541, the Portuguese answered the cries of the beleaguered Christian nation and came to the rescue of the Abyssinians, killing Gran and defeating his forces in 1543. The Muslim army scattered and eventually returned to the Hararghe plateau and the eastern low-lands.[20] Over the next century, the fortunes and territorial gains of the Adal state gradually deteriorated, as Oromo tribes advanced north seeking new grazing lands and water sources. For survival, the Adal people built a stone wall (that still exists) around the city of Harar, but they soon retreated to the stronghold of Awsa. There the remnants of the once-proud sultanate dissipated in the desert amid infighting among its leaders and conflicts with the Afar.[21]

By the mid-seventeenth century, however, Adal once again became a recognized independent Muslim state, though confined to the Hararghe plateau, while Awsa became an independent Afar sultanate. Both remained so until they were brought into the orbit of the modern Ethiopian nation a century ago by Emperor Menelik II and his compelling join-the-empire-or-be-exterminated policy.[22] As for the Oromo, including the Kariyu, they occupied portions of the middle and lower Awash Valley, except for Awsa, for nearly 300 years, until they too were beaten down by Menelik's forces and eventually driven to the southern fringes of the Depression by the Afar.[23]

Among the first Muslim émigrés to land on the shores of Eritrea (c. A.D. 620), was one Ja'far ibn Abi Talib, the first cousin of Mohammad, probably the same Ja'far that our guide Ali Axinum believed was the first Afar man. Subsequent Arab patriarchs emigrated across the Red Sea and married into leading Afar families, linking many Afar genealogies with the Prophet. One of these Arabs is said to be Har al Mahis, common ancestor of the Afar "nobles" living in the fertile areas of Awsa and Beadu in the area of Gewani. Afar tradition also has it that the Issa Somali, whose language is distantly related to Afar, are descendants of one of the sons of Har al Mahis.[24]

The first known reference to the Afar region as the "Danakil" was recorded by an Arab geographer in the thirteenth century.[25] The name *Dangali* appeared on a Venetian map printed in 1541, the same year the Portuguese went to the rescue of the Abyssinians from the Adal invaders.[26] A Venetian map published in 1555, four years before the stone wall was built around Harar, shows Adal confined to the Hararghe plateau. Ruins of stone buildings in Awsa today, near Aysaita, report-edly date to the 1570s, when the seat of the Adal sultanate moved to the security of the Awash lowlands.[27] Afar oral tradition deals very harshly with the Adal lead-ership in Awsa during this period, a time of great internal strife that was surely exacerbated by continuous raids on the sultanate by the Kariyu.[28] This is probably the time, or the last time, when the Kariyu Oromo occupied the splendid plain south of Tendaho that bears their name, from which the Kariyu could easily have conducted raids into Awsa. By the mid-1600s, however, they were driven back into the southern Awash Valley at the time the Adal state was overthrown.[29] The descendants of the Kariyu, as well as other Oromo tribes that may have invaded the middle and lower Awash Valley in the sixteenth century, can be found today along the Hararghe escarpment.

By mid-July 1974, Liz Oswald and I had mapped 23 faults at Geraru, in an area of just 2 square kilometers (see Map V on page 30). The many faults are due to the site's location squarely at the intersection of the Red Sea and the East African rifts. By comparison, the immediate Hadar area has half as many faults in an area of 15 square kilometers but lies on the margin of the two rift trends.[30] Despite the complexity of Geraru, Liz and I were slowly getting a sense of its stratigraphy. Using eight tuff layers and several sandstones as marker beds for correlation, we pieced together the chopped-up sequence as best we could.[31] The lower Geraru beds were similar to the middle lake beds at Hadar, and the middle Geraru beds were similar to Hadar's upper riverine beds. The possibility of Hadar-equivalent lake beds at Geraru, located 40 kilometers northeast of Hadar itself, suggested a greatly expanded Pliocene paleolake filling the central Afar—one, moreover, that had migrated north and east-to lower elevations into a subsiding triple junction (see Appendix II on page 320).

At the northeast end of the Geraru basin we found numerous Oldowan stone tools in a shallow depression eroding from a clay layer unconformably overlying the middle beds. Just to the west of Geraru in the next drainage were thick tuffs and gravels, apparently of a much younger age, that were warped downward toward the Kariyu basin. The strata had been deposited during a period of volcanism at the same time the basin was subsiding, causing paleostream flow to be directed to the east, to lower elevations, no doubt draining into a proto-Lake Kariyu. We also found some fairly fresh Acheulean tools on the surface of the tuffs, which suggested this tectono-volcanic activity may have taken place during the middle Pleistocene, perhaps a million or more years after the Oldowan toolmakers roamed the area.

Late one afternoon in the third week of July, Liz and I came dragging into camp and found Hebah there waiting for us. He had just returned from seeing the sultan in Aysaita and was touching up his hairdo in the mirror of the Land Rover. As we approached, he quickly cast an eye in Liz's direction.

"Say, Hebah, did you find the sultan, or were you off chasing those Amhara girls again?" I asked.

Hebah was preoccupied.

"*Hebah!* Any luck with the sultan, or were you off chasing those Kariyu girls again?"

"Yes, yes, the sultan will see you in two days, midmorning."

"Great! You did well, my gorgeous friend. Now listen, about Liz . . ."

Hebah was well connected in Afarland. Some days before, he had gone to Aysaita to arrange a meeting for me with the sultan, Ali Mirah. Since the night raid on our camp at Hadar by thieves the previous year, it had been clear that we needed to work more closely with the Afar. Going into areas without first contacting Afar chiefs was not the way to win friends. Yes, we had permission from Addis Ababa to go where we wanted, but the government had limited authority over the Afar people—and little to none off the roads. During our visits to Hadar, we had just "appeared" and busied ourselves doing whatever we wanted. We had invited trouble. Because Hadar was also near the boundary of two Afar clans, I wanted to be able to work freely on both sides of the Awash; I did not want to return to the area and wake up one morning with a knife in my back. This is what I wanted to discuss with Ali Mirah: long-term health insurance.

The sultan had his hands full that year. Famine, political turmoil in Addis Ababa, the opening of the Trapp road—all were bringing an invasion of outsiders: relief workers, political agitators, more truckers. There was a great reshuffling in the works, and Ali Mirah must have been concerned about the future. On top of it all, his subjects were getting bad press. In late 1973 British television aired a documentary entitled "The Forbidden Desert of the Danakil" that referred to the Afar as "some of the most . . . savage people on earth."[32] The film featured Thesiger's exploration through the Awash Valley 40 years earlier, although by his own account he witnessed little that was "savage."[33] After all, during the course of his travels, he was able to collect 876 specimens of birds for the British Museum of Natural History.[34] During his visit with the sultan, Thesiger was "astonished by the orderly conditions which prevailed throughout his land."[35] Apparently the worst that Thesiger suffered—and I do not belittle its severity—was an attack by a scorpion that crawled inside his shorts and stung him near his private parts.[36]

Whatever the truth about the Afar people's capacity to kill and plunder, relative to "savages" the world over, their fierce reputation has long served the interests of highlander Ethiopians. From the time of the Axumites, the nomads occupying the eastern lowlands—whether they were Ophirians, Afan, Ausan, Ankala, or Afar—have effectively served as the "border guards" or "watchdogs" of the Ethiopian plateau, as one scholar (who is Afar) has put it.[37] And it is true that the Afar have traditionally controlled, sometimes ruthlessly, the caravan routes to and from the coast as well as the salt basins, which are so vitally important to the highlanders.[38] Levying tolls for protection against brigands, however, and wreaking havoc on those who failed to pay for this protection, was a time-honored business to the Afar.

The Afar *are* also known for emasculating their enemies.[39] I recall being offered a pinch of snuff—it's terrific for flushing out desert dust—from an Afar who claimed the pouch he carried it in was human. Historically, however, the Afar are not the only tribe in Ethiopia known to carry out the ritual of castration.[40]

Much of the unfavorable reputation of the nomads in years past stems from their attacks on politically minded agents, adventurers, and explorers whose goals were at cross-purposes, either real or perceived, with the livelihood of the Afar.[41] It is apparent that some of these attacks were encouraged by the schemes of highlanders, most notably King Menelik and Emperor Yohannes in the late nineteenth century. Both used the Afar as pawns in their imperial intrigues, particularly in their efforts to obtain shipments of firearms from the coast.[42] During the ensuing scramble for African colonies, the Afar were drawn into the machinations of foreign powers and were divided among themselves as to the degree of their cooperation with these elements.[43] This was also the period during which the politically motivated expeditions of Munzinger, Giulietti, and Bianchi were destroyed by the "barbarous Dankali." The Afar in Awsa in particular feared that the opening of trade routes and the shipments of firearms to the insatiable King Menelik would ultimately be used against them. These fears were realized in 1895, when the recently enthroned Emperor Menelik sent 30,000 troops to Awsa to discipline the sultan for becoming too friendly with Italian arms merchants. The punishment was contradictory, because Menelik had repeatedly encouraged the Afar to cooperate with the Italians after the destruction of the Giulietti expedition in 1881, when he was desperate for modern weapons.[44]

Relations were not helped when another imperial army punished the Afar in 1913, this time because they had reportedly slaughtered 300 Oromo Kariyu who had occupied parts of the middle Awash Valley. The leader of the attack on the Afar, Lij Iyasu, the teenage pretender to the throne of Menelik, was then courting the larger Muslim population of the country, the Oromo included. Iyasu's soldiers attacked several Afar encampments, killing numerous people and confiscating livestock. Thesiger reports that Iyasu castrated the dead and dying and, after watching his soldiers rape an Afar girl, chopped off her breasts when she refused his advances. Another account states that the encampment Iyasu's soldiers attacked was not the one responsible for the assault on the Kariyu.[45]

Iyasu's reign was a bizarre time in the country's history because the young prince seemed hell-bent on turning Ethiopia into a Muslim state, at least in part by procreation. Among his many (Muslim) wives was the 14-year-old daughter of an Afar sultan, by whom he had a son named Menelik III. After a brief reign that caused much political chaos in the country, Iyasu was overthrown and eventually died in prison, where he was kept in an iron cage. Some say T. E. Lawrence (of Arabia) secretly came to Ethiopia during this period and helped bring about Iyasu's downfall, because the British feared his leanings toward the Turks and Germans in World War I. The fate of Iyasu's Afar wife, to whom he was especially attached, and their son, the alleged heir to the throne, is uncertain. One report maintains that she and her son fled to Tadjoura in present-day Djibouti.[46]

By the time Nesbitt (1928), Baron Franchetti (1929), and Thesiger (1934) traveled to Awsa, Ethiopia was seething with intrigue, and everyone—the Amharas, Tigrayans, French, English, and especially the Italians—sought favor with the Afar in anticipation of an Italian assault. There is no doubt that Franchetti spied for the Mussolini government, as did Nesbitt's companion, Tullio Pastori, as he later confided, and we can assume that British intelligence scrutinized the published maps and journals of Nesbitt himself and Thesiger.[47] It is also certain that Italian intelligence made use of Nesbitt's knowledge of the Afar, particularly since he published a full account of his explorations in Italian. This is not surprising, because Nesbitt was born and raised in Italy (although he is known by the first name of Lewis, his full name was actually Ludovico Mariano Nesbitt). Thesiger is less circumspect, saying that in 1935 the Italian army used information provided by Nesbitt and Pastori for its march through Awsa to the highlands.[48]*

In September 1935, when Mussolini was poised to invade Ethiopia, the League of Nations made a last-minute offer of appeasement, proposing that Italy occupy the Afar Depression and part of the Ogaden instead of the whole country. With sarcasm heard around the world, the Fascist leader replied that Italy was not "a collector of deserts."[49] Two weeks later, the Italian armies began their invasion of Ethiopia on three sides: south from Eritrea, north from Somaliland, and west from the port of Assab. By April 1936, the Dancali Irregular Unit, supported by 25 aircraft, had punched straight across the Depression along the partially completed Bati-Assab road until it reached Sardo, then the seat of the Awsa sultanate. There, its efforts uncontested by the Afar, the invaders hastily built an airfield that served as a stepping-stone to the highlands. The Afar sultan at the time, the cooperative Mohammed Yayo, was later received in Rome by an appreciative Mussolini.[50]

Given the contentious relationship that the Afar from Awsa had had with the highlanders, it is not surprising that they sided with the Italians during the war, nor that Haile Sellassie ordered another punitive expedition against them afterwards. In 1944 a new sultan was installed in Mohammed Yayo's stead, the strapping young Afar chief Ali Mirah Hanfare II, thirteenth sultan of Awsa, known far and wide simply as Ali Mirah.[51]

Along with their well-deserved fame as a hardened desert tribe, it is certain that much of the reputation of the nomads as "savages" comes from literary travelers. In 1856 Richard Burton described them as "wild as ourang-outangs, and the women fit only to flog cattle,"[52] which suggests that Burton probably chewed more *chat* than he revealed. Nesbitt, who lived to tell about it, said, "the Danakil kill any stranger on sight,"[53] and Evelyn Waugh concluded from reading Nesbitt that the Afar were "naked homicidal lunatics."[54] A contemporary of Waugh, Byron de Prorok, described the Afar as "devils out of hell . . . like starved jackals, dipped in tar and perfumed with brimstone."[55] A propagandist for Mussolini, Louise Diel, evoked diabolical images of the Afar by describing the Depression as "the African chamber of horrors."[56] The great Islamic scholar J. Spencer Trimingham, writing in 1965, offered a more balanced view, describing the Afar as "viva-

---

*Nesbitt died in a plane crash on July 21, 1935, while flying from Rome to Berlin; Franchetti died in a plane crash two weeks later, while flying over Egypt. Thesiger is still living.[47,48]

cious, bold, and courageous, but living in a country where life is so precious [that it has] made them suspicious, treacherous, and bloodthirsty."[57] Loren Bliese, a missionary, educator, and linguist who has worked with the Afar for the last 40 years, recently characterized them in this matter-of-fact fashion:

> The Afar have pride in their nomadic freedom and their ability to care for their animals. They are loyal to their clan, and friendly to outsiders who are accepted by the clan leaders. The men are volatile when they or their herds are threatened. Afar women are hard-working in building and maintaining their homes and providing food, water and firewood.[58]

Liz, Yesahak, Hebah, and I piled into the Land Rover and took off from Geraru for Aysaita to meet with Ali Mirah, leaving the rest of our crew to look after the camp. We spent that night at the Tendaho Plantation. There Liz and I socialized with the English manager of the plantation, Michael Quimby, and his wife Peggy at their new house with its view of the Awash. I had known the Quimbys since 1971, when Maurice and I first met them on our way to Lake Abhé. Michael was then in charge of a remote part of the cotton plantation down the valley, where Peggy nearly went berserk from the isolation. On this occasion, they both were distracted because the Awash had risen all day, threatening to flood their home. That evening, as we sipped drinks in the garden surrounded by flame trees and bougainvillea, we watched Michael direct workers placing sandbags around the back perimeter of the lawn.

The next morning I took the dirt track from the plantation to Aysaita. I crossed a nearly flooded bridge to the south bank of the Awash and drove a short way beyond town to a beautifully forested area where Ali Mirah lived in a modest, white wooden house. About 30 Afar men were gathered beneath enormous trees, giving the impression that a meeting was about to take place. Another four or five men stood next to the house, all armed, muscular, and well groomed. I assumed they were Ali Mirah's bodyguards.

As Hebah, Yesahak, and I walked up to the house, one of the guards came up to ask our business. After a brief exchange, he disappeared into the house and then reappeared with a spectacled man wearing brown pants and shirt and sandals, who introduced himself as the sultan's aide. I introduced myself. I was wearing shorts, sandals, and a khaki safari shirt. The aide knew Hebah and our business and showed Yesahak and me into the house. We were led into a small, spartan, spotlessly clean room to chairs in front of a small divan upholstered with royal blue cloth. Shortly the sultan entered. Ali Mirah was an impressive man, larger than any Afar I knew: heavy-set, with broad shoulders, thick forearms, and a full trimmed beard. He wore a long waistcloth with a *shamma* draped over his shoulders and an embroidered, royal blue skullcap. In his left hand he held a fly whisk. I shook hands with him in the western manner, although the traditional Afar handshake is like a gentle slap of the right hands, which then slide past one another into a firm grip. There are endless variations of this handshake to reflect status and circumstances. The most respectful ends when one draws the hand of the other to one's lips, which I started to do, but the sultan firmly held my hand down. I guessed his action was either a sign of respect or an indication that this was not a way he was accustomed to being greeted by non-Afar.

His aide led the conversation, speaking alternately in English and Amharic, while translating what we said into Afar, even though Ali Mirah was known to speak perfect Amharic and probably spoke several other languages as well, including at least some English. I explained who I was, what I was doing in the Afar, where we were camped, and the nature of the problem we had had at Hadar the previous year. After I spoke, Ali Mirah asked a number of questions.

"Why are the stone remains of animals important?"

"Because they tell us the nature of life many thousands of years ago."

"Why is this important?"

"Because we learn how our ancestors lived."

"How did they live?"

"They lived around a series of large lakes that over time moved from the southern Awash to Lake Abhé."

"Why is it important that we know this?"

"Because understanding the past is the key to finding resources that we can use in the present."

"What resources have you found that we can use for the present?"

"None, yet."

"Where have you looked?"

"I've looked at Leadu, Agàbu Koma, Weranso, Malob Koma, Ledi, Howoona, Amado, Ahdaitoli, Badena, As Koma, Denen Dora, Sidiha Koma, Ahdi D'ar, Kada Da Moumu Koma, Gona, Hourda, Meshellu, Busidima . . ."

When I paused he asked, "Is that all?"

"No," I said, "I've also looked at Geraru, Korkora Koma, Amare, Deneba, Abaco, Kariyu, Sana-as, Layagili, Sagale, Korkora Koma, Kohilta Koma, Asboli, Kasuri, Adisgura, Dereleka, Dalu Ali, Arissa . . ."

He interrupted, "Fanta?"

"Pardon?"

"Fanta?"

"Oh, *Fanta!* Sure, yes, that would be very nice." Fanta is a bottled orange drink. A servant then brought one for everyone. As I sipped the drink, I began to realize that on this particular day, I probably would not be castrated and murdered.

Ali Mirah then asked, "If you did find some resources useful for my people, will you then inform me about this?"

"Absolutely."

"So, how can I make your task easier?"

I explained that it would be very useful if the chiefs on both sides of the Awash at Ahdi D'ar (Hadar) knew that I had spoken with him and that my colleagues and I had his approval to work there.

"Tell me exactly where Ahdi D'ar is located."

After I did this ("It's a four-hour walk upriver on the north bank of the Awash from the former German camp near Millé"), the sultan said that he would have his son and an assistant accompany our team in September to help us deliberate with the local people. He would also introduce me to some Afar chiefs in the area whom it would be useful for me to know. I told him I would be very grateful. With that, Ali Mirah's aide told us that the sultan was always willing to help whomever he could. He then made a special point of telling us that Ali Mirah's authority extended across the *entire* Afar area, from Massawa to Assab to Awash Station.

As I was wondering when the sultan would introduce me to the chiefs he mentioned, he abruptly rose and we again shook hands. He motioned us to the door. I assumed the audience was over, but he then followed us outside and led me over to the men standing beneath the trees. He said they were chiefs and introduced them to me one by one. He told them all that if I were to visit their area, they were to assist me as best they could. He made a special point of telling me which chiefs were from south of Hadar, because I had told him I already knew the chiefs to the north.

I thanked him again. We shook hands one more time, and Yesahak, Hebah, and I walked back to the Land Rover to an excited Liz, who had watched the proceedings.

We drove into Aysaita, weaving around clusters of camels and goats, to do some shopping and celebrate the successful meeting. Over lunch, in a restaurant on the lattice-covered roof of a small hotel, we discussed the assistance the sultan offered. Certainly I was pleased, but that still left the politicized nature of the Hadar research; also, major political changes were taking place in Addis Ababa. From listening to Yesahak's transistor radio at night, we learned that a group of junior military officers were effectively running the government through a secretive committee soon known as the *Derg*, meaning "a group of equals."[59] Haile Sellassie was powerless, and many in the ruling class were being arrested. I wondered how Ali Mirah, essentially a Haile Sellassie appointee, fit in with all this.

Back at the Tendaho Plantation, Michael Quimby told me that the plantation airplane would be returning from Addis Ababa later that day and that he would ask the pilot to fly me over Geraru so I could photograph the site. I had told Michael of the complexity of its geology and of my need for better-resolution photos to use for mapping the site. The next morning I met the pilot at the dirt airstrip, as instructed. The mechanic servicing the plane told me the pilot was a former major in the Ethiopian air force who had flown jet fighters in the Korean War. I thought, *Oh great, that's just what I need, a top gun to fly me around the Afar.* But the major turned out to be a mild-mannered man who inspired confidence.

We took off at 8:00 A.M., swooping off the dirt runway in the usual cloud of dust. With the side windows open and wind buffeting the cockpit, we headed south. Soon we were over the Magenta Mountains and then the glistening Kariyu plain, where Urs Carol had flown me (and at me) two years previously. As before, the landscape was speckled with herds of animals. Flocks of birds shuffled back and forth below us. And then,

WHOOM!

A blast of wind blowing through Magenta Gap rocked the plane as the pilot struggled to maintain our course. A head wind jerked us along westward like a kite in a storm.

"Sorry, Major, I should have warned you about that!" I yelled over the roar of the wind and the engine.

"Not to worry, not to worry!"

As we approached the Awash the wind slackened off, but then we were caught in the thermals rising off Kariyu. The combination of wind and swirling heat seemed to move the plane in any direction it chose, like those bumper cars at carnivals that get pushed in one direction when turning in another. While readying my camera, I directed the major to the hill cut by the Geraru stream. There below was our modest wind-battered camp, perched on the edge of the gravel plateau. With our first pass, my crew remaining in camp came running out from under the work tent yelling and waving. I assumed they were saying, "Praise God, come rescue us!" As the major flew the airplane back and forth over the site, I fired away with my camera, taking close-ups of the main fossil areas, the downwarped tuffs, the Oldowan artifact locality, and the fault scarps. But I needed some shots with my camera lens pointing perpendicular to the ground, so I could enlarge the photographs to make a base map with as little angular distortion as possible.

"You mean you want to take photos of the ground with the camera pointing *straight* down?" the major asked.

"Yes, that's what I want. I want to take photos with the camera pointing *straight* down."

It was only a few minutes later that I bombed camp Geraru with my breakfast of English muffins, sliced fruit, and coffee. The major flew over the site with the wing tips 90 degrees to the ground, or so it seemed, while turning the plane in a pinwheel pattern. I lay with all my weight on top of the copilot's door with the window open, both hands holding my camera jammed to my face and pointing straight down into space, certain that the door would burst open, my safety harness would snap, and the major would have to return to the plantation without me. Millions of years later an anthropologist of a higher species would find my remains with a camera embedded in the skull and announce the discovery of . . . Camera Man!

"Major, can we do that one more time" I asked. "I still have a couple of English muffins to go."

I took 101 photographs of Geraru that morning.

Back in camp at the end of July, we had another four days before returning to Addis Ababa, or so I had planned. Some people in camp who shall remain nameless—well, *all* the people—were complaining about the food. Yesahak, always the young scientist, asked me how much money I had spent so far on food for the camp. He then calculated that all the lentils, peas, beans, rice, sardines, tuna, and flour I had bought came to just under 50 cents a day per person.

"Oh yeah," I said, "and what about the jar of jam and the mustard? And the onions and peppers we bought in Millé? You didn't figure those in, did you? See, you Ethiopians always exaggerate!"

This exchange was followed by a long discussion about how we could get more food without going over my budget.

"*Budget?*" Yesahak asked.

We talked about going into Millé and asking for relief food, because on our way back from the plantation we had seen large sacks of grain with US AID stamped on them being unloaded at a hastily built relief camp on the edge of town.

"Look, Yesahak, who the hell needs food when you're doing science? How are you going to be a famous physicist if all you do is eat?"

As a compromise, we decided that even though hunting was illegal without a permit, considering the circumstances, we could not prevent Hebah from shooting an antelope.

"I mean, these Afar *are* savages, right?"

"Yes. *Yes,* YES!" everyone screamed," in their various languages.

The next day about noon, Liz, Yesahak, and I were looking over some red sediments baked beneath a basalt flow in the company of Hebah, when we flushed a gazelle out of a shallow tree-lined ravine.

"Take him, Hebah!" I yelled.

The gazelle dashed up a hill. By the time Hebah jerked the Mannlicher from his shoulder, chambered a cartridge and aimed, the gazelle was still no more than 20 meters away.

BAM!

Our fresh meat disappeared over the hill. He missed the animal by a camel length.

"Jesus, Hebah, if that had been an Issa you would be in dreamland! Hand me that rifle and let me give it a try."

He agreed after some fussing, not because he did not want me to use his rifle, but because cartridges were expensive. Firearms and ammunition were the real gold of the Depression. I was told that the sale of both was strictly controlled by Ali Mirah, at least in the central Afar.

After I had agreed to pay for two cartridges, at three *birr* each (U.S. $1.50), I took the rifle, slid a cartridge into the chamber, and planted the barrel in the fork of a tree at eye level. I aimed at the knot on a tree stump about 20 yards away, took a deep breath, and let it out ever so slowly while pulling back on the trigger . . .

BAM!

The shot ripped through a bush well to the right of the stump and plowed into the dirt 10 meters beyond that. The bush exploded into flames. As the fire crackled in the heat of high noon, Hebah dashed forward, raced up the hill to the spot where the bullet had dug into the ground, fell to his knees, and after a few seconds raised an arm and shouted something like, "Eureka!" The copper-jacketed bullet was barely scratched, ready to be used another day.

I gave the extra cartridge back to Hebah and advised him either to invest in a few cartridges to sight his rifle properly or to move to Switzerland, where there are no Issa.

That night, after a repast of "famine food," a surly and hungry crew lounged around on their foam mattresses as I made a case for spending just two more days at Geraru.

"Now, look, it's not my fault that Hebah couldn't hit a fuel truck with that rifle. Besides we still have some flour and an onion . . ."

No response.

". . . and this afternoon Mohammed found some ants. Not those tiny brown ones you're use to, but the fleshy, tangy red ones."

Not funny.

"All right, an Ethiopian compromise: We'll break camp in the morning, but we won't leave until noon."

Everyone lightened up. Kebede put on a pot of water for tea while confessing that we still had enough sugar for one spoonful for each of us.

Liz yelled, "Party!"

As the moon began peering over the eastern sky, I played a tape of Hebah and some of his friends singing with the rhythmic stamping of feet, which I had recorded days before. Later, Yesahak turned on his radio to listen to the news as we all dragged our mattresses into a closer circle on the gravel surface. There was a strong breeze, and the sky was free of rain clouds.

Listening to the news that night was as surreal as it had been on previous nights that month. The Voice of America and the BBC told us about the tailspin of the Nixon administration over release of the Watergate tapes, and the Radio Voice of the Gospel and the Amharic stations told us of more arrests in the imperial government. Both Nixon and Haile Sellassie were being consumed by wolves, and it was only a matter of weeks or days before nothing was left of them. The only question was which would go first. Nixon's ouster, however, would lead only to the demise of a rogue presidency; the Emperor's would end 2000 years of tradition, and feudalism. Yesahak and I talked about these matters night after night. He learned about the U.S. system of government, and I learned about the grievances he and many of his fellow students had against their overlords. I was ambivalent about returning to Addis Ababa and the current events, but it was time to catch up on the changes and get back to my family.

Everyone began packing up camp at dawn. By midmorning we had bid our farewells to Sheik Mohammed and Hebah in Millé, stored much of our field equipment with Epfrem (the restaurateur and former operator of the "Amoeba Bar" at Camp 270), and were on our way to Bati and Addis Ababa. Not surprisingly, the Land Rover was impounded at the customs checkpoint at Bati, because Johanson had failed to pay the duty on the vehicle. "Yes, officer, you're absolutely correct, the duty must be paid, so *please*, take the keys." The customs chief did agree to let me drive the Land Rover back to Addis Ababa the next day, accompanied by his sergeant (who had to attend a wedding in the capital). Once there, the vehicle would be locked up in the customs yard at the railroad station until the duty was paid.

That night in Bati, Liz, Yesahak, and I stayed at the 1930s Italian-built Saba Hotel on the heights west of town and savored the crisp mountain air on the balcony while sharing a bottle of wine. The town below us was bathed in the light of the Afar moon rising over the escarpment through a curtain of cobalt blue sky draped across the Depression. I felt some satisfaction knowing what mysteries lay behind that curtain: another curtain, then another, and another.

The next day was market day, held in the square on the south side of town. Afar with their camels packed with wares lumbered in from the escarpment canyons, and farmers and merchants came with donkeys laden with goods. People milled about, setting up plots and displays, talking and hawking their products in whichever language and dialect they spoke. To encourage good behavior, an iron gallows—with a rope attached to a pulley for efficient operation—stood prominently on one side of the square. The highlanders sold millet, barley, teff, sorghum, peas, beans, and yams; and the Afar sold livestock, salt, sandals, rope, palm mats, and cloth from Djibouti. Then there were farm implements for sale by local merchants, and spices, sugar, coffee, tea, soap, candles, lanterns, pots, knives, ladles, combs, brushes, cigarettes, beads, clothes, incense, and perfume.

After roaming around for a couple of hours while waiting for our customs passenger, we headed south from Bati to Addis Ababa on the road that follows the eastern edge of the plateau. To the west was the captivating scenery of the Blue Nile basin, to the east the rugged terrain of the escarpment and the Awash lowlands. As I pointed out to Liz along one stretch, when it rains all the water on the right side of the road ultimately drains into the Mediterranean, and all the water on the left side goes directly to Geraru. Liz understood.

CHAPTER 11

Dawn of Humanity

Back in Addis Ababa I received a letter from the director of FORGE, Alfred Kelleher, telling me that Johanson and Taieb had phoned from Cleveland attempting to discourage FORGE from awarding me more funds. Kelleher told them that he had no plans to discontinue my funding. Nevertheless, it was apparent that somewhere in the infighting between Johanson and me, I had lost. I later learned that an associate of Johanson's at Case Western Reserve University—a geologist—and Taieb were applying jointly with Taieb to NSF for funds for the Hadar work. The connection would eventually bring substantial funds to the IARE for radiometric dating purposes, an opportunity Taieb would not pass up, and understandably so. Shortly, the National Geographic Society wrote me to say that the grant application I had submitted to it several months earlier had been declined. The secretary of their grants committee said that Taieb had informed him that the IARE had all the funds it needed for the pending field season.

It was time to start thinking of forming my own team.

In early August 1974, I used $200 of my nearly depleted FORGE money to charter an airplane for a reconnaissance survey of the northern middle Awash Valley, which was essentially unexplored. This was the area between the Gewani and the Wallo Province boundary. The pilot was an American friend, Fred Lundahl, who worked for a small air charter service. I called him Flying Fred.

We took off in the cool morning air and headed northeast. After reaching the edge of the Ethiopian plateau at Ankober, the birthplace of Menelik II, we flew over the escarpment, dropping into the increasingly warm lowlands, and then over a series of small crater lakes on the west side of the Awash. Afdem volcano stood in the distance, where Johanson and I had chewed *chat* into the night two years earlier.

After picking up the Awash River, we headed north, crossing to the west of the Mataka hot springs, and then Gewani (see Maps IV and VII on page 140). To the east was Ayelu volcano and the infamous "Issa Graben," where Maurice and I had rustled that herd of livestock in 1971. Thirty kilometers north of Gewani, in the center of the valley, we flew over a large, heavily cross-faulted complex of multicolored, sedimentary exposures and volcanic-capped hills that certainly looked promising for fossils. I snapped some photos. To the east were brown and white banded sediments hugging a low basalt plateau; far to the west were brilliant white exposures, surely tuffs or diatomites or both. Damn, I thought; these sediments must be old: According to my hypothesis, sediments get older *up* the Awash Valley, because the lakes migrated *down* into the unfolding triple junction.

Beyond the faulted complex I had Fred fly northeast, following the Meshellu stream. Soon we reached the magnificent badlands south of Hadar that I had come so close to two years earlier. Between Meshellu and the Busidima River to the west I shot two rolls of film. The panorama of exposures looked every bit as extraordinary as those north of the river, perhaps more so. On the remote chance that I would continue working with the IARE, I also took numerous photos of the greater Hadar area and the area to the west at Gona. Next we flew north to the Ledi River, where I photographed the fossil localities in the central Ledi basin, "Elephant Park."

By this time Flying Fred was saying our fuel supply required that we head back to Addis. On the return trip I got a second look at what lay between Meshellu and Gewani. We overflew a sea of sedimentary exposures for 50 kilometers on both sides of the Awash next the sediments on either side of the valley that I had seen earlier, and then the complexly faulted area in the middle. Who needed the IARE, I thought? There was certainly room for more than one tribe in this prehistoric wilderness.

After a while we again reached the escarpment east of Ankober and began our ascent to the plateau.

"Fred, don't you think we're getting a bit low on fuel," I said, pointing at the fuel gauges.

"Nah, these planes have three gas tanks," he said, pointing to three fuel gauges, "one tank on the right side, one on the left, and one in the center. After a while the remaining fuel in the two side tanks will drain into the central tank and you'll see the needle on that gauge go back to nearly full."

"Aha." I said, marveling at the technology.

Fifteen minutes later we were crossing the escarpment approaching the 1500-meter-deep Kasem River gorge east of Addis Ababa. "Fred, I don't see the needle on the central gauge going anywhere but down, in fact, like the other two needles. They're all showing nearly empty as far as I can tell, unless you have a secret fourth gauge somewhere."

"Jon, do I argue with you about rocks? If you say something is a rock, I believe it's a rock. If you say it's a black rock, it's a black rock. If you say it's a pink rock, it's a pink rock."

"Yeah, I know Fred—and I appreciate that—but what worries me . . ."

Just then the engine sputtered. And died.

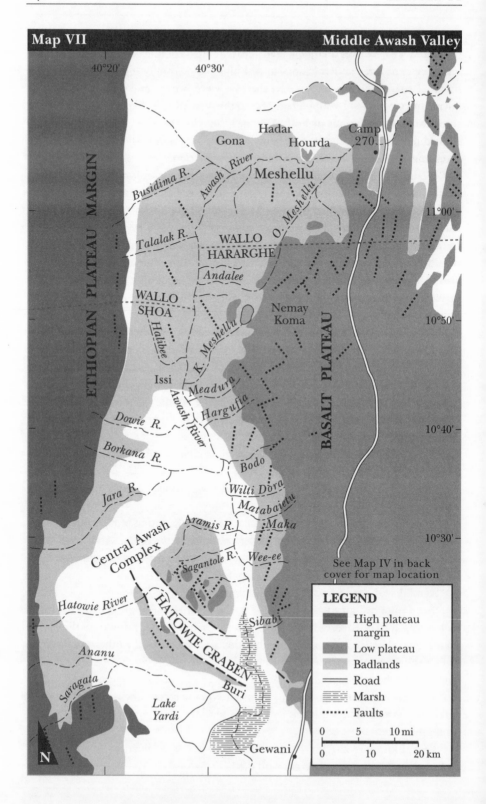

Map VII   **Middle Awash Valley**

40°20'   40°30'

Hadar   Camp 270

Gona   Hourda

ETHIOPIAN PLATEAU MARGIN

Busidima R.   Awash River   Meshellu

11°00'

Talalak R.   WALLO HARARGHE   O. Meshellu

Andalee

WALLO SHOA   Nemay Koma

10°50'

Halibee   K. Meshellu

BASALT PLATEAU

Issi   Meadura

Dowie R.   Awash River   Hargufia

10°40'

Borkana R.

Bodo

Jara R.   Wilti Dora

Matabaietu

Aramis R.   Maka

10°30'

Central Awash Complex   Sagantole R.   Wee-ee

See Map IV in back cover for map location

Hatowie River   HATOWIE GRABEN   Sibabi

LEGEND

Ananu

High plateau margin

Low plateau

Saragata   Buri

Badlands

Road

Lake Yardi

Marsh

Faults

N

Gewani

0   5   10 mi

0   10   20 km

"What the hell's going on?" I asked.

"I think we're running out of gas."

"*Terrific!*"

The engine sputtered again, then died again. By now we were gliding over the Kasem River with the walls of the gorge rising on three sides. I was not having fun. Seven years earlier, while working for the U.S. Army Corps of Engineers in Central America, two fellow geologists and I had been in two near crashes in light planes on the same day. The first time was over the Gulf of Panama when the engine caught fire. The cockpit filling with smoke, the pilot frantically radioed our position to the Canal Zone seconds before the engine died. We barely made it to the only strip of open beach within miles. When we landed the pilot jumped out of the plane and fell to his knees, crying and kissing the ground. Later that day, after the rescue plane had picked us up, we got lost in a thunderstorm over the Darien jungle on the way to our base camp across the border in Colombia. As darkness closed in, and low on fuel, our new pilot began his descent while explaining how he was going to land the plane on the tree tops, "like a bird." Just then one of my colleagues caught a glimmer of light in the far distance. The generators that powered our electricity in camp had just kicked on. After we reached and buzzed the camp, two trucks drove over to the grass airstrip and shone their lights at either end for our landing.

"Maybe there's something wrong with the tanks," Flying Fred said as he tried to start the engine again—it sputtered. He tried again; it started. Then it died again.

"What do you mean, 'Maybe there's something wrong with the tanks?' How the hell could there be anything wrong with the tanks?"

"The plane was recently overhauled; maybe some flakes of dried paint got into the fuel line, blocking flow to the central tank."

Fred tried the engine again.

It sputtered, but kicked on.

"Fred, surely you're not telling me that dried paint could be in the fuel line because they painted the *inside* of the tanks?"

Just then the engine died.

"Don't worry." Fred said. "This time I cut the engine intentionally to save fuel. I think we'll have enough gas to climb out of the gorge after we go a little further. Then I'll cut the engine back on and we'll be outta here." ·

"Yeah, but what if . . . ?"

The engine started up; we made it over the lip of the gorge—to my reckoning, by the skin of our eyeballs. Fred thought we could "probably make the remaining 40 kilometers to Addis Ababa."

"*Probably* make it to Addis, Fred? Are you outta your mind? I want out of this death trap *now!* What if we 'almost' make it to the airport and we land on a schoolhouse?"

We landed on a gravel road just west of the gorge. It was filled with people coming or going to market. They all waved as we made our approach. Then when they realized we were landing where they were, they scattered into the fields like grain birds. From a nearby village Fred called the Addis Ababa airport and they sent a Jeep out, bringing us two jerricans of fuel. Fred was insulted when I rode back to Addis in the Jeep.

That afternoon I filled out a research authorization form given to me by the director of the Antiquities Administration, Bekele Negussie.[1] A former high school teacher from western Ethiopia, Bekele had followed the growing dispute within the IARE throughout the summer and had encouraged me to assemble a new team, as long as the study area I requested did not conflict with the IARE's. Nevertheless, attached to my permit application was a map depicting my proposed permit area, which lay in the center of the IARE's 33,000-square-kilometer concession (see Map VIII, page 143). It contained no area where the expedition had previously worked. I noted in a cover letter to Bekele—with searing insight—that "Taieb and Johanson will no doubt complain bitterly" at my request. In the event that the Antiquities Administration refused to break up the IARE permit, I also requested three smaller areas completely outside of its boundaries: in the southwest Afar against the Ethiopian escarpment; to the northwest in the upper Millé River valley; and between the Magenta Gap and Lake Abhé. Each sampled a different arm of the triple junction, and I hoped that together they might produce sediments and fossils ranging from late Miocene to late Pleistocene in age.

The following morning when I gave Bekele the permit application, he told me that the IARE permit was now under both Taieb's and Johanson's names and that Johanson was now codirector of the IARE, despite the fact that the protocol we all signed named Taieb as the sole head of the expedition. Bekele said the two had omitted my name from the list of participants in the permit application for the pending field season.

Instead of giving my own application serious consideration, Bekele wrote Taieb and Johanson, telling them that "Mr. Jon Kalb contests the manner in which the scientific research [of the IARE] is being organized under your direction" and that until those differences were resolved, the Antiquities Administration "sincerely regret[s] not being able to give our approval for this year's field season."[2] Furthermore, he required the signing of a new protocol "by the signatories of the [original] agreement." Bekele ended his letter by saying, "Should it be impossible to come to an agreement, it will be up to the Antiquities Administration . . . to reconsider the applications for research."

Bekele was immediately inundated with letters, telegrams, and telephone calls from Johanson, Taieb, and the French embassy protesting the hold on the IARE field permit. Johanson started the ball rolling by writing, in his inimitable style, "I

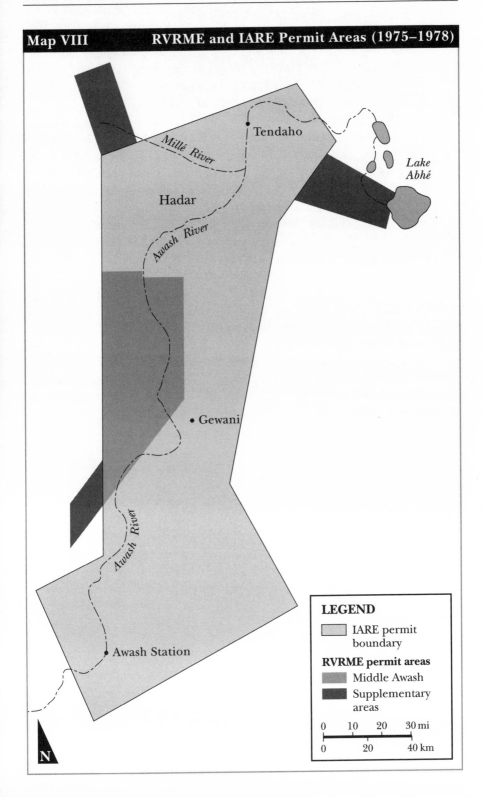

Map VIII   **RVRME and IARE Permit Areas (1975–1978)**

Millé River

Tendaho

Lake Abhé

Hadar

Awash River

Gewani

Awash River

Awash Station

LEGEND

IARE permit boundary

RVRME permit areas

Middle Awash

Supplementary areas

0   10   20   30 mi

0   20   40 km

N

hope this letter finds you in good health and that your duties at this time are not too overwhelming."[3] He was "most surprised by [Bekele's] comments concerning Mr. Kalb" because he had no idea that I was contesting anything. "So, you can see, I have been quite surprised by your letter." Bekele next received a letter from Taieb, saying that the National Geographic Society, NSF, and FORGE informed him they were not supporting my research because I lacked adequate degrees.[4] Taieb copied his letter to the Ethiopian minister in charge of the Antiquities Administration and to the French, American, and German ambassadors. In a follow-up letter to the minister, he listed the names of individuals he said disapproved of my participation with the IARE.[5] At the head of the list was Mary Leakey, who soon wrote from Olduvai, bless her, denying having said any such thing.[6]

With the French embassy bombarding the Antiquities Administration with protests, on August 28 Bekele again wrote Taieb denying authorization for field work until a new agreement was drawn up "duly signed by the signatories of the 1972 agreement."[7] Soon Bekele received a letter from the scientific director of the Centre National de la Recherche Scientifique,[8] sent through the French ambassador, and another letter from Taieb, both letters insisting that field work proceed under the terms of the original agreement.[9] Taieb added that he would "take into consideration" a supplementary agreement.

What Taieb and Johanson did not appreciate was that the old guard in Ethiopia no longer held power. The minister to whom Taieb had previously appealed was now minister in name only. The shift in the government had been underscored in recent days by jets streaking over the rooftops of the capital and tanks rumbling through the city in a show of unity among the armed forces. Helicopters dropped leaflets promising a new order in the making.

On September 10, Bekele sent Taieb both a telegram and a letter once again denying authorization for field work, this time pending a request to CNRS, NSF, and FORGE to help arbitrate the dispute.[10] That same day Taieb sent Bekele an agreement, written in English and French and signed by himself, Johanson, and other members of the IARE.[11] In apparent reference to Johanson's handling of the hominid discoveries the previous year, the two-sentence document stated, "We the undersigned . . . are in full agreement to work together as an integrated team devoted to scientific research. In addition, we agree to freely share any and all results of this work."

The next evening two films were shown on Ethiopian television, now under the control of the *Derg*, the secretive military committee that had effectively run the country for the last few months.[12] The first film contrasted the lives of the underprivileged masses with the comforts Haile Sellassie afforded his dogs, including his beloved Lulu, a Chihuahua that had traveled with him overseas. The second film was a doctored version of Jonathan Dimbleby's 1973 film about the Wallo famine, showing people starving, interspersed with scenes of the Emperor feeding his dogs from a silver tray.[13]

The following day at dawn, September 12, 1974, a small group of army officers went to the Jubilee Palace and summoned the Emperor to the library, where he soon appeared in full uniform. As Haile Sellassie stood erect, a nervous officer read a proclamation informing him that he was deposed. The scene was described by Marina and David Ottaway in their book *Ethiopia: Empire in Revolution:*

> It is said the police officer's hand shook uncontrollably while he read the proclamation and that some of the soldiers in the room wept so profusely that they had to leave. The Emperor refused at first to leave the palace. But then *Ras* Imru [then Prime Minister] walked up to him, kissed him on the cheek, and said, "Go." The Emperor, dazed and apparently still uncomprehending, walked with the *Ras* to the palace steps looking for one of his six different-colored Mercedes limousines. Instead, he was driven in a small Volkswagen the short distance to the Fourth Division Headquarters and imprisoned in a small wattle-and-mud building.[14]

The King of Kings, said to be the 225th descendent of the biblical union of the Queen of Sheba and King Solomon, was finished. He had outlasted President Nixon by 35 days.[15]

As the news raced across the city, there was great rejoicing, although in many homes, wealthy and poor alike, there was also a great sense of tragedy and fear. Haile Sellassie, after all, was no mere mortal, but "Elect of God." In midday, with Judy at home tending to four-month-old Spring, I piled Justine into our vintage Hillman Husky station wagon, drove down our lane to Teodros Circle, then east to Menelik II Avenue and to the parking lot of the Hilton Hotel. From there we walked the half-block to the palace, where I had heard that a large, festive crowd was gathering. Armored tanks were parked on either side of the front gates. Joyous people were throwing Meskel daisies, symbolic of good fortune, on the vehicles and handing them out to soldiers. Justine, who now spoke Amharic as well as any four-year-old Ethiopian, generally knew the significance of the occasion, having heard our servants talking excitedly about it all morning. I explained to her that it was not for us to celebrate the overthrow of the Emperor, but to hope that this was the beginning of a prosperous time in the country's history.

Although the city was peaceful, the ruling class—those not already arrested—had every reason to be fearful. Over the preceding month the press had vilified the Emperor, the royal family, and the entire aristocracy, accusing them of profiteering and corruption, even blaming them for the many prostitutes in the capital.[12] The Emperor himself was accused of being a "thief" who had amassed billions of dollars in private wealth from the sweat of the masses.[13] And he was accused of personally sanctioning the cover-up of the famine in Wallo. There were cries of "hang the Emperor," although the leaders of the coup insisted that the revolution would proceed without bloodshed.[16] By the time Haile Sellassie was deposed, all of his assets not in foreign banks had been seized, the Constitution suspended, and the

Crown Council, Crown Court, and Parliament abolished. In the place of the monarchy was the snowballing power of the military, increasingly under the influence of radicals from their own ranks, those released from prisons, and those returning from overseas.

On Revolution Day, as September 12 was subsequently known, the leaders of the "bloodless coup" announced that the country would be run by a provisional military council—still referred to as the Derg—until a "people's government" had been established.[14] Four days later, several thousand students demonstrated at Addis Ababa University demanding a civilian government, as did the national labor unions.

The struggle between the civilian population and the military had begun.

That same day, September 16, Johanson and Taieb arrived in Addis Ababa and promptly confronted Bekele, demanding that the Antiquities Administration give them a permit for the field season. Bekele again refused, pending a new protocol agreed upon by "the signatories of the 1972 agreement." In a letter to Bekele I told him that I would not sign anything that "makes me beholden to the control of people who do not even reside in this country. . . . I did not come to Ethiopia four years ago to see my work directed from Paris."[17] At the same time I insisted that Johanson write a letter to all IARE members apologizing for withholding information about the hominid discoveries the previous year. Finally, I appealed to the Antiquities Administration not to renew the permit that gave Taieb and Johanson exclusive control of the entire middle and lower Awash Valley.

After eight days of deadlock, Bekele came by my office at lunch time and told me that Johanson claimed I was a CIA agent. Piecing together Bekele's account, Johanson told him that I received covert money from a secret CIA front organization, FORGE, and that I had a "connection" with the U.S. embassy through Dennis Peak's father, who worked in the political section. Johanson argued that FORGE was not listed in any foundation directories and that his own inquiries had produced no information about what FORGE was or where its money came from. No reputable organization, he said, including the National Geographic Society and NSF, would support me because I lacked proper academic credentials.[18] Johanson made it clear to Bekele that unless he gave the IARE a field permit immediately, he would take his charges elsewhere in the newly established military government.

I explained to Bekele that FORGE was a small but reputable foundation, whose president was a former president of the American Association for the Advancement of Science, and that its board of advisors read like the Who's Who of American business and science.[18] I produced copies of FORGE correspondence to and from Johanson that showed that he knew exactly what FORGE was. Included were two letters he had written to its director endorsing my early grant applications. As for whatever Dennis's father did with the embassy, I said, that was no concern of mine.

Bekele said he believed me, but he also said that these were dangerous times in Ethiopia and that he could not have Johanson going from office to office in the government until he found someone who would listen to his story. He asked me to sign the new agreement drawn up by Taieb and Johanson—and by Coppens, who was also then in Addis Ababa. I signed the document two days later. By then, however, Bekele had already granted the field permit, and shortly thereafter the IARE left for the field. I remained in Addis Ababa.

Needless to say, my relationship with the expedition was over, although in fact it had ended months previously. The new agreement, which we all signed, was my going-away present to my former colleagues, and to Hadar paleoanthropology. In most respects it was similar to the 1972 agreement, except for one major departure. The new agreement contained the following stipulations:

The precise location of all hominid localities will be fully documented under the control of M. Taieb, Y. Coppens, D.C. Johanson, and J. Kalb, as follows:

— informing all concerned expedition participants of new finds,
— having neutral parties view specimens in situ,
— properly entering all hominid fossils into the Expedition's fauna catalog,
— allowing all new hominid localities to be geographically and geologically documented according to established procedures,
— allowing the Expedition's archaeologist the opportunity to investigate hominid localities for habitation evidence or other paleobehavioral information.[19]

There was one more addition to the agreement. After signing the document, I scribbled across the top in bold letters: *ETHIOPIA TIKDEM*. "Ethiopia First," the slogan of the Ethiopian Revolution.

On that day I joined the masses.

The IARE had been at Hadar for scarcely three weeks when four hominid jaws were found.[20] Remarkably, three of the jaws were found over a two-day period, October 16 and 17, at three different localities by the same person, Alemayehu Asfaw, the antiquities representative who had worked with us at Hadar the previous year. All specimens came from near the base of the section. Johanson identified the fossils as early *Homo* and rushed back to Addis Ababa to announce the discoveries at a press conference. In a written statement, he described the finds as "the most complete remains of this genus from anywhere in the world" and estimated their geological age as "approaching 4.0 million years."

He elaborated:

This find is unparalleled. Discovery of the genus *Homo* older than 3.0 million years is a major step forward in our understanding of early man's evolution and represents a major revolution in all previous thinking concerning the origin of the group representing modern man. The completeness of the specimens refutes any controversy regarding their identification. . . .

Previous discoveries from Olduvai and even Lake Rudolf [Turkana] have only taken the origins of the genus *Homo* back to slightly over 2.0 million years. We have in a matter of merely two days extended our knowledge of the genus *Homo* by nearly 1.5 million years. All previous theories of the origin of the lineage which led to modern man must now be totally revised. We must throw out many existing theories and consider the possibility that Man's (*Homo*) origins go back well over 4.0 million years. The discovery of specimens assigned to *Homo* in this geographical region further suggests the possibility that the origins of man may have been outside of Africa. This fact is a revolutionary postulate which will meet with extreme skepticism but must now, with the Afar discoveries be considered. . . .

These discoveries of the earliest specimens of the genus *Homo* anywhere in the world will provide perhaps the most provocative human fossils ever discovered on the African continent. It is certain that anthropologists from all over the world will meet these discoveries with extreme controversy and amazement.[21]

The announcement was carried widely in newspapers. The finds were truly significant, except that D. Carl would soon retract everything he had said and then spend years arguing that only australopithecines were recovered from Hadar.[22] As for his estimate that the fossils were to "close to 4.0 million years," this too would be refuted; the fossils were closer to 3 million years old. Much of the ensuing "debate" about the fossils was with Richard Leakey, who was delighted with the identification of early *Homo* at Hadar.[23] Like his father, he believed in the great antiquity of humans and believed that *Homo* and the ape-like australopithecines diverged from a true ape some 4 to 5 million years ago.

So how did Taieb and Coppens fit into this new burst of media-science? Although in our newly signed agreement Taieb was reinstated as the single head of the IARE, and Coppens was still its senior anthropologist, the roles of both would become increasingly subordinate to Johanson's. Given the extraordinary richness of Hadar, monumental discoveries were forthcoming, to be sure, but despite decades of scientific work to be done at the site, incredibly, the IARE lasted only two more field seasons. By then the expedition had been entirely subsumed by Johanson's ambitions and, ultimately, by the ramifications of the CIA allegations about me that would be spread far and wide. As the Ethiopian Revolution progressed, many would learn that spreading false accusations of subversion was a treacherous game to play.

During the first few weeks of October 1974, the positions of the military and those demanding civilian rule became increasingly adversarial. There was ample opposition within the military itself to the Derg, and confrontations with students and labor unions led to more demonstrations, arrests, and shootings. Open opposition to the Derg continued into November, involving military officers of higher and higher rank. Then, in mid-November, special courts established to prosecute corruption and other abuses returned indictments against 87 former officials for their roles in the cover-up of the Wallo famine, an offense proclaimed punishable by death.[24]

Prominent among those who opposed the most radical elements of the Derg was General Aman Andom, called the "Lion of the Ogaden" for his leadership role in the 1964 border war with Somalia. After Haile Sellassie was overthrown, Aman was named head of state—but not of the Derg, where the real power resided.[25] Aman, an Eritrean, favored a negotiated settlement with the northern separatists and opposed the execution of former officials and aristocrats, a stance that put him directly at odds with many in the Derg. In late November, Aman made the mistake of discussing anti-Derg measures over the telephone, which was tapped.[26]

On the evening of November 23, a Saturday, Judy and I left the girls in the care of our housekeeper and went to a movie at the U.S. commissary compound on the south side of the city. Nonofficial Americans could go to the small cinema there for a small fee that included the opportunity to buy popcorn and eat American candy bars. On our way home, we drove past the Ministry of Mines at Mexico Square, then east to the intersection of Churchill Road and Johannas Street. There we were halted by a soldier standing in the road with his arms spread as a convoy of military vehicles rolled by, moving north. Judy and I were struck by the eerie silence of the convoy, passing in the night as it did so quickly and purposefully. Like a snake in the grass: it's there, now it's gone. When we got home, it was not too late to call a journalist friend and ask him what was happening. David Ottaway covered eastern Africa for the *Washington Post* and made Addis Ababa his base. His wife, Marina, taught sociology at the university.

David knew nothing about what was going on that night but said he would try to find out. And he did. The story, filed early the next morning, appeared on the front page of the *Post*:

Ethiopian Officials Executed

The lead was: "Ethiopia's military rulers announced today that 60 former Cabinet ministers and government officials were executed last night."[27] Those killed included General Aman, 2 former prime ministers, the former admiral of the navy (who was the grandson of Haile Sellassie), 15 other generals, 11 other officers of high rank, 8 former provincial governors, 5 former cabinet ministers, 5 former vice-ministers, 5 former officials of the Imperial Court, a former judge of the Supreme Court, and 2 former security chiefs.[28] The Ethiopian revolution had entered a new and bloody phase.

Unlike many former officials who were arrested without a fight, General Aman was not executed. The "Lion of the Ogaden" lived up to his name. He died that night in a furious gun battle with the men who had come to arrest him under orders from Major Mengistu, one of the top leaders of the Derg.[29] Soon a story circulated around Addis Ababa, particularly among Aman's Eritrean admirers, putting more and more soldiers in the party that Major Mengistu had taken to arrest the Eritrean hero. First it was said that a platoon of the Derg's men had been killed, then a company, then a regiment, and finally an entire division. It was also said that when Major Mengistu went to view the carnage after the battle, he realized that he had been tricked. Not only did he find Aman's bullet-riddled body, but lying next to him was a *second* Eritrean.

It is apparent that at the time I had spoken with Ottaway (about 9:00 P.M.), the political prisoners had not yet been killed. He later learned from eyewitnesses that at about 9:30 P.M. the prisoners were taken from the basement of the old Menelik Palace in the north part of the city to the central prison in Akaki, south of town. Because Aman lived in the south part of the city, when Judy and I saw the convoy heading north on Churchill Avenue, it must have been on its way to pick up the prisoners at Menelik Palace to transport them to Akaki. And there, according to reports, over the next few hours they were machine-gunned to death two at a time, the hideous job apparently being completed in the very early hours of November 24. The slain were buried in a mass grave.[29]

Later that same morning at Hadar, Johanson found "Lucy," the most complete skeleton of an early hominid known.[30] Thus, on the day that Addis Ababa awoke to the news of the end of humanity, at least as the families of the slain understood it, the IARE celebrated the discovery of humanity's beginning.

One of the ironies of these respective events is that a reason given for the execution of many of Ethiopia's elite was their cover-up of the famine in Wallo, where Lucy was found, and where tens of thousands of Afar nomads had died of government neglect. The two monumental events on that November day in 1974 may have marked the first time in Ethiopia's history that the Afar people and their unique land were given so much attention.

Lucy was a great discovery. Announced at another press conference in Addis Ababa on December 20, the find was described as a 40-percent-complete skeleton of a diminutive, bipedal, adult female about one meter high.[31] The expanded pelvis indicated its sex, the erupted wisdom teeth revealed its maturity, and the articulations of the leg bones proved it had walked upright. The array of 63 pieces of the skeleton were found the day after Richard Leakey, his wife Meave, and Mary Leakey had visited Hadar, and there was great celebration in camp that evening. The hominid got its name when the Beatles' song, "Lucy in the Sky with Diamonds," was played on a tape recorder.[32]

Lucy was found high in the stratigraphic section in the northern part of the site, just near the spot (Artifact Locality 12) where Corvinus had found a single basalt flake. The Lucy locality, L288, was surrounded by a cluster of seven other fossil localities mapped by Dennis Peak and myself.[33] At one time or another in 1973, probably everyone in camp had walked across L288, Johanson included.

The skeleton was discovered eroding from a sand layer stratigraphically above a tuff that thickens into a massive stream channel to the far west. At the time Lucy was found, however, Taieb still had not sorted out the basic stratigraphy of Hadar, with respect either to the tuff layer or to the basalt flow east of L288.[34] Nevertheless, two days after Lucy was discovered, Johanson sent the NSF program director of anthropology, Iwao Ishino, a glowing progress report that included a generalized geological section of Hadar.[35] The figure showed the basalt located 30 meters above L 288, instead of at its correct position 50 meters below L288, a difference equivalent to the height of a 20-story office building. Because the basalt was later dated radiometrically at close to 3 million years old, its position was critical to Lucy's age. Lucy was either "approaching 4.0 million years old" (with the basalt above L288) or closer to 3 million years old (with the basalt below L288), as was the case.[36] By the end of the 1974 field season, however, Taieb had finally determined the correct position of the basalt, which led Johanson to tell reporters with confidence at his next press conference, two months later in Cleveland, that Lucy was nearly 3 million years old.[37] Nevertheless, it would in fact take many years and research dollars before researchers reached any kind of consensus on Lucy's exact age.*

As for the identity of Lucy, Johanson must have immediately recognized its affinities with the australopithecines by its ape-like, V-shaped lower jaw, which contrasts with the U-shaped mandible of *Homo*. In a 1976 *Nature* article, he would describe how Lucy's jaw resembled those of the gracile australopithecines of South Africa, although certain features of the jaw, such as the greater compression of the dental arch, suggested a more primitive condition similar to chimpanzees.[39] In the same article, Johanson compared a skull fragment and partial femur from the lower part of the Hadar section to robust australopithecines from Olduvai and South Africa. Finally, he continued to maintain that the jaws found by Alemayehu in October 1974 were *Homo*, although in November 1974 he must have been struck by their similarity to the Lucy jaw. I suspect, however, that Johanson was reluctant to contradict this identification any time soon, because only five weeks previously he had proclaimed to the world that "[a]ll previous theories of the origin of the lineage which led to modern man must now be totally revised."

---

*Current estimates of the age of tuffs above and below Lucy are 2.9 and 3.2 million years, respectively; the underlying basalt itself is now dated at 3.3 million years. The "*Homo*" jaws from near the base of the Hadar section, which are australopithecine, are dated at approximately 3.4 million years.[38]

CHAPTER 12

A New Mission

Between late 1974 and early 1975 I struggled to stay afloat in revolutionary Ethiopia. While Johanson and the IARE were making discoveries and holding press conferences, I renewed my attempt to establish a research center of Afar studies. My efforts were now directed to the newly created Ministry of Culture, established after a restructuring of government offices, which included the Antiquities Administration, the National Library, and offices concerned with the arts and sports affairs. In meetings with ministry officials, I promoted my research center proposal, arguing against the "reactionary" expedition style of research that allows "hordes" of foreign-based scientists to routinely "invade" the country to "exploit" Ethiopia's heritage. I argued that a permanent research center devoted to studies in the Afar Depression would be aligned with the revolutionary government's demands for change and progress by advancing institutional development and the training of Ethiopians. Further, I reasoned, that unless the government stopped the flow *en masse* of fossils and artifacts overseas, the Ministry of Culture had no chance of attracting international monies to develop its own institutions for the study of antiquities. With the advice of people from USAID and the director of FORGE, I also charted routes that the ministry could pursue to begin raising modest sums to fund such a center. In other words, I sought to convince the ministry that *Ethiopians* could conduct prehistory research and that the likes of those running the IARE were unnecessary.

Of course the research center proposal never had a chance—certainly not without the support of foreign researchers who were essential for raising funds. And Johanson, Taieb, and Coppens were not about to support efforts they surely viewed as a threat to their own research.

In January 1975, I again sought permission to form my own research team, which I insisted would not be an "expedition," because it would be based in Addis Ababa and operated year-round. And once again I tried to convince Bekele Negussie to break up the IARE concession in the Awash Valley. As in August, I sought permission to work in the unexplored Middle Awash area south of Hadar, bolstering my appeal with maps showing that the size of the Taieb–Johanson permit was as large as the neighboring French Territory of Afars and Issas (see Map VI on page 91.)

"Look, Ato Bekele," I said, pointing at one of my maps, "there's already one French colony in the Depression, you don't need another!"*

At the same time, I ran around Addis Ababa talking to more government officials and giving slide presentations about the "limitless" potential for prehistory research in the Afar, particularly as it applied to unexplored areas in the IARE "monopoly." In an article I wrote for the *Ethiopian Herald*, I said that "the future may well show the Afar [to be] the richest single fossil area in Africa,"[1] and in reports for the Ministry of Culture, Dennis, Liz, and I submitted 36 maps depicting the fossil localities established by the IARE, illustrating the "vastness" of prehistoric deposits in the central Afar alone.[2]

I argued that the entire 33,000-square-kilometer IARE concession was 10 times larger than Lake Tana and 550 times larger than Hadar, which itself was twice the size of Olduvai Gorge, "where generations of Leakeys have worked for decades." Anticipating that the ministry might refuse my request, I continued to include in my permit application the three smaller areas peripheral to the IARE area. But I considered the Middle Awash critical to jump-starting a new team.

These efforts and deliberations went on for six excruciating months. Because Bekele and other ministry officials were exercising caution under the military government, and anticipating the protests that would follow from the IARE should my team be approved, I was forced to make almost daily trips to the ministry to provide officials with documentation about one formality or another. These mostly concerned proof of funding and the qualifications of specialists I was recruiting. I recall becoming so frustrated with some new formality during a meeting with Bekele in his office that I slammed my fist down so hard on the armrest of my chair that a pencil in my hand remained stuck in the wood. Fortunately, as tough as he was, Bekele never stopped being amused by the "abnormalities" of foreigners, and his scornful laugh would roll from his mouth like marbles.

While I was preoccupied with these matters, Judy was still teaching English in the Muslim Abadir School to supplement our income and was also picking up the odd job here and there. On one occasion she helped the former minister of tourism prepare an English text for his defense against corruption (he went to prison anyway), and later she presented a weekly classical music program for the Radio Voice of the Gospel that was broadcast across eastern Africa. "Good evening, this is Judy Kalb bringing you new selections of classical music for your listening pleasure." Later still, she weekly cleared all the furniture in our dining room and gave country dance lessons to children of a dozen nationalities.

Ato in Amharic is the equivalent of "Mister"; by custom, Ethiopians are addressed formally by their first names.

In October 1974, Justine began attending the French school on Churchill Avenue just down the hill from Teodros Circle and our house. A few weeks later soldiers went to the school and closed it down, pending the government's evaluation of how the curricula should best meet the revolutionary goals of the country. When the school reopened, classes were taught in Amharic as well as French. Henceforth she would write her French lessons in one notebook and Amharic script, with its 44-symbol alphabet, in another. That was the only time the school was closed except on holidays. It remained open even when bombs rocked public buildings in late November in protest to the mass executions of the nobles. When one explosion blew up part of the city hall just up the hill from the school, the vibrations rattled the windows of the classroom as well as of my office. The school also remained open over the following few months, when hundreds of thousands of people marched down Churchill Avenue in support of land reform and the government's opposition to Eritrean independence. During one of these parades, I walked with Justine as far as my office and then let our zabanya walk her the rest of the way to school. To this day I can still see her in a pretty green dress weaving her way through the chanting throng. She seemed unaffected by all this, but on Christmas day blanched with surprise when she saw her first Santa Claus at a party given in the home of an American friend.

After 34 formalities had been completed, on April 2, 1975, the vice-minister of culture, Solomon Tekalign, and Bekele Negussie both signed all eight pages of my permit application. This included the permit boundary maps and authorization to work in the Middle Awash, as well as in the three peripheral areas I had specified (see Map VIII on page 143). I named my new organization the Rift Valley Research Mission in Ethiopia (RVRME). The Ministry insisted that the word *Afar* not be used in the name lest it be confused with the IARE or with an "Afar" mission, and I insisted that the term *mission* be used, because we were not an expedition. The RVRME was to be a permanent organization with a full-time staff. Our offices were those that I occupied at Teodros Circle.

As I had promised Bekele, I began recruiting Ethiopian students to work with the mission. I was able to do so through a government program called the Development through Cooperation Campaign. The *Zemacha*, as it was called—"campaign" in Amharic—mobilized students for a year or more to work largely in the countryside organizing peasant associations, attending to land reform matters, and working with famine relief, literacy, health, and agricultural programs. They were also allowed to work with organizations that promised the students practical experience in research or data gathering that served the "common good" of the country.[3] The Zemacha headquarters allowed me to retain Dejene Aneme, still a senior at the recently closed Addis Ababa University, as our administrative officer and, over the coming months, to recruit four more senior undergraduates to work as research assistants in anthropology, paleontology, and geology. I also hired a secretary, a part-time draftsman, and several other students part-time. As determined by the Zemacha, the salary of the full-time students was 50 *birr* (U.S. $25) per month, which included my purchasing a brown uniform for each and a pair of thick-soled shoes. It turned out that six undergraduates, including Liz, formed the core of the RVRME scientific staff, a fact that was looked upon with disdain by the high-powered research teams working in the country.

For specialists I was able to recruit Fred Wendorf, an archeologist with Southern Methodist University; Clifford Jolly, Glenn Conroy, and Doug Cramer, anthropologists from New York University; and a complement of Ph.D. students in archeology, anthropology, paleontology, and geology from SMU, Yale, Harvard, and Princeton. Most had had field experience in Africa, and Wendorf, Jolly, and Cramer had worked previously in the Main Ethiopian Rift. Other specialists agreed to work with the mission but would later back out for security, scheduling, funding, or "political" reasons. I assumed some of the latter had to do with Johanson, who fought the RVRME authorization in his usual manner. According to Wendorf, Johanson flew to Dallas to try to convince him that I was with the CIA.[4] He made similar overtures to others, and soon stories began circulating—from unknown sources—that Bekele had been bribed.[5]

For funding, RVRME participants received grants from 16 organizations and institutions over the ensuing years.* As it happened, I also contributed substantial funds to the mission from my own pocket. This included the salaries of the Ethiopian students, although their field expenses were shared from grants given to individual team members. Per diem expenses provided in grants were used to contribute to the office expenses when mission members crashed at the office. Much of the field equipment came from Clifford Jolly, who in recent years had maintained a large, well-equipped research camp in Awash Park to study baboons. For field vehicles, we began by renting two Land Rovers in 1975, but later that year I bought a used, long-wheel-base Land Rover from a one-armed English mechanic. The next year Wendorf and I purchased four Toyota field vehicles from Tenneco Inc. at $500 each, using money Fred raised in Dallas. I immediately sold one of the vehicles to pay the duties on the two best Toyotas; the fourth, a junker, we kept for spare parts.

Tenneco was closing up shop in Ethiopia after failing to find commercial amounts of oil and gas in the Ogaden, while enduring more than its share of problems in the country. In late 1972 a Tenneco vice-president was shot in the back during the same attempted skyjacking of an Ethiopian Airlines flight to Europe that I described on pages 126–127.[6] Three of his colleagues were injured by a concussion grenade set off by one of the hijackers during the desperate midair

---

*The Bache Fund of the National Academy of Sciences, the Boise Fund of Oxford University, the Explorer's Club, FORGE, the Harvard Museum of Comparative Zoology, the L.S.B. Leakey Foundation, the National Institutes of Health, New York University, Providence College, Rutgers University, the Sigma Xi Fund, Southern Methodist University, the Spencer Foundation, Tenneco Inc., the Wenner-Gren Foundation for Anthropological Research, and Yale University.

shoot-out with security guards. After the grenade blew a hole in the cabin floor, the jet went into a dive, and the stricken oil men were sure they faced certain death. However, the pilot pulled the plane out of the fall, and they safely returned to Addis Ababa. Then four geologists working with Tenneco were kidnapped by rebels in Eritrea in March 1974 and kept in chains for six months before being released for a modest ransom on the Sudanese border.[7]

We were ready to begin explorations in the Middle Awash in early April 1975. I was joined by a graduate student of Wendorf's, Herb Mosca, a bearded, gangly fellow Texan. Just prior to our leaving for the field, however, Judy became seriously ill with hepatitis and was hospitalized at Saint Paulos Hospital for a month. While she was there, I put Herb to work with aerial photographs making a 1:60,000 drainage map of the northern sector of the Middle Awash beginning at Hadar, as an extension of the base maps that Dennis and Liz had made farther north.[8]

In early 1975 the geology of the Middle Awash Valley from Gewani to Hadar was virtually unknown (see Map VII on page 140). Although Ludovico Nesbitt traversed part of the western Middle Awash in 1928, and Wilfred Thesiger the eastern Middle Awash in 1934, neither made any geological observations to speak of.[9] The first geological reconnaissance of the area, in early 1938, was carried out by an expedition sponsored by the national oil company of Italy, AGIP.[10] Led by geologists Michele Gortani and Angelo Bianchi, the purpose of the expedition was to look for oil seeps or "asphalt lakes" ostensibly found 50 years earlier by Emilio Dulio, a former administrator of Italian Somaliland. Dulio had passed through the area in 1888 accompanying a caravan traveling to Awsa and was convinced he had seen hydrocarbons in the region, principally at the Teo hot springs just southwest of the Kariyu basin. Although Gortani and Bianchi viewed the prospect of finding oil with skepticism, Dulio apparently held enough sway with the Occupation government to see the exploration carried out, and to accompany the expedition himself, even though he was 78 years old.

The expedition was no ordinary affair. It consisted of 85 men, including a platoon of 30 Somali soldiers and an assortment of drivers and mechanics to operate 6 heavy-duty Fiat trucks, 2 smaller Fiat trucks, 2 Ford trucks, and a Caterpillar tractor. The expedition also included a medical doctor, a sanitary aide, radio operators, a carpenter, Ethiopian workman, 2 Afar chiefs, 2 interpreters, 2 guides, 30 camels, and 8 mules.

As described in the official log of the expedition, on January 24, 1938, the exploration was preceded by an aerial reconnaissance of the area from Gewani to "the big elbow" of the Awash, almost certainly the eastward bend of the river in the area of Gona and Hadar. Thus the pilot and anyone who accompanied him may have been the first to set eyes on the fossil deposits, at least from the air. By early February the ground expedition was under way. Two days were spent traversing the area between

Gewani and Hargufia 60 kilometers to the north; an additional three days were spent going to and from Teo, and another day was spent returning to Gewani. No oil was found. Two trucks and the tractor broke down during the week-long trip. On February 11 Bianchi wrote to his wife, "The legend of oil in Dancalia is finished."[10]

Despite the brevity of their survey, Gortani and Bianchi made a respectable regional map of the southeastern Middle Awash, which was published in 1941 and reissued in 1973.[11] They distinguished several sedimentary and volcanic units, but in their report they made no mention of vertebrate fossils or artifacts.

Paul Mohr's 1963 map of Ethiopia did little more than distinguish broad areas of volcanic rock in the Middle Awash from sedimentary areas.[12] A geomorphology map, made largely from aerial photos, was included with a 1965 UN FAO study to develop the Awash Valley. This map distinguished areas suitable for agriculture from volcanic rocks and "badlands," deeply eroded terrain nearly devoid of vegetation.[13] Taieb was aware of this map, which may well have called his attention to Hadar. His own geomorphology map of the Awash Valley, contained in his 1974 Ph.D. dissertation, is virtually identical to the FAO map, except that he failed to depict the badlands.[14]

In addition to these sources available to Taieb when he compiled his 1972 geological map of the Awash Valley, an unpublished photogeological map of the southern Afar had been made by geologists Getahun Demissie and Ceri James with the UN Geothermal Project and the Geological Survey.[15] Although Taieb added almost nothing to his own map of the Middle Awash area that differed from the combined Italian and UN maps, he did distinguish the Pleistocene gravels in the Hadar area from the underlying richly fossiliferous Pliocene sediments. The older, badland beds—exquisitely clear on aerial photographs of the western Afar used by Demissie and James—were curiously omitted from their map. This was almost certainly done at the urging of Taieb, who proffered the geologists his advice on the sedimentology of the Awash Valley, for which he is credited in the legend of the UN map. We can assume that in late 1971, Taieb was desperate to keep knowledge of Hadar under wraps and to be the first to publish a geological map of the region. He did just that in the proceedings of the French Academy of Sciences in August 1972, in an article co-authored with Coppens, Johanson, and myself. But the map itself, included as an insert, was published as compiled by Taieb only.[15] The map by Demissie and James was not officially released with the UN geothermal report until 1973. On their map, the Pliocene exposures centered at Hadar do not exist; in their place is depicted a monotonous surface of "Upper and Middle Pleistocene conglomerates."[15] On Taieb's map, however, the extraordinary, fossiliferous sediments surrounding Hadar are clearly shown.*

---

*While at the Ethiopian Geological Survey in 1971–1972, I unwittingly served as the go-between in the exchange of maps between Demissie and James and Taieb that resulted in the two versions of the geology in the greater Hadar area. Taieb was later reprimanded by the Survey for liberal use of the UN map before it was officially released, even though on his map he credited the work of the UN team. Taieb also credited the FAO for use of the geomorphology map.[15]

Although these various ground and aerial surveys conducted between the 1930s and the early 1970s drew attention to the geology of the Middle Awash, as of mid-1975 not a single artifact or vertebrate fossil had been described from the area. For the purposes of the RVRME, therefore, our entire Middle Awash permit area, 4475 square kilometers, was unexplored.

That would soon change.

At the end of April Judy was released from the hospital, and our team was ready to go to the field. The government saw fit to celebrate the occasion by having a full-dress May Day parade that pitched the city into a delirium of political activity. Before loading up the Land Rovers, I hoisted Justine on my shoulders and marched off to watch a half-million people flow down Churchill Avenue shouting political slogans and waving placards of every persuasion. Included were regiment after regiment of soldiers, soon to be the largest standing army in Africa, peasant associations, labor union groups, contending political parties, "Red Guards," and Zemacha students. Many carried Mao's Little Red Book, while extolling universal freedom, class struggle, and land to the tiller. The spirit of Mao, Marx, and Lenin had surfaced in the Ethiopian highlands.

We had stayed in Addis Ababa as long as I dared, because during the time that Herb worked with me, he had fallen savagely in love with a beautiful young woman from a prominent Armenian family in the city. Increasingly, he was talking about scuttling his career and spending the rest of his life raising Armenians. During his courtship, however, he had held his beloved's hand only once, in the sitting room of her home, while one aunt was being replaced with another. Such chastity was a phenomenon Herb had never experienced in Dallas, and he was entirely befuddled. After listening to him tell me over and over what it was like to hold that precious hand for those fleeting seconds, I realized he was becoming truly witless. I had read accounts of how brutal Henry Stanley was to his men while dragging them through the Congo, and I knew that if we were to make great discoveries, I had to act now or never.

At dawn on May 13, 1975, we left the city to explore the Middle Awash.

CHAPTER 13

An Acheulean "City"

Over the next two months Herb Mosca and I traveled 5000 kilometers, nearly a third of that in the bush. My plan was first to survey the eastern Middle Awash by going south from Meshellu and then to cross into the western area by going north from Awash Station. The first stop was Geraru, where we dropped Liz. She was to spend a month there with several of my crew, including Kelati Abraham, who was last with me when I went into the Kariyu basin. I had hoped to have Hebah join her as well, but we were saddened to learn that our congenial colleague and consort to the sultan had recently been killed in a shoot-out with Issa. Geraru was not in my permit area, but Bekele agreed to let Liz continue working there on the stratigraphy as long as she did not collect fossils. Herb and I planned to pick her up in mid-June when she would join us in the Middle Awash along with Fred Wendorf and Glenn Conroy. Wendorf wanted to bring two potential financial backers with him from Dallas. I was confident we would have something to show them.

There were easier ways to get into the eastern Middle Awash than from the north, but I considered Meshellu unfinished business, having twice failed to make it into the main basin in early 1972—once with Kelati and Ali and later with Taieb and Johanson. This time I had picked out a route to follow on aerial photos, and three days after leaving Addis Ababa, Herb and I stood on the south rim of the plateau overlooking the Meshellu badlands.

The vista looking north was a jagged sedimentary expanse raked by wadis leading to the Awash. Reliefs seemed higher and steeper than those across the river at Hadar. Consisting of an immense U-shaped basin, the sedimentary exposures were also twice the breadth of those at Hadar. After taking photos and marveling at the panorama, we climbed back into the vehicles and followed a dry streambed down to the Awash. We saw no people in the area, but we did flush two leopards out of some craggy shadows. I wondered whether they were the same as those I had seen in 1972, or perhaps their offspring.

After reaching the Awash, we drove westward next to the river and its envelope of deep green forest, stopping periodically to look at sediments or fossils. Soon we reached an expanse of flat alluvium south of Denen Dora, the small wadi on the north bank that had been the subject of Johanson's first press conference. Denen Dora was also the area where Corvinus had found her stash of Acheulean handaxes. The alluvium, at Meshellu, covers a broad area between outcrops of Pliocene sediments and is long enough for hominid hunters to use as an airstrip.

Because of the northward slope of the Pliocene strata on both banks of the Awash, there are 10 to 15 meters or more of basal section exposed at Meshellu than across the Awash. At least one undated tuff lies below the lowest tuff at Hadar (itself dated at 3.4 million years), which could provide an age for any pre-Lucy fossils discovered in the lowest Meshellu strata.

The next morning we moved south. A careful study of the basin will someday produce many great discoveries. Johanson reportedly crossed the Awash in a rubber dinghy in 1976 and briefly looked for any obvious hominid loot, but the area remains basically untouched today. One of the few far-sighted things the Ministry of Culture has done in recent years is to set this area aside for future Ethiopian scholars. At a minimum, the Pliocene sediments in Meshellu should be the best sample of strata that are equivalent to the middle levels of Hadar, from which most of the hominids have been recovered.

For the next two days we dug our way south through Pleistocene badlands until we crossed the Wallo–Hararghe provincial border, which was also the boundary between the IARE and the RVRME concessions. On one occasion we spent hours building a dirt ramp for my Land Rover after I had driven it off the bank of a stream. "No, I'm not 'goddamn blind'"! I yelled at Herb. When we moved near the Awash, we were forced to chop our way through dense forest. It was not the "quivering green hell" of the Ituri forest in the Congo that Stanley described, but it took us some hours to pass through. The adjacent terrain consisted of riverine sediments containing thick tuffs, fossils, and artifacts of obvious Pleistocene age. Herb judged the stone tools to be Middle Stone Age (MSA), like those described by Wendorf in 1972 in sites he excavated in the Main Ethiopian Rift.[1] As Herb explained it, MSA toolmakers—early *Homo sapiens*—lived between about 35,000 and 185,000 years ago (although these time period estimates differ among experts). They are known for manufacturing a diverse toolkit of light-duty flaked tools, a distinct shift from the large handaxes and cleavers used for butchering large animals that characterized the Acheulean.

On the afternoon of the second day, we stopped among some low hills just east of the Awash to take a breather and look at some interesting stone tools. As we passed around some canvas waterbags, I pulled out my tape recorder to have our local guide give us the name of the stream passing through the site.

"*Andalee*," the guide said into the microphone, "*Andalee*."

"*Endegana*," I said in Amharic. "Repeat that."

"*Anda lee. Andalee. An da lee.*"

"Again."

"*Andalee. An Da Lee. Andalee.*"

"Okay, what does *Andalee* mean?" I asked through an interpreter.

"Nothing, it's just a name."

"It's got to mean something."

"No, it means nothing."

"Is he saying *ounda,* meaning 'small,' and *lee,* meaning 'water'—*small water?*"

"No, he is saying *Anda lee.* He is not saying *Ounda lee.* He is saying *Andalee.*"

"So, what does *anda* mean?"

"It doesn't mean anything. It's just a name."

"Could *Anda* be the name of a person, maybe a person who lived here years ago?"

"Who knows?"

I snapped the stop button on my tape recorder.

What *Andalee* means or originally stood for I never found out. But that is the name for the stream that I put on our maps of the Middle Awash, which we later published in a dozen scientific journals, along with the names of other landmarks we mapped up and down the valley.[2]

Herb and I examined the Andalee stone tools. Among the artifacts were crudely made core axes, small bifacial handaxes, flakes, and abundant rock cores and debris. The handaxes were made of volcanic rock and were typically Acheulean, but small—no larger than my hand—and more triangular than those I had seen at Hadar. Herb thought they looked like a very late phase of the Acheulean tradition. The most striking stone tools, however, were large, crude picks made from stream cobbles. At first glance they looked like the early Acheulean handaxes described from the middle levels of Olduvai. A second look, however, showed them to be shaped to a point on three sides, whereas the surface on one end, the cortex, remained unworked. Herb said these trihedral tools were *pickaxes* in archeology terminology and looked like those described from "Sangoan" sites in East Africa. A poorly understood late Pleistocene tradition, the Sangoan is named for a site in Uganda that follows a succession of Acheulean tool levels.[3] He explained that some archeologists speculated that the picks were made for woodworking or for some use in a forested environment.

Herb had barely finished this explanation of the tools when one of our crew, Meles, reached down and picked up the fossil jaw of a small animal. It was a monkey. And yes, monkeys live in forests. It was perfectly preserved for such a delicate fossil, and soon everywhere we looked we found monkey jaws and postcranial bones, including a complete face and maxilla. The teeth were coated with a shiny oxide that made them glisten in the afternoon sun. The site was *primatiferous!* We found other fossils, mostly small animals such as small bovids and pigs. We had found similar large numbers of monkeys at Amado in late 1973, the Pliocene site north of Hadar, but those fossils were mostly isolated teeth associated with well-broken-up bones and teeth from large animals, reflecting deposition from high-energy streams passing through a combination of forested and savanna paleohabitats. The excellent preservation of the Andalee monkeys could have been possible only where animal remains were buried in a low-energy environment with minimal physical transport and damage by streams. This fit with the geology of the site, which suggested overbank deposition removed from the main stream channel. Perhaps the environment was next to or within a gallery forest inhabited by monkeys, such as the gallery forest surrounding the Awash at Hadar, where the chatter of monkeys could not be mistaken at sunrise and sunset.

Herb also said the Sangoan was a "terminal" Acheulean tradition that apparently preceded or coincided with the early MSA. If the Andalee tools were post-mainstream Acheulean, or very late Acheulean, how much younger was Andalee than the Acheulean levels at Hadar? From letters I had received from Corvinus, I knew that in late 1974 she found the source of the bulk of the Acheulean tools at Hadar beneath the plateau gravels. This made the gravels at Hadar post-Acheulean in age. I recalled that Corvinus reported concentrations of MSA tools on the surface of the gravels, which suggested the gravels were pre-MSA in age as well. What, then, was between the Acheulean and the MSA? The "terminal" Acheulean and the Sangoan? And where were the gravels at Andalee? Over the previous two days we had found plenty along the banks of the Awash. . . .

I walked over to the Land Rover and grabbed my topographical map. I thought again of the areas we had passed through in the previous few days. In areas where we found massive gravels or conglomerates next to the river, even in the river, the Awash flowed in nearly a straight line. It did so for 8 kilometers beginning just north of Andalee. Nearly everywhere else in the Awash Valley the river flowed in sweeping meanders, through soft sediments. I recalled that beginning just east of Hadar there is a 5-kilometer stretch of the river that also flows in a nearly straight line all the way to Camp 270. I knew that because one afternoon, while waiting for my Land Rover to be repaired, I had followed that part of the river by foot high on the bluffs overlooking the Awash. I was curious about its course because my topo map showed the river flowing in nearly an east–west line, and I knew of no such trending faults in the region. My survey revealed that the straight stretch of the river was not due to a fault. Instead, there the Awash sliced through tens of meters of basalts. In the absence of faults, a river can surely cut through solid rock simply by erosion, but it will do so along the shortest and least resistant path, even if that is through igneous rock, marble, or cemented gravels.

That straight 8-kilometer length of the Awash north of Andalee had to be flowing through massive gravel conglomerates. From the monkey site, I walked downhill toward the Awash and almost immediately found gravels eroding from beneath a veneer of alluvium. The fossil beds appeared to be lying at most a few meters above the gravels. This meant that within a relatively short period of time geologically, Andalee had been transformed from a sterile cobblestone desert pavement into an environment supporting a thriving population of monkeys, and humans. But could the gravels be a localized stream channel? Herb and I walked east of the site—finding more areas covered with monkey fossils—until we reached a steep streambed. Its banks were made of several meters of massive gravels, including large cobbles, surely the source material of the pickaxes and probably of the small handaxes as well. These gravels had to be the same as those in the river. But were they also the same as those overlying the Acheulean tools at Hadar? From within and on top of the gravels Corvinus had described a "plateau stone tool industry" at Hadar. What was its relationship to the industry on top of the gravels at Andalee? Could they have been contemporaneous? No idea. The last place we had seen plateau gravels before moving south from Meshellu was in the center of the basin, where they overlay Pliocene strata. If the artifacts between Meshellu and Andalee were MSA, or Sangoan, was it possible that all the intervening sediments and artifacts between the two areas lay above the gravels?

As we sat on the streambank looking at the gravels, Herb was picking up one cobble after another to see what they were made of and then slamming them into the streambed as though he were trying to kill a snake. "For Christ's sake, Herb, what if they are made of goddamn plutonium?"

Unless I had missed something in my haste to move south from Meshellu, it was apparent that the gravels formed an old erosional surface that sloped from the higher plateau in central Meshellu to the Awash north of Andalee, just as it did from the plateau at Hadar to the river. The sediments above the gravels, the gravels themselves, and the sediments immediately below the gravels bearing Acheulean tools all must thicken toward lower elevations, toward the Awash. Sedimentary deposits thicken downhill. Was this supposed to be profound? Had I been in the sun too long? What about the gravels? What *about* the gravels?

That night we camped at Andalee, and over a dinner of lentil stew Herb and I discussed it all in scholarly terms.

"Kalb," Herb said, "I'm sick to death hearing about your frigging gravels. Gravels don't mean squat to archeologists. It doesn't mean piss all whether you have artifacts *on* gravels or *in* gravels—they're all transported from the moon and back. They have as much value as this horse manure you call food."

"Oh yeah! You archeologists don't know *squat* then! I love gravels. I like to sleep on them, lick them, and *eat* them. They mean *everything*."

"You're sick, you know that, *sick*. You've got gravels in your brain and coming out your ass!"

"Well, I'll tell you this, Herb," I said matter-of-factly, "I'll get more out of those gravels than you'll ever get out of your *precious* girlfriend in Addis Ababa."

Herb immediately went for his knife. I could see the glint of the blade in the moonlight as he lunged for my throat. Fortunately, it was just one of those small plastic picnic knives, and it snapped harmlessly against my neck. He then grabbed me and we rolled into the campfire, slugging and punching. We fought and rolled under the Land Rover, into thorn bushes, into the fire again, and down the hill. All the while everyone else in the camp surrounded us, yelling and betting on which one of us would kill the other.

"*Ullet birr* [two dollars in Amharic] on *Yon* [Jon]!" one of them yelled out.

"*Sidiha dollare* [three dollars in Afar] on the Bearded One!" another yelled.

"*Arat* [four] *birr* on *Yon!*"

"*Konoy* [five] *dollare* on the Bearded One!"

After a furious fight that seemed to last for an eternity, we both lay battered, bloodied, and exhausted. Following more angry words, the party broke up and everyone turned in for the night.

I had been afraid something like this would happen. The tension and hardship of the expedition. The dense forest, the badlands. One insufferable day after another. Push onward, push, push, push. Herb was finally losing it. He couldn't take the pressure. The same thing had happened on Stanley's last expedition across the Congo. Some of his men degenerated into madness and depravity in the jungle. That was happening here. I had no choice. I would have to shoot Herb. I would have to. . . .

"Come on, princess, time to get up," Herb said, as he kicked me. I looked up. It was daylight and, yes, it was the same old Herb. My God, another horrible nightmare!

After breakfast we photographed the Andalee site, the fossils, and the artifacts. We collected nothing, neither there nor anywhere else on that field trip. At that stage we were just looking and mapping. We packed up camp and headed south.

Much of the day was again spent grinding and digging our way through badlands until we finally climbed out of the drainages southeast of Andalee. In midafternoon, we reached the stream our guides called the *Kada* (large) Meshellu to distinguish it from the *Ounda* (small) Meshellu stream on the other side of the ridge, even though the Ounda Meshellu is significantly greater in length than the Kada Meshellu. The gradient of the Kada Meshellu is greater, however, which could mean a stronger stream flow during heavy rains.

We dropped into the dry riverbed and headed upstream to look at a small, oval basin lying in its uppermost catchment. Light-colored sediments in the basin showed up brightly on aerial photos, and I wanted to take a look. The basin is about 6 square kilometers in size and lies high on the western margin of the plateau in a graben about 100 meters above the Awash.

We drove up the narrow streambed, where basalt and some Pliocene sediments formed high bluffs on either side, until we reached two large basalt knobs called Nemay Koma, Two Hills, that stand like gateposts at the entrance of the basin. We passed between the hills into a beautiful, secluded landscape of low rolling terrain drained by a score of small streams feeding into the basin. Instead of the fossils of Pliocene animals that I had fully expected to find, we found abundant Late Stone Age (LSA) artifacts scattered among the remains of fish, turtle, and mollusks. The fossils and stone tools lay along the margins of what had been a small lake perched in the basalt in the late Pleistocene or Holocene. As the lake had filled with sed-

iment and had become shallower and shallower, its waters must have risen and
breached its southern end, leaving it simply as the stream catchment we see today.
During its short history, the lake must have offered an idyllic setting for its inhabitants, one both hidden and readily defensible. Although I did not climb the hills at
the basin's entrance, they surely offered a splendid view of the region. It was probably
a great place to be on a Saturday afternoon 10,000 years ago with your Stone Age
sweetheart.

The LSA artifacts were multicolored microliths, mostly very small blades and
scrapers, struck from cores of cryptocrystalline quartz. As Herb explained it, blades
were the great innovation of the LSA and were believed to have been devised about
40,000 years ago by anatomically modern humans. The razor-sharp rock slivers
were hafted into wood or bone handles, like X-acto knives, and used to cut meat,
hides, or other materials.[4] Scrapers (blades broken transversely at one end) could be
similarly socketed and used to scrape hides or to fashion wood into spear points.
The Nemay Koma LSA tools differed from those in Kariyu by being made from
stream cobbles, not from the more easily worked obsidian, the raw material most
often used for toolmaking throughout the Afar during more recent times.

While our field crew made camp, Herb and I walked around the basin until the
sun started fading. Herb was worried. "Jon, all this is very interesting. *Very* interesting. The broken rocks you showed me at Geraru that *you* call Oldowan,
Meshellu basin—neither of which is in your permit area—Andalee, this site; but
Fred Wendorf is paying the bills for my part in this survey, and Fred wants old
stuff, not LSA, not MSA, not Sangoan. I mean, he wants *old*—the oldest stone
tools on earth by made by human beings."

"Yeah, Herb, I know all that, but we've just started. Sometimes you have to be
circumspect in science, inductive, piece by piece, the little to the big. We have to get
the lay of the land to figure out what is where and why—the whole picture, the big
picture."

"That's your great philosophy? That's bullshit. I'm talking m-o-n-e-y. Grant dollars. Fred is coming here in three weeks with two big shots from Dallas. We have
to produce the goods, *old* stuff, or we're screwed."

As we approached camp with darkness quickly closing in on the Nemay Koma
basin, in 1975 as it had for thousands of years, I said, "Herb, you have to relax.
God will take care of us. If God wants us to find old, we will find old; if God does
not want us to find old, we will not find old."

"Who is talking about God here? I'm talking about *Fred!*" The wind shifted and
the intoxicating smell of dinner wafted across our path. "Oh no," Herb shouted,
"not lentils again!"

———

The next day, May 21, 1975, we discovered what may be the largest known Acheulean stone tool complex in Africa.

That morning we drove back down the Kada Meshellu stream to the Awash, where we refilled our jerricans; then we headed east, following a small stream called Meadura, until we reached some low sedimentary foothills abutting plateau basalts. Immediately we began finding basalt handaxes and cleavers. First we found three or four. Then several dozen. Then hundreds. Then thousands! The sediments were overlain by plateau gravels that were surely contemporaneous with those overlying Hadar and underlying Andalee. At least three or four tuff layers were readily visible, interlayered with what we later determined to be deltaic and nearshore lake deposits. The bulk of the fossils and artifacts came from coarser channel layers.

A rough count of a single cache of handaxes and cleavers numbered more than a thousand tools, which enabled me to later say in lectures that "the artifacts are so dense in places that they can be seen from an airplane at 2000 feet." Both heavy-duty and light-duty stone tools and rock debris were probably present originally in several levels in that area, but differential removal of the surrounding sediments and the lighter rock material by erosion had left the heavier tools concentrated at their present level in what is called a "lag" deposit. Most bifaces on the surface were abraded and patinated to various shades of brown, depending on how long they had been exposed to the elements. However, we also began finding smaller concentrations of undisturbed tools, chippings, and debris, along with the remains of hippos, suids, bovids, rhinos, and elephants. This included artifacts in situ lodged in streambanks, or freshly eroding to the surface, that retained sharp, unweathered edges and their original basalt-gray color. There was no doubt that the local basalt was the source of the tools; we found a partially buried factory site on a basalt flow where broken or imperfectly made handaxes and cleavers and rock debris lay scattered in abundance.[5]

Herb was incredulous and for the first few minutes refused to believe the handaxes were man-made. Finding three of something is one thing; finding hundreds, or thousands, can be unbelievable.

"Herb, these are not only bona fide Acheulean tools, but you're standing in the middle of an Acheulean *village!*" The site proved to be 2 square kilometers, but it was soon clear that we had misjudged the magnitude of our discovery.

The next morning while our crew was packing up camp, Herb and I took a Land Rover and blitzed back north a few kilometers to the next stream, Buyelle. There were more tools in abundance, made out of basalt as well as stream cobbles. Both artifacts and fossils were eroding from red, iron-rich sediments that appeared to be at a higher level than Meadura. We drove further north a few more kilometers to Garsalee. More red beds, more Acheulean tools and fossils. We crossed the Kada Meshellu. More red beds, more Acheulean tools, more fossils. To As Koma. More of the same.

We went back to camp and loaded up. For the next five hours we drove south 5, 10, 15 kilometers, zigzagging our way up and down streambed after streambed and areas in between. Guenelea stream. More Acheulean tools and fossils. Kada

Hargufia stream. More Acheulean tools and fossils; enormous lance-shaped hand-axes up to 30 centimeters in length; thick crystalline tuffs. Ounda Hargufia. More Acheulean tools, more fossils. Koba-ah. More Acheulean, more fossils. Kada Bodo. More Acheulean, more fossils. Ounda Bodo. More. More.

We came to a halt at Bodo. It was one o'clock.

"Herb, this isn't an Acheulean village; it's an Acheulean *city!* Do you realize we've found 20 kilometers of Acheulean sites? We've probably seen 100,000 hand-axes this morning."

Herb responded with his uncanny version of Marlon Brando imitating James Brown, bopping around, flinging his arms about, and singing, "I feeeel good . . . *Stellaaa!*" More bopping around, then James Brown imitating Brando, "*Stellaaa . . . I feeeel good.*"

We grabbed some canvas water bags off the trucks and all huddled under an acacia tree hanging over the streambed to eat our tuna, bread, onions, and limes that had turned brown and were hard as walnuts. While everyone else rested, Herb and I walked around Ounda Bodo. The site was rich in fossils—the most concentrated we had seen—eroding from several levels of paleostream deposits. Abundant large-bladed flakes, bifaces, and tool-manufacturing debris were everywhere. We also found large basalt boulders that must have weighed 100 kilograms (220 pounds). They had been shaped and transported 2 or 3 kilometers from the nearest basalt flow to be used as an on-site source of tools. "I would like to see the fellas that did that," Herb said. Or their robust mates, I thought. The boulders showed scars where large flakes had been struck off to be used as blanks for the manufacturing of bifaces. I wondered if they pounded them off with their fists? But we also found round hammerstones that were battered and pitted, presumably from striking flakes off other rocks, among other uses. With such numbers of fossils and artifacts, Bodo seemed a likely site to find a toolmaker. The wisdom of the time gave *Homo erectus* credit for inventing the Acheulean tradition during the early Pleistocene, some 1.5 million or more years ago (Appendix I). The tools we found that morning were more refined than the earliest documented Acheulean industries, but still old. The question was just how old?

We were having constant fuel pump problems with Herb's Land Rover, and it had a broken mainspring. My plan was to go to one of the plantations in the southwest Afar, where there were mechanics and spare parts for Land Rovers. From there we could cross the bridge near Awash Station and then head north to begin surveys in the western part of our Middle Awash permit area. I figured we had enough Early Stone Age sites in the eastern Awash to make Wendorf happy and keep him busy well into his next life. If we did not find Oldowan-age sites in the western Middle Awash, and if this was still a major priority with Wendorf, then on our return to the eastern Awash we still had promising areas to survey.

Before leaving Bodo, I dashed to the area a couple of kilometers east of the Acheulean site to look at some elevated white exposures visible from a distance—the same deposits that I had first seen on that reconnaissance flight with Flying Fred the previous year. I ran up the Ounda Bodo stream, then cut across some sandstone hills to a thick sequence of banded white and tan sediments. Nice. They were bounded to the east by basalts and continued into the distance north and south. Lake sediments, and tuffs. I looked around for fossils. I found two broken elephant molars eroding from some channel sands. Catching my breath, I knelt down to look at them carefully. I drank from my canteen. Not Miocene as I had hoped. Looked like *Elephas*. The genus does not appear in the fossil record until about 4 million years ago. These looked like *Elephas* molars from lower Hadar, maybe lower-crowned and more primitive. Middle Pliocene?

I ran back.

En route to Gewani our guides directed us onto a track south of Bodo that the Awash Valley Authority maintained for a nearby hydrology station on the river. After making a few more stops to look at sediments, we pulled up to a waterhole in the basalt at Sibabi not far from the village, which Taieb, Schönfeld, and I had visited 4 years earlier, as had Gortani and Bianchi 33 years before that. We were caked with dirt and sweat after our Acheulean roadtrip and took the time to bathe. After a wide detour around a swamp, we arrived in Gewani after dark.

Ludovico Mariano Nesbitt, approximately age 40, in Rome in 1930, two years after completing his Afar expedition.

Wilfred Thesiger, age 24, at the Gulf of Tadjura in 1930, after completing his trek through the Awash Valley. Courtesy of © Sir Wilfred Thesiger. Permission granted by Curtis Brown Group, Ltd., London.

Maurice Taieb (left), Kelati Abraham, and Don Johanson in May 1972.

Hadar badlands, looking northwest. Photo: D. Peak.

Author with Afar chief near Bati,
November, 1973. Photo: D. Peak.

Liz Oswald, near Awash National Park,
July 1974.

Hebah at Geraru, July, 1974. Photo: D. Peak.

Afar woman assembling a hut.

Afar men and boys watering animals at a well dug in river bed.

Hadar badlands, with Dennis Peak (foreground) and the author, in November 1973. Photo: G. Corvinus.

Aerial view of eastern Hadar looking northwest with the adjacent riverine forest surrounding Awash River, Kada damoumou basalt-capped hill is on the right. Photo taken by author during reconnaissance flight in August 1974.

The mood of the fall 1973 expedition. Maurice Taieb (top), Yves Coppens (bottom). Photo: D. Peak.

Don Johanson (top) and the author (bottom).
Photo: D. Peak.

Herb Mosca (right) and the author in the Middle Awash, May 1975.
Photo: Ali Mosa.

Acheulean handaxes and cleavers in "lag"
deposit at Hargufia. With the author are Ale-
mayehu Asfaw (right) and Selati Alemma Ali,
October 1976. Photo: C. Wood.

Oldowan chopper, or "pebble tool," with one end flaked. It was found in situ at Matabaietu by Paul Larson, eroding from tuff, in September 1976. Note the sharp edges, and the tuff impregnating the stone tool. This chopper is probably 2.5 million years old. With centimeter scale. Photo: S. Tebedge.

The author with a Land Rover crossing the swollen Awash River on Dobel plantation ferry, July 1976. Photo: D. Cramer.

*Afar teenage girl in a market at Kure Beret in
May 1975, with guide Ali Koka.*

Afar teenage boy at Bodo in October 1976.

Early Pliocene sedimentary exposures in the Central Awash Complex at Sagantole, with Aramis in the background. Photo: C. Wood.

Upper Miocene white diatomite overlain by stratified volcanic tuffs just north of Ananu. Note the tilt of the strata that has resulted from faulting.

The Bodo cranium. These photos were taken in March 1978 while the fossil was being reconstructed. The face view shows the broad nasal area still impregnated with sand. The horizontal crack immediately below the brow ridge is where the lower and upper face pieces fit together. The top view of the cranium shows much of the assembled braincase and the prognathic lower face protruding outward from the massive brow ridge. Photos: D.Cramer. Note centimeter scale.

Sleshi Tebedge (top) with Afar child. Alemayehu Asfaw (bottom) at Bodo in October 1976.

Dejene Aneme at Addis Ababa in November 1976.
Photo: S. Tebedge.

Doug Cramer at Bodo in March 1978.

The RVRME, around a fig tree at Andalee. Front, left to right: Alemayehu Asfaw, Assefa Mebrate, the author, Paul Larson, Selati Alemma Ali (lying down), Tsrha Adefris, Sleshi Tebedge, and Ado Ali. Back, left to right: Bill Singleton, Paul Whitehead, Menda Wordofa, Makonnen, and Craig Wood (on the ladder).

Justine and Maleeka at Bodo with Selati, chief Ismail Mohammed, and Ado Ali in October 1976.

Judy, Justine, and Spring on a campout in February 1975.

CHAPTER 14

Rebellion

After some vehicle repairs at the Italian-run plantation near Arba, on May 24, 1975, we crossed the bridge near Awash Station into Shoa Province and headed north on the west side of the Awash River (see Map VII on page 140). On the first day, Herb and I and our crew passed herds of Kariyu cattle heading into Awash National Park, forded the Kasem and Kebena tributaries of the Awash River, and met with the Afar chief who presided over a large, prosperous encampment spread across the escarpment foothills. The chief assigned his son, Ali Koka, a proud young aristocrat with elaborately rich tribal scarring on his chest and back, as our guide.

That night we camped on a plain nearby, just west of a small, brackish lake called Leado. Ali Koka tried to convince me to put the camp elsewhere, but I told him I liked the view. After dinner and discussion about the events of the day with Herb, we turned in. As usual, we spread our bedding on the ground without tents, because it was still the dry season. We enjoyed a steady mosquito-banishing breeze. I drifted off to sleep thinking about plans for the next day just as the moon was starting to rise in the sky. Some time around midnight, however, I was awakened by the low groans of animals. *Lots* of animals. I quickly distinguished the shouts and whistles of Afar herdsmen and the slaps of long sticks on animal hides. It was a camel drive! In the diffuse moonlight I saw the shapes of large animals fading in and out of a rapidly advancing cloud of dust.

I glanced quickly at Herb, who was between me and the animals. He was sleeping with a grin on his face as though he were indulging in unspeakable fantasies. I picked up my canteen and threw it at his head. He shot up into a sitting position, staring at me wild-eyed. "Hey, cowboy," I said, "don't look at me, look *behind* you!"

There were 500 camels within seconds of turning Herb into chicken scrap.

He screamed.

We both leaped to our feet, scooped up our bedding, and with the rest of our crew dived into, onto, and under the Land Rovers just as the herd rolled by, obliterating our camp.

Could that have been laughter I heard coming from the herdsmen?

The next morning I surveyed the scene as we removed camel droppings from our midst. Okay, my mistake: no big deal. At my insistence we had camped in the middle of the Afar Interstate, the main north–south trail on the west side of the Awash.

From 1959 to 1972, Italian geologists from the University of Padua mapped the thick succession of volcanic rocks in eastern-central Ethiopia and determined that basalts and associated rocks along the western escarpment—the direction we were heading—were lower to upper Miocene in age, 10 to 25 million years old.[1] During periods of rifting, lower and middle Miocene rocks were faulted and exposed at higher elevations, and upper Miocene rocks were exposed on the lower escarpment. While Herb was wasting his dreams at night on depraved reveries, I was dreaming of finding middle Miocene sediments exposed between volcanic flows that contained fossils of *"Ramapithecus."*[2] In the mid-1970s, many still thought this putative taxon to be the earliest hominid, dating from the divergence of hominids from apes, which was believed, arguably, to have occurred some 15 million years ago.[3] On the basis of the Italian work, I believed the volcanic rocks northwest of Lake Leado could be of this age. Aerial photos showed the area to consist of enormous blocks of down-faulted volcanic rocks beveled and deeply scarred by streams flowing off the escarpment. The streams commonly flowed along narrow strips of sediments exposed by faulting. Finding fossils was a long shot, but that was why I had asked the Antiquities Administration to grant this area to the RVRME as one of the three smaller permit areas on the periphery of the original IARE concession (see Map VIII on page 143).

As we headed north from the lake that morning, I hoped to find an easy way into this extremely rugged area without losing a lot of time that could otherwise be devoted to finding Oldowan sites. After fording the Hawadi River, we followed a rocky track leading up the escarpment to the village of Kure Beret, where I planned to find a guide. On reaching the village, however, I learned that vehicle access to the localities that interested me would be difficult and that it would require several days to conduct any reasonable reconnaissance of the area. I decided to abort the escarpment survey for the time being and continue north into the western Middle Awash. As I inquired about these matters, Herb and the rest of our crew were busy socializing in the village; it was market day. Oromo and Amhara traders and farmers were down from the highlands doing a lively business with the Afar up from the lowlands. Soon Herb was following our Afar guide, Ali Koka, who was hard at work rounding up a bevy of beautiful Afar girls who wanted a lift back down the escarpment.

All were probably in their middle to late teens and looked every inch the desert princesses they were. Adorning their necks were simple strings of small brass and ivory beads, contrasting with longer necklaces, made of brightly colored glass, aluminum, copper beads, buttons, and pierced silver coins, draped over and between their bare breasts. They also wore large earrings, bracelets, and multiple rings on their fingers. Their dusky brown skin and dark eyes contrasted sharply with the whiteness of their teeth and the sheen of butter rubbed over their upper bodies. One girl had braids tied beneath her chin, of some particular significance; another had a red ribbon wrapped around her head; and all wore baubles or large metal medallions on their foreheads. Tightly woven plaits of glistening hair parted in the center of their heads fell thickly and evenly to their shoulders and wrapped around their narrow waists were long linen skirts dyed the color of cinnamon from tree bark. Their feet were clad with thick-soled sandals made from camel hide. Hanging loosely about their hips were thin handmade ropes of hemp that had been used to bring goats to the market. Several of the girls carried goatskin water bags draped over their arms or shoulders.

Giggling, laughing, and flirtatious, several of the girls piled into my Land Rover, and several into Herb's. It was probably their first ride in a motor vehicle, and as we left Kure Beret and headed down the escarpment, our new passengers were flushed with excitement, talking loudly, holding hands, and shrieking with each bump in the road. Herb and I had no idea where we were taking them, but I did know that *Ramapithecus* was history. After a short distance, I saw in my rearview mirror that Herb's Land Rover had come to a halt and that he was peering at the undercarriage. I stopped and jumped out. Herb stood up, looked over at me with disaster on his face, and made a snapping motion with his fists, as though he were breaking a stick.

Another broken mainspring.

We had to tell the girls the party was over while we made some repairs. To our collective dismay, they piled out of the vehicles amid more giggling and laughing and went running and skipping down the road, turning around again and again to wave good-bye to us.

Over the next week we continued north. After fording the Nejeso River, we found abundant Late Stone Age microliths along former lake terraces that were probably contemporaneous with the Holocene terraces that Taieb found near Gewani. Around the margins of Yardi Lake, we found promising Middle Stone Age sites and, nearby, ceramic shards dating from more recent times, the first I had seen in the Afar. On a ridge northeast of the lake at Adgabula and Buri, amid large Afar settlements, were small, triangular handaxes and flake tools of a late Acheulean or Middle Stone Age type. The artifacts were associated with vertebrate remains, some of which were charred, suggesting the use of fire.

At the end of May, while we were camped northwest of the lake and Herb was tied up working on the yet-again-faulty fuel pump of his Land Rover, I followed the Ananu stream west into the foothills. I knew abundant white sediments were exposed there; I had seen them twice on reconnaissance flights. They also showed up like a drive-in movie screen on conventional aerial photos and to a lesser degree on space photos. In view of the pattern of paleolakes becoming older up the valley, I felt these deposits were our best shot at finding Miocene fossils. Ananu is 110 kilometers southwest of the middle Pliocene deposits at Hadar and is located in the foothills of the escarpment.

Getting into the Ananu area by vehicle was no problem. Wide foot trails leading through the area and up the escarpment were as worn as the paths to Mecca. In the main streambed there was only a trickle of water, mineralized and warm from hot springs, but enough to form algal soup along the way for livestock and thirsty travelers who were indiscriminate about what they drank. It was immediately apparent that because the area was so heavily traveled, any fossils that had eroded to the surface would soon have been pulverized. I also realized that there would be few vertebrate fossils at Ananu in any case, because most of the sediments consisted of deep- to shallow-water lake deposits. The few vertebrates I did find were casts of fish, isolated crocodile teeth, turtle remains, and the occasional hippo tooth. But I did find *billions* of plant fossils: silicic, microscopic skeletons of Bacillariophyceae algae—diatoms. Pure, brilliant, white diatomite in the greater Ananu area cropped out over some 25 square kilometers, as we later determined, in layers up to 3 or more meters thick, indicating a lake of long duration. I guessed it was one of the richest deposits of the industrial mineral in Ethiopia.*

In the few hours I spent in the area, I found and collected only one vertebrate fossil that I thought might reveal something about the age of the deposits. It was a small *Hipparion* molar. At the first opportunity I would send photos of the equid to a specialist for identification, as I would a sample of the diatomite to another. There was much more to be dated at Ananu, however: Overlaying the diatomite were massive, water-lain tuffs, some of which had been folded and distorted along fault planes as the tuff was still being laid down. It was apparent that volcano-tectonic activity had drained Lake Ananu.

I was back at camp in midafternoon. Herb had repaired the fuel pump, so we loaded up and headed south around Lake Yardi to a ferry crossing on the Awash at the Dobel Plantation near Mataka. Wendorf and his associates were due in a week, and I planned to set up camp in the eastern Awash amid the Acheulean sites.

The ferry consisted of 20 or so 50-gallon drums lashed to a wooden platform. It allowed just enough room to take one vehicle across the river at a time. A cable, suspended between the banks of the river, was used to pull the ferry across the river by hand, helped by the current. Heavy chains attached to trees on the riverbank and to the ferry kept it from slipping away while it was being loaded. Once the vehicle was on board and the chains were removed, getting it across the river atop

---

*Among its many uses, diatomite is used as a stabilizer in dynamite, as a filter in the production of beer, and as an agent for spreading airborne insecticides.

the ferry was like balancing a cannon ball on the head of a camel. As the ferry drifted into the current and across the river, we constantly had to run from one side or end of the ferry to the other to keep the wooden platform level and to prevent the vehicle from tipping over. Once we reached the other riverbank, chains again attached the ferry to trees so that the vehicle could roll onto the bank without the ferry backing away.

Herb and I reached Gewani in the late afternoon. There we learned that the previous night, at full moon, the Afar had attacked several military targets in the central Depression. The Afar sultan Ali Mirah was out of favor with the new regime in Addis Ababa, and rebellion was his response.

The next morning, June 1, we left for Geraru. Liz had been there for two weeks with Kelati and the others. She still had two weeks to go to complete her map work of the site, but perhaps it was a prudent time to move camp. We arrived in mid-morning and everyone was fine. After Liz and I went over her progress on the geology of the site, we broke camp. We then went to the Tendaho Plantation at Dubti, where we put Herb's Land Rover in the shop for replacement of the fuel pump. We left early the next morning. Later I learned that the Afar went on a rampage that same day, attacking the plantation and massacring several hundred highlanders. I had gone to precisely the wrong spot to avoid the rebellion.

Back in Gewani, we retrieved our field equipment from the Red Cross clinic, picked up our guides for the eastern Middle Awash, and drove to the Acheulean site of Meadura. We were deep in the recesses of the valley, well removed from any trouble spots. After making as comfortable a camp as we could, we marked out an airstrip nearby for Wendorf and his colleagues to fly in on. The next morning Liz and I went over a work plan for her to pursue while Herb and I went to Addis Ababa to meet the SMU dignitaries. I left Liz with my Land Rover and a full crew of assistants and and told her I would fly back with our visitors in five days. Herb and an assistant would return in four days by Land Rover with new provisions.

On arriving back in Addis Ababa the next day, I walked into my office just as the phone was ringing. It was someone with the U.S. embassy following up on a report that "Jon Kalb was missing in a rebellion in the Danakil." No, he's here, I told him.

———

Following his installation as the Sultan of Awsa by Haile Sellassie in 1944, Ali Mirah Hanfare II claimed to be leader of and spokesman for all the Afar.[4] In fact, his authority was largely restricted to the central part of the Depression, to probably fewer than 100,000 subjects of perhaps a million Afar then estimated to be in Ethiopia, Eritrea, and the French Territory.[5] Nevertheless, the sultanate was strategically located astride the Bati-Assab road, a major artery to the sea and to the Soviet-built oil refinery at Assab (completed in 1970). Awsa was also strategically located on the border of the French Territory, which was nearing independence (obtained in 1977), and was essential to the security of the region, including the Djibouti–Addis Ababa railroad. Finally, Awsa was centered in the extremely fertile floodplains and delta of the lower Awash, and with the creation of the Tendaho Plantation in the 1960s by the Mitchel Cotts Company, it had become a prototype for large-scale agribusiness in the country.[6]

The successful production of cotton, "white gold," as a major cash crop in the Awash Valley ultimately diminished Awsa's autonomy from Addis Ababa. Throughout the late 1960s and early 1970s, more and more plantations were carved out of the Afar grazing lands in the Awash by the imperial government and handed over to foreign companies as profit-sharing concessions. The Awash Valley Authority, established in 1954, was the official organization to oversee this development.[7] By the end of the Emperor's reign in 1974 much of the Awash had been appropriated by the government. Because of the extreme pressures caused by the loss of grazing lands in the 1960s and early 1970s, the Afar people along the valley became desperate for pasturage for their livestock, which led to more and more territorial clashes with the Issa, Kariyu, and other tribes marginal to the Awash.[8] When the drought struck eastern Ethiopia in 1973, the displaced Afar animal herds were already suffering from the loss of grazing areas, leaving the nomads without livestock reserves to see them through hard times.[9] Although the government was later accused of indifference during the ensuing famine, which reportedly claimed the lives of 25 percent or more of the entire Afar population, in fact the origins of the problem—in the Awash Valley at least—went back a decade or more to developmental planning by the UN FAO, the Mitchel Cotts Company, and others. The causes of the suffering accompanying the famine of 1973–1974 were thus more complicated than just the failure to distribute relief grain in a timely manner.[10] Agribusiness, and government support of agribusiness, had irreparably damaged a centuries-old way of life.

For years Ali Mirah led delegations to Addis Ababa, protesting the evictions of the Afar from the Awash. At the same time—ostensibly to check the spread of the Tendaho Plantation across Awsa—the Sultan himself claimed large pieces of fertile land at the periphery of the Mitchel Cotts concession, which he opened to cultivation by Afar, becoming a major landowner himself.[11] Although members of his own clan prospered, much of the remaining Afar population in the Awash became increasingly impoverished. Resettlement schemes and efforts by the government to get the Afar to become field hands offered little if any real relief, particularly because migrant workers from the highlands were enlisted to plant and harvest crops. As a result, many displaced Afar drifted to towns such as Aysaita, Millé, and Gewani, where relief grain, and chat, were in high demand, and where increasing numbers of Afar women turned to prostitution.[12] Although some valuable efforts were made to open schools, training centers, and clinics in these growing urban areas, such as the agricultural trade school and Red Cross clinic in Gewani, the investment by the government in such endeavors was minimal.

With the overthrow of Haile Sellassie in late 1974, Ali Mirah soon became a target of the reformist military leaders because of his past alliances with the Emperor, his semi-autonomy, and his role as a feudal landowner. When the land reform proclamation announced in March 1975 nationalized all rural lands, including the Tendaho Plantation and Ali Mirah's holdings, it was only a matter of time before there would be a showdown between the Sultan and the new government.[13] Efforts by the military to come to some kind of terms with Ali Mirah through intermediaries went nowhere, and in late April the Sultan was quoted as saying, "So long as they do not touch our land or our religion, there will be no problems."[14]

Although accounts differ on the timing of the events that followed, near the end of May the military tried to lure Ali Mirah to Addis Ababa for direct negotiations, sending an airplane to Aysaita to bring him to the capital.[15] Feeling that his chances of surviving such a visit were remote, Ali Mirah instead fled to Djibouti, where his brother-in-law was president of the French Territory.[16] The Sultan and a heavily armed entourage reportedly used a full moon to escape across the border in a convoy of vehicles driven through the Dobbi graben north of Aysaita. Apparently the flight coincided with a preemptive strike by the Afar on several key bridges and military garrisons to aid the Sultan's midnight escape. The ensuing attack on the Tendaho Plantation by the Afar, and the killing of the several hundred highlanders, may have also been a spontaneous reaction by the Awsa population to the news that the Sultan had been forced to flee. Whatever the case, the retaliation by the military was predictably brutal. A battalion of troops was sent to Awsa, ostensibly to capture the long-gone Sultan; instead, a two-day battle ensued, during which a reported 1000 Afar were killed in and around Aysaita. The punitive attack was supported by armored cars and air force jet fighters that bombed and strafed targets around the Afar town.[17]

Among those caught in the fighting was the British anthropologist Glynn Flood, a Ph.D. student from the London School of Economics who was studying the kinship system and political organization of the Afar. He had spent months living with them, spoke their language, and had taken to dressing like them. We had spoken on the telephone several times in Addis Ababa but had failed to meet. At the time of the attack on the plantation, Flood was in Aysaita; reportedly he had been warned to leave the area immediately, because the Afar fully expected the Ethiopian military to retaliate, but chose to stay. Apparently he thought his British passport would protect him. According to reports, when the military did go to Aysaita, they rounded up and executed some 400 Afar, Flood among them. All were buried in a mass grave. The number of soldiers killed is unknown, because the Mengistu government ordered an information blackout on the rebellion; however, the Sultan had surely anticipated this conflict for some time. He was reported to have smuggled truckloads of arms from the French Territory into Awsa, including factory-new, Soviet-made AK-47s.[17]

Following the military strike on Aysaita, the Afar immediately began hit-and-run attacks on trucks carrying fuel and merchandise from Assab, which coincided with the formation of the Afar Liberation Front by Ali Mirah and his sons.[11] Before long a fuel shortage was felt throughout the country.

As to my concern with these events, the task ahead of me, as Herb and I pulled into Addis Ababa from the Middle Awash in the late afternoon of June 5, was to use the next 72 hours to find a plane, locate fuel for our vehicles, and secure a permit from the authorities to allow me to escort a group of science-minded tourists into a region that had just exploded into war.

CHAPTER 15

The Oldowan

F red Wendorf arrived in Addis Ababa the next day. His two financial supporters, Owen Henderson and Russell Morrison, arrived the day after. Both
were successful businessmen and trustees of Southern Methodist University, and
both had donated large sums to its Department of Anthropology, making it one of
the best in the United States. Fred was the Henderson–Morrison Professor of Prehistory.

Then 51 years old, Fred was tall and lean, with a boyish face; he came from the
East Texas backslapper school of public relations. Raised in Terrell, originally a railroad town, he was lucky to be pursuing his childhood dream of archeology. He had
served in a combat unit in Italy at the end of World War II and was hit by shrapnel,
leaving him with only partial use of his right arm. After the war he received a bach-
elor's degree in archeology from the University of Arizona on the GI Bill and a Ph.D.
from Harvard.[1]

Fred soon established a reputation in the archeology of the American Southwest
Indians. This included work on the Midland skull site discovered in a West Texas
oilfield, which helped push human origins in North America to the end of the last
Ice Age, some 10,000 years ago. In 1961 he shifted his interest to African archeology when he joined the Combined Prehistoric Expedition to Nubia, a multinational project devoted to locating and salvaging sites soon to be flooded by the
Aswan Dam. Much of his work since had been concentrated in Egypt, where he
had pursued investigations on the origins of food production.[2]

I had met Fred in early 1971 in Dallas when I was making inquiries about Ethiopia. That year he was leading a team in the Main Ethiopian Rift excavating MSA
and LSA sites.[3] The next year I had tried unsuccessfully to get him involved with
the IARE.

On June 9, 1975, Fred, Henderson, Morrison, and I piled into a twin-engine airplane and headed to the Middle Awash (see Map VII on page 140). After an hour
or so we reached Meadura, and on the first pass I could see the huge cache of

Acheulean bifaces Herb and I had found a few weeks earlier. We circled around over the Kada Meshellu stream and came in low to buzz the camp. Liz had been off in the exposures but was already running back to camp to drive the Land Rover over to our airstrip. Herb was also there. So he had made it back okay.

I yelled to the pilot, "Make your approach *there,*" pointing to an area I had previously selected, "and land along this area!" But misunderstanding my instructions, the pilot started landing on the intended airstrip *sideways,* giving us one-third the proper landing distance.

"Sweet Jesus!"

We came in with full flaps and landed low, but not low enough, hitting the ground with a shattering jolt. The pilot yanked back the throttle and slammed on the brakes as hard as he could without flipping the plane over. We rumbled and bounced along with the wind rushing through the cabin as he fought to keep control of the plane. Amid a churning cloud of dust, we came to a full stop within a few feet of a bank of acacia trees.

"*Damn,*" Fred said, "that wasn't a landing, that was *an arrival!*"

After unloading some gear, we thanked the pilot and bid him good-bye. He wanted to leave immediately; he was nervous about the Afar. If he knew something about the rebellion, I didn't ask. He wheeled the plane around, taxied to the proper end of the "runway," and took off. By then I was introducing our guests to Liz.

Fred's first impression of Meadura was not good, because most of the fossils were well broken up and most of the stone tools were abraded and patinated. But then he saw the several tuff layers that would make dating possible—a major problem in most Acheulean sites—and fresh, sharp artifacts eroding from at least several levels. Herb and I also showed him isolated concentrations of bifaces, flakes, rock cores, and debris associated with animal remains. Fred was certain they were discrete butchering sites where tools had been fashioned and used on the spot. Elsewhere he identified a large area of artifacts and bones that he thought might be a living floor, where people had camped. We found the remains of a baby elephant in this same area; it might have served as a snack in the middle of a long day of slaughtering and maiming. Blackened bone suggested fire use, long suspected of Acheulean toolmakers in Europe and Asia, but unknown in Africa.

Fred thought Meadura might be an upper Acheulean site because of the presence of some advanced-looking artifacts: large, sharp flakes struck from specially faceted rock cores—a tool-manufacturing technique that is a relatively late innovation, about 300,000 to 400,000 years old (Appendix I on page 319).

The next morning, while Henderson and Morrison went with Liz and Herb to see the Awash River and replenish our water supply, Fred and I walked north to the next stream drainage, called Buyelle. There were quartzitic as well as basalt tools amid red sediments that looked somewhat higher stratigraphically, and therefore possibly younger. Fred again identified the predominantly bifacial tools as upper Acheulean. En route to the site we nearly stumbled over a broken orange ceramic pot with a basalt handaxe sitting in the middle of it. I gasped. Fred looked at me and said, "Don't even think about it!"—there was no way the two were contemporaneous. We walked on. I have always regretted that incident, because I took no photos or notes, nor did I later return to that spot. Who made that pottery?

The visit to Buyelle was quick, and we were back in camp by midmorning. Fred was not satisfied with another Acheulean site, even though it and Meadura alone represented a major discovery. Back in camp we had a group meeting. Fred reminded me that his priority was the Oldowan not Acheulean and that this was the agreement he had made with Henderson and Morrison for their continued support. Herb looked at me and raised his eyebrows: *I told you so!*

Why all the fuss about Oldowan tools? And how do the earliest stone tools fit into the grand scheme of human evolution?

When Louis Leakey first found crude artifacts made from stream pebbles—"pebble tools"—at the base of Olduvai Gorge in 1931, naming them the "Oldowan Culture," no one knew who had made the tools. When *"Zinjanthropus" (Australopithecus boisei)* was found in Bed I at Olduvai in 1959 at the same level as the tools (Appendix I), Louis thought he had finally found the toolmaker, and he immediately announced his conclusion to the world. A year later, however, after a partial cranium of early *Homo* was also found in Bed I, with hand bones very similar to our own, Louis gave our lineage the credit for making the tools. He named the new hominid fossil *Homo habilis,* which means "handy man." What took Louis a mere press conference to conclude, however, took Mary Leakey ten years of meticulous excavations to document. Radiometric dates placed the age of the tools and toolmaker at 1.75 million years.[4]*

Although a number of other sites with Oldowan-like tools have been reported over the years, it was not until the discoveries by Richard Leakey's team at East Turkana that the magic combination of early artifacts, hominids, associated mammalian fauna, and datable volcanic layers again came together. The blend was highlighted by the discovery of the presumed toolmaker, represented by the "1470 skull" found in 1972, which was initially identified as *H. habilis* but later described as a new species, H. *rudolfensis.* Both the artifacts and the skull were eventually dated at about 1.85 million years.[6]

Older stone tools were quickly discovered by both the French and the American teams working in the Omo. After the inevitable claims and counterclaims about which were the oldest, all eventually clustered between radiometric dates of 2.3 and 2.4 million years (Appendix I).[7] None of these finds, however, was associated with a toolmaker.

By 1975 the compelling questions that interested Wendorf and other archeologists were exactly how old the oldest manufactured human tools would prove to

---

*Current dates place the age of Bed I at 1.75 to 1.80 million years.[5]

be, who had made them, and what had the tools been used for. Most archeologists expected stone tool use to coincide with the first appearance of *Homo,* some 2.0 to 2.5 million years ago.* Many also assumed a causal relationship between the use of stone tools and the origin of a large-brained *Homo.*

If the earliest appearance of stone tools did have a direct bearing on the origin of our genus, in what form might these early tools have been, and when did they *first* appear? Did the use of a stone tool such as a hammerstone, used to manufacture other stone tools, originate with *H. habilis* or with an earlier, yet unrecognized, species of *Homo?* What dietary changes, such as meat eating, accompanied tool use? And did climatic changes, such as drier conditions, force these dietary changes? If *H. habilis* used Oldowan tools, what led to the demise of the species and its replacement by *H. erectus,* the presumed inventor of Acheulean tools?

———

As we discussed our itinerary for the few days that Fred and his associates would spend with us, his position on the long-standing mysteries of African archeology boiled down to one central issue: hard cash. As he explained it, there were Acheulean sites from South Africa to northern Europe, but only a handful of well-dated Oldowan sites. If we came up with an *early* Early Stone Age site that was datable and contained fossils, then he could confidently go to NSF and ask for funds for a major field effort. We could tie into that project funds to investigate the Acheulean sites and some of the younger sites, such as Andalee. Henderson and Morrison might be convinced to keep the archeological surveys going until then on a limited basis. Fred emphasized that without an Oldowan site we would be competing for money from the same NSF pot as the teams working in the Omo, East Turkana, and now Hadar that were awash in either hominids or "the oldest" tool sites. Fred estimated that NSF had already poured close to a million dollars into the Omo Valley alone.

Still, I persisted. "But, Fred, what Herb and I have seen during one *brief* survey is not another Melka Kontouré. Meadura and Buyelle are only the tip of the Acheulean iceberg. I'm guessing that when we finish surveying we will have mapped hundreds of individual archeological localities that blanket a 50-square-kilometer area. There may be dozens of artifact-bearing levels. We may have entire Acheulean encampments, villages, paleo-shopping malls!"

"Jon, I'm not getting through to you. NSF will *not* buy into this, unless you find a fat hominid sitting right in the middle of it."

I then suggested a compromise. "Okay, the sites seem to get older from north to south and from west to east on this [the eastern] side of the Awash. Let's go to our farthest site south, Bodo, which Herb and I think might be earlier Acheulean. It's

———

*Richard Leakey, like his father, proposed that the origins of *Homo* would prove to be at least several million years earlier.[8]

also our richest site so far in fossils. We'll see what you think of it; then we can survey further east and south for Oldowan tools."

Fred agreed.

We all took off from Meadura for Bodo, arriving by noon. This time I got a better look at the site. There were two prominent artifact levels. The lowest, consisting of coarse sands and gravels, contained large numbers of moderately abraded handaxes and concentrations of broken-up vertebrate fossils; the second, a thick overlying sand, contained artifacts of all sizes, whole skulls of large animals, and minimally disturbed skeletons. The second layer showed great promise for finding intact butchering sites and occupation floors. In one area, for instance, we found several hippo skeletons lying side by side, associated with heavy- and light-duty stone tools. The association suggested that the huge animals either had been fortuitously killed and butchered at the same place or had been killed and somehow dragged to the spot where they were butchered. If the latter was the case, then we can guess the hunting party required at least 15 adults to drag a 3000-kilogram hippo carcass to or along the water's edge. With that many individuals, there may have been 40 to 50 people in the group, including women and children, living at or near the site. The several basalt core boulders Herb and I had found in May also suggested an encampment of some size and duration, because it is doubtful that the 150-kilogram rocks were transported several kilometers just to accommodate a few people bent on spending a quiet weekend of carnage.

Fred did not think Bodo was early Acheulean. He thought the stone tools represented a middle Acheulean tradition, on the order of, say, 400,000 to 500,000 years (see Appendix I on page 319). On the basis of the regional geology, I still suspected the site was older than Meadura. Nevertheless, there seemed to be enough tuffs at both sites to give us, eventually, reliable dates and correlations of strata betwen sites.

The next day we drove south from Bodo about 5 kilometers, turned up a wadi called Wilti Dora, and found a small site exceedingly rich in fossils, eroding from fluvial sands and tuffs. The *Elephas* molars suggested an age of about 2 million years. We spent some hours searching the area. We found a few fairly fresh Oldowan choppers made from cobbles and a beautiful polyhedral hammerstone with flat surfaces formed by repeated, random blows on one side of the stone or the other. However, we were unable to trace the tools to an in situ source. We guessed they originated from overlying layers that were now eroded, leaving only the stone tools behind. We also guessed that these sources had been nearby.

Two days later we headed back to Addis Ababa via Gewani, where we spent the night at the AGIP station. There were military vehicles about and soldiers in the village, looking as though they were waiting for something to happen. The word was that Ali Mirah's Afar Liberation Front had now started blowing up trucks along Millé-Awash Station road. The next morning on the way to Awash Station I had to believe the stories were true when we passed the smoldering wreckage of a big tanker truck. Then we passed a long column of armored vehicles filled with soldiers, probably on their way to Aysaita. At the front and rear of the column were the familiar Jeeps with machine guns mounted on the hoods, manned by gunners wearing helmets and goggles. When we stopped at the Awash Buffet at the train station for breakfast and Owen Henderson asked me what all the troops were for, I confessed that we had just left a combat zone.

Over fried eggs, fresh fruit from a nearby plantation, hot bread, and strong Harar coffee, Fred told Herb and me that he thought the prospect of finding an Oldowan site in the Middle Awash was worth pursuing and that Henderson and Morrison had agreed to provide continued support for Herb for another month of fieldwork.

Herb slapped me on the back, "Lucky you!"

The followers of Ali Mirah were not the only people in revolt in mid-1975. Ethnic brush fires were flaring up across the provinces as one group after another sought independence from the military government. There were rebellions in the highlands of the Blue Nile and Tigray, in the semi-arid lowlands of Bale, and in the desert terrains of Sidamo and southern Hararghe. The Eritrean rebels held much of the countryside in the north, and Somali-backed "agents" were infiltrating the Ogaden. Nevertheless, the military continued to strengthen its hold on the country, sending troops here and there to quell conflicts big and small.[9] At the same time, provincial power bases long controlled by royalty and chieftains appointed by the imperial government were being replaced by peasant associations intent on realizing the entitlements of land reform the single issue that brought the Derg the most popularity in the rural population.[10]

There was much resistance in Addis Ababa to the socialist military leadership, although it was largely ineffectual, because civilian ideologs were outflanked by the progressive elements of the military government.[11] There were demands for a democratic government, but civilian groups opposed to the Derg were internally divided, and they advocated Marxist–Leninist reforms similar to those of the Derg. When labor and student outcries for a "people's government" became overly aggressive, the military resorted to force. In early June the military arrested hundreds of protesting high school students in Addis Ababa, releasing them the next day to grateful parents.[12]

On our return from the field, everyone went their separate ways. Wendorf went to Dallas, Herb traveled to Kenya to see the game parks with Henderson and Morrison, Liz went to Djibouti, and I took my family camping with friends at Lake Shalla, in the rift south of Addis Ababa.

A week later, Herb, Liz, and I were all back at it, getting ready for our return to the field: going over aerial photos, buying supplies, repairing Land Rovers, patching up field equipment, and chasing down enough fuel to sustain two vehicles for another month in the bush. Because of the attacks by the Afar on tanker trucks, gasoline was being strictly rationed by the government, and there were long lines at filling stations. The ever-resourceful Dejene, however, always had ways to expedite paperwork for extra fuel rations.

At the end of June we were off, joined by two new participants. The first was a senior biology student from Addis Ababa University, Assefa Mebrate, whom I had recruited to be a paleontology research assistant. Short and thin, Assefa was from Dessie, on the edge of the Ethiopian plateau just west of Bati. He had previously worked as a research assistant in ethnobotany in Ilubabor Province in western Ethiopia, and on a malaria control project in Goma Gofa Province in southern Ethiopia. Most recently he had served as a relief worker distributing grain and medicine in Wallo. In high school he was president of the biology club, and his hobby was collecting tropical fish. In other words, Assefa's background was perfect for becoming the RVRME's expert on fossil elephants. On his first day on the job I gave him a copy of Vincent Maglio's monograph, *The Origin and Evolution of the Elephantidae*, published by the American Philosophical Society, and told him that he had four days before we left for the field to become a paleontologist.

Also joining us was Glenn Conroy from New York University, who taught physical anthropology in the Department of Anthropology and Human Anatomy in the Medical School. The year before, he had received his Ph.D. from Yale. Boston-bred, of medium height with long, dark-brown curly hair, he had field experience in Egypt, East Africa, and Pakistan. His dissertation was on postcranial remains of early anthropoids from the Oligocene Fayum Depression of Egypt. He had also published two papers on "*Ramapithecus.*" Just as Herb had helped me search for these putative hominoid fossils, it certainly made sense that Conroy should help us look for Oldowan stone tools.

We set up our first camp at Wilti Dora, where Wendorf and I had found Oldowan tools on the surface. From the site we recovered 99 fossils representing 18 taxa, all recovered from thick fluvial sands and tuff layers. The fauna still suggested an age of about 2 million years.

Along the drainages of a nearby stream, Matabaietu, we found another site exceedingly rich in fossils similar or identical to those at Wilti Dora. Fossils were so abundant that the Afar had used them for constructing stone graves; we even found a hearth built out of an elephant mandible.

And almost immediately we began finding Oldowan tools. We found them spilling across a gentle slope at the base of a small ridge at the north end of the site. They were made from basalt cobbles and were moderately to lightly patinated. Some retained gray basalt coloring on unexposed surfaces, indicating a nearby source; many were fresh with relatively minor edge damage. The artifacts included large choppers, heavy- and light-duty scrapers, polyhedral hammerstones, and numerous chips and flakes. Nearby we found a large serrated scraper lying amid the remains of an elephant.

The day after we found Matabaietu, Herb and I made a side trip back to Bodo to photograph artifacts and collect a sampling of fossils for dating purposes. Near midday we were standing in the middle of the site, inert, watching the heat waves ripple skyward through the haze. Dust devils jetted around in the distance. The sky was a very pale blue, as though it were liberally mixed with white paint. No clouds. A jackal cried in the distance. After a moment Herb looked at his watch and said, "Guess what?"

"What?"

"It's noon, the 4th of July." Independence Day in the U.S.A.

That night we slaughtered a goat and feasted.

The field conditions in July were not optimal for newcomers, and I sensed that Conroy was not particularly happy. The temperatures during the day were blazing, and it rained almost every night. I still had not picked up a field thermometer, so I had no idea how hot it climbed to during the day, but I guessed the temperature was somewhere above 43°C (110°F) in that part of the Awash, with high humidity in the mornings. But it only took a couple of hours for the sun to burn off the night's rain, turning it into steam, before we were blessed with another clear, sunny day. In the evenings we had to climb into our mosquito nets early or be devoured by the merchants of malaria. Also present were lions, which could clearly be heard at night, although so far as we know, none came into our camp.

To Glenn's credit I never heard a word of complaint, although the political climate must have been at least as great a discomfort as the meteorological climate. On the way from Awash Station to Gewani we passed another military convoy, and there was an even larger military presence in the village than before: soldiers carrying rifles or machine guns with bandoliers of cartridges wrapped around their chests. Ali Mirah's men were obviously still on the prowl. Also, the Issa were taking advantage of the instability and attacking more Afar encampments along the Awash, no doubt encouraged by the Somali freedom fighters moving up through the Ogaden. In the Middle Awash, the Afar were agitated, and when we went to the river on our first day for water, we encountered some armed men who kept looking at Glenn as though they wanted to plant a spear in his brain. I could not understand why they disliked Glenn. Perhaps they suspected he was an Issa because of his bushy hair. Then, over several days, gunfire erupted around Wilti Dora, climaxing one morning when armed Afar burst through our camp, leaping over cots and boxes in pursuit of the real Issa. Afar women followed a short time later carrying goatskins of water to nourish the front lines, which seemed to extend from Gewani to Glenn's tent. I told him that "this sort of thing happens all the time in the Afar and has nothing to do with us," but he lived in New York City and was nobody's fool.

After a week, Glenn told us that he had dysentery and thought it best he return to the United States. Earlier he had told us he had found two hominid fossils at

Wilti Dora, a complete upper arm bone and a partial upper leg, and therefore his primary mission—finding hominids—was accomplished. I offered Glenn a fail-safe method that Ethiopians use for getting rid of dysentery : two heaping tablespoons of ground red pepper mixed with a glassful of cooking oil, but he felt the "cure" could be worse than the malady.

Herb volunteered to take him back to Addis Ababa, which he did.

It turned out that both of Glenn's "hominids" were monkeys. I never faulted Glenn for the misidentifications, although they later caused me some embarrassment. I fully appreciated that had he plundered an Afar grave for comparative bone material, this never would have happened.

One afternoon Liz and I had spent some time following a marker bed south from Maka, near Matabaietu. It was a distinctive limestone layer about 15 centimeters thick, packed with fossil snails that had lived along a lake margin. The layer appeared to be the same limestone that I had seen along the eastern margin of Bodo, because both were overlain by a coarse, black tuff. And if it turned out to be the same, then the limestone indicated a lake basin at least 20 kilometers long. From Maka we traced it south to Belodolie.

By late afternoon it was time to call it quits. The mauve colors of the descending sun were starting to appear in the west, and soon the landscape would turn deep purple and gray and blur over, making it difficult for us to distinguish one streambed from another when returning to camp. But Liz had disappeared. I hit the horn of the Land Rover and waited a while. No Liz. I laid on the horn and waited. Still no Liz. More waiting. *Damn.*

I grabbed a canteen and took off on foot to the east, periodically calling out her name. "Liz . . . Liz . . . Lizzz!" Suddenly she appeared behind a distant hill, like a pop-up at a shooting gallery.

All breathless she yelled, "I've found a very primitive elephant molar . . . *pant pant* . . . very primitive . . . *pant pant* . . . I think . . . It has only eight plates . . . *pant pant*. You really should see it!"

By then Assefa, our new fossil elephant expert, had joined me, and we followed Liz double-time to the next stream drainage, called Wee-ee.

The molar was definitely primitive and belonged to a skeleton, much or all of which was eroding from a crystalline tuff. The bones and teeth were completely silicified. The teeth were broken up, except for the molar that Liz handed me with eight complete plates. In length it was equal to my outstretched hand, and it was completely intact except that the roots were sheared off. There were growth layers (similar in principle to the growth rings in a tree) encircling the plates, and the underside of the molar, which we later learned indicated that the animal was about 30 to 35 years old at the time of death. Perhaps it had suffocated in the volcanic ash fall.

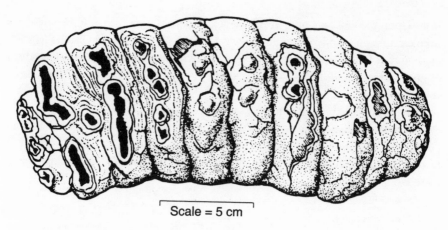

Scale = 5 cm

FIGURE 4.   EARLY PLIOCENE PRIMITIVE ELEPHANT MOLAR ("Mammuthus")
from Wee-ee, found by Liz Oswald in July 1975.

From studying Maglio's monograph, Assefa thought the molar was a third
molar, although he was uncertain whether it was an upper or a lower. The difference is important, because the plate number and other metric features that vary
over time also vary between the upper and lower molars, as they do among the
first, second, and third molars. In this case, however, the distinction did not matter,
because a plate number of eight on either the upper or the lower third molar was
at least three or four plates fewer than I had seen on any elephant third molar, from
Hadar or anywhere else. Also the plates were wide, thick, widely spaced, and low-
crowned, and the plate enamel was thick—all criteria for primitiveness.

Nice. Very nice.

"Liz, I take back all the vile and disgusting things I've said about you."

"Screw you."

We collected the molar and made it back to camp as the sky flooded with blood-
red clouds. More rain for sure.

The next morning after breakfast, we all crowded around Assefa as he sat on a
wooden food box, with the Wee-ee molar propped between his knees and Maglio's
monograph in his hands. Maglio had compiled a number of tables giving the vital
statistics of the molars of each species of the recognized elephant genera—*Elephas,
Mammuthus, Loxodonta,* and *Primelephas.* The last he described as the common
ancestor of the other three. As with the stages of *Elephas recki,* these data included
plate number, relative height, plate thickness, and enamel thickness. Maglio plotted
these features on charts that showed their variation with time. Beginning 6 or more
million years ago, with the first known appearance of *Primelephas,* the relative
height of elephant molars has steadily increased up to the present, as has the plate
number, while the relative thickness of plates and tooth enamel has decreased. As
is the case with the stages of *E. recki,* these trends reflect an increase in molar dura-
bility as vegetation became coarser and the climate more arid over time. If a tooth is
higher-crowned, for instance, it will last longer. Collectively, these changes increased

the total amount of enamel in the tooth, which is essential for increasing its "molar life" and thereby guaranteeing successive generations of elephants a long, happy life chewing increasingly coarse grass, leaves, bananas, and peanuts.

After taking the necessary measurements of the Wee-ee molar with some calipers that Conroy had left us, and after studying Maglio's tables and charts and his numerous figures of molars, Assefa concluded that the molar was *Primelephas* or one of its immediate successors, very early *Mammuthus* or *Elephas*. Either way, he estimated that Liz's find was at least 4 million years old. *Early* Pliocene.*

We knew we had made a great find. The Wee-ee molar turned out to be the linchpin of much of our subsequent paleontological research in the region, because it told us that we had found fossils at least half a million years older than those at Hadar. The find held out the promise of finding human ancestors in the eastern Middle Awash that much older than Lucy.

Among our other great discoveries that month was a trail of several dozen coprolites (fossil feces) found at Matabaietu that could not be mistaken for any animal other than horse. There may be a coprolite expert somewhere who would argue with me over this, but having stepped in a number of modern counterparts of these fossils in my youth, I feel confident they were horse, and *Hipparion* at that, because that was the only equid we found at Matabaietu. Make no mistake, coprolites are serious science business, because if they are well preserved, they can be loaded with fibrous and pollen plant remains, which are invaluable for both habitat reconstruction and dating purposes. I later sent samples of the Matabaietu fossils to a specialist at the University of Washington for pollen analysis, but the specimens were too altered chemically through fossilization, and no paleoflora was preserved. Years later I sent the remaining sample of coprolites to a *Hipparion* expert at the Florida State Museum, hoping that new technology or observations would reveal more about the specimens, but I received no information.

If anything, the rains increased while we were camped at Matabaietu. The clouds started to build up in the late afternoon and commonly unloaded about midnight or in the early morning hours. Afterwards a deathly stillness would settle over the camp. The slightest cough or murmur, or the sound of someone slapping a mosquito, would reverberate off the saturated ground or bounce off raindrops hanging onto the thorns of an acacia tree, causing them to patter across a tent or the hood of a Land Rover. Seemingly, sounds several kilometers up the valley could be heard just as plainly as those a few meters away.

For several days running we were visited by lions. When the rains ended, we could hear them far off to the south, bellowing to one another in the darkness. Perhaps they came from Maka, from Belodolie, or from Wee-ee, where we found abundant recent bone remains of antelopes. We could plainly hear the lions getting closer and closer. I would be in my tent, falling in and out of sleep, but always listening to the sounds: 200 meters away, 150 meters, 50. By then the camp was roused and my

---

*The tuff layer containing the molar was later dated at 3.89 million years.[13]

colleagues would start murmuring to one another. When the roaring stopped, we really perked up, because we knew that the lions had picked up our scent, like a dinner bell. When the Afar with us got up and jumped into the vehicles, we knew we were about to have some serious guests. One night in particular this happened three or four times over a period of hours, the lions wandering in and out of our camp and people going for the vehicles, as though the big cats enjoyed agitating the bipedal apes. Because of existing clouds or the lack of a moon, we never saw them at night; in the mornings, however, we found their tracks everywhere.

In the latter part of July our supplies started running low, and it was time to pack up. When I saw Liz poking a stick in a termite mound searching for tidbits, I got the message. No problem. I figured that we had more than completed our mission. I was certain that future excavations would reveal in situ Oldowan tools and living floors at Matabaietu. Moreover, I was confident that the deposits there were stratigraphically very similar to those at Wilti Dora, which would enhance our prospects of finding more artifact localities. Just north of Wilti Dora, at Gemeda, we found another site of similar lithology that was as large as Matabaietu, or larger, and strewn with fossils, but we found no artifacts during our brief survey. It was essentially another elephant graveyard, like "Elephant Park" in the central Ledi area, and probably was of a similar age.

We found no hominids that July, but I told Herb that I would kiss the backside of every anthropologist in Africa if we did not eventually find early *Homo* in those deposits.

The day before we broke camp we procured another goat. Kebede went the limit and used our last can of sliced pineapples to put over the meat as he grilled it on an open fire, basting it with a sauce made from pineapple juice, vinegar, and red pepper. Herb entertained us with James Brown music on his tape player while he danced it up:

> I feel good. I knew that I would now.
> I feel good. I knew that I would now.
> So good, so good, I got you.
>
> I feel nice, like sugar and spice.
> I feel nice, like sugar and spice.
> So nice, so nice, I got you.

CHAPTER 16

The "First Movie"

O n August 28, 1975, a brief notice appeared in the *Ethiopian Herald:* "Yesterday, Haile Sellassie I, the former Emperor of Ethiopia, died. The cause of death was circulatory failure." Of course there were rumors that he had been murdered—smothered in his sleep, or garroted. Another account said that he had become ill and no effort had been made to give him medical treatment.[1] He was buried in a secret grave.[2] There was no official mourning period; rather, life went on as usual in Addis Ababa, at least on the streets. The deposed leader had spent his last days imprisoned in the old Menelik Palace, rising at dawn and attending morning mass, followed by a day of reading and meditation. It was said that he read without eyeglasses in spite of his 83 years.[2]

Not long after his death, I was poking around in the large tin shed behind the National Museum, looking for extra storage space for fossils. At the back of the building in a heap were stone, plaster, and bronze busts of the former emperor. Apparently even the royal lions had been discarded. Shortly after Haile Sellassie was deposed, an anthropologist visiting a taxidermist in Addis Ababa saw a pile of freshly killed lions said to have come from the palace, the very symbol of the Conquering Lion of the Tribe of Judah.[3]

Haile Sellassie is reported to have said, until the end, "If the revolution is good for the people, then I am for the revolution."[2] How good was still a matter of debate. During a parade on September 12 celebrating the first anniversary of the revolution, a number of students and workers who passed the reviewing stand shouted anti-government slogans: *"Down with the military." "People's government!"* and *"Free the political prisioners."*[1] Over the next two weeks, more provocations followed, increasingly taunting. A confederation of labor unions released a number of demands, including civilian rule, the formation of political parties, establishment of a minimum wage, and the right to strike. When a worker was caught by police at the airport distributing leaflets supporting these demands, a riot ensued, resulting in the deaths of seven people.[1] Following a strike a few days later, the Derg declared a state of emergency and placed Addis Ababa under martial law. Over the next few weeks the windowless, gray Mercedes military trucks swept the city, hauling off union officials, teachers, and students in mass arrests.

———————

At this time the IARE came back to town, its ranks now swollen by new members, all hanging on to Lucy's apron strings. D. Carl was riding high, flushed with media attention and $100,000 in new grant money over the last year.[4] The first order of business of the IARE was to petition the Ministry of Culture to reinstate the Middle Awash to Taieb's original concession. In doing so, they had a new ally.

Following a reorganization of the Ministry, Dr. Berhanu Abebe became head of a new department that included the National Museum and the Institute of Archaeology. The Antiquities Administration had been absorbed into this department. Despite the rumor that surfaced in the United States, it was not true that its former director, Bekele Negussie (who had authorized the RVRME) had been arrested; rather, he now headed the Institute of Archaeology. It was true, however, that Bekele no longer had the final say about issuing research permits.

Berhanu was a curious relic of the imperial government. He had received his Ph.D. from the University of Paris in anthropology and took great pride in having studied under Claude Lévi-Strauss, renowned for his studies of kinship, primitivism, and myths. Berhanu's own ample contributions to these subjects are contained in his dissertation, which traces Shoan land ownership policies from the days of Emperor Menelik to the Ethiopian Constitution of 1931, promulgated under Haile Sellassie.[5] Shoa Province is the ancient heartland of the Amhara tribe, which claimed both Menelik II and Haile Sellassie, as well as Berhanu, as its own. The 1931 Constitution was progressive in its day but above all reaffirmed the Emperor's absolute constitutional right to rule the country absolutely.[6] A central tenet of this document—with profound religious and political implications—was the claim that Haile Sellassie was a direct descendant of Solomon and the Queen of Sheba, an assertion long since discredited by Ethiopian scholars.[7]

Berhanu's studies of the Shoan monarchy prepared him well to become director of the Haile Sellassie I Prize Trust, which had been created to dispense scholarships and awards for meritorious contributions to humanity.[8] The gold medals and cash awards handed out by the Trust were Ethiopia's equivalent of the Nobel Prize, and they were given to such recipients as the president of Senegal for his poetry, Louis Leakey for his life's work, and various scholars for studies of Ethiopian history, culture, and other subjects. The trustees were members of the royal family and had included the former prime minister and the Emperor's grandson, both of whom were executed by the Derg in November 1974. The assets of the Trust were among the first to be seized by the military government.[9]

How Berhanu had moved so quickly from the imperial family to the revolutionary family was a mystery, but he was no fool and he managed to prosper in both worlds. After the Trust was dissolved, his appointment to the Ministry of Culture followed because of his credentials as Ethiopia's first social anthropologist and secretary of the 1971 Pan-African Congress of Prehistory, over which he had presided as the only active Ethiopian participant. Berhanu was handsome, and he charmed the delegates with his flashing smile and excellent command of English and French, which included recitations of French poetry.

As an ardent Francophile with entrenched connections in the royal family, Berhanu had long enjoyed access to cocktail parties and luncheons at the French embassy and was very close to the French archeologists and advisors working with the Ministry of Culture. I first met him at the 1971 Congress through Taieb, who considered him an important door-opener in the imperial government. When Berhanu moved to the Ministry in late 1975, Taieb saw him as a means of getting the Middle Awash reinstated into his original concession. On the face of it, Berhanu was the man for the job, because he considered his work at the Ministry an extension of his position as a giver of awards at the Trust. Handing out prizes to foreign scholars, whether gold medals or concessions to prehistoric sites, was a means of currying favor and gaining influence, something Berhanu understood very well.

In early October 1975, I was summoned to the Ministry by Berhanu to discuss the RVRME's research authorization. I knew before Berhanu knew how the meeting would end.

"Mr. Kalb, you do understand, do you not, that just because there is a new regime in Ethiopia, there is no reason why agreements established with the previous government should not be honored by the present government."

"Meaning?"

"Meaning, Mr. Kalb, the agreement Dr. Taieb made with this Ministry in 1972—the one that gave him the permit in the Awash Valley."

"Oh, you mean the permit *for* the Awash Valley," I said, "that was engineered by a French advisor with the Antiquities Administration? That would be the same permit that established a French colony in the Afar region that is larger than the neighboring French Territory . . .?"

"Mr. Kalb, there is such a thing as international law, and international law states that agreements must always be kept from one government to the next. Otherwise, there would be no diplomatic continuity between nations."

"Dr. Berhanu, I am surprised. I had no idea that you were so well versed in legal matters. I thought you were an anthropologist."

"Well, as a matter of fact, I studied under Claude Lévi-Stra . . ."

"*Really?*"

". . . but I have also studied law." The law under Emperor Menelik and the Constitution of 1931.

"I had no idea."

"Yes, Mr. Kalb, that is why I am quite sure that agreements with the previous government must be adhered to."

"This is all very interesting, Dr. Berhanu. And have you told the Derg your theories about how the policies of the former regime should be continued under the revolution? I feel certain the Derg would find this fascinating."

". . . Mr. Kalb, that is all for now, but let's keep in touch."

A few weeks later, the IARE made an epic discovery that in many ways was even more important than Lucy. Dubbed the "First Family" by Johanson, the discovery consisted of 214 fossil hominid bones and teeth from a single locality, L333.[10] The fossils represented a minimum of 13 individuals. Because one complete adult hominid skeleton contains 238 bones and teeth, and 13 would comprise 3094 bones and teeth, clearly the L333 hominid fossils were a remnant of a much larger number of skeletal elements lost through erosion or other means. Nevertheless, for the first time ever, a sample of early hominids of both sexes and different ages were found in a single locality, almost certainly representing a single species. Among the 13 individuals were at least 9 adults, 3 juveniles, and an infant. It may be that the fossils came from the same group, or band, given that the proportion of adults to juveniles is similar to that expected from a single hominid social unit.[11]

Marked sexual dimorphism—differences in body size and shape between males and females—was also apparent. A constant source of confusion for anthropologists, sexual dimorphism has frequently led to fossils of different sexes of the same species being identified as representing more than one species. That would not happen to this trove of associated fossils from L333 that were traced up a slope to a single, thin sandy clay layer. A number of fossils, all of the same apparent population, were eventually excavated from a 33-square-meter area of this layer.[12]

The first hominid from L333 was discovered by a colleague of Johanson's in midmorning.[13] When Johanson arrived on the scene shortly afterward, he immediately appraised the situation and postponed recovery of the fossil until early the next morning. The following day at 8:00 A.M. a movie crew and a *National Geographic* photographer were gathered to focus on D. Carl retrieving the fossil. While the shoot was in progress, a cry rang out from another team member who had found another hominid. Soon others joined in finding yet more hominid fossils! As Johanson euphorically described the occasion in *Lucy*, it "was the first time ever that professional photographers have taken pictures of [hominid] fossils actually being found."[13] He recounted how "A near-frenzy seized us as we scrambled madly to pick them up." The problem with this photogenic moment was that unless the exact location of each specimen was carefully marked, valuable information that might be useful for associating one bone or one fragment of a bone with another was surely lost.

Nevertheless, once again a major find was made. L333 came from the middle of the section just below the level of Lucy (L288), making the hominids featured in the "First Movie" and Lucy essentially the same age, about 3.2 million years.[14]

So what explains this potpourri of hominid remains? In the December 1976 issue of *National Geographic,* Johanson recreated the death scene of the band of hominids: "Crashing through the stillness of a primeval dawn, a flash flood races down a dry channel, overwhelming a terrified group who have no place to run. Triggered by highland rainstorms, the flow drowns and then buries them under sediment."[15] Geologist James Aronson and Taieb later suggested that a flood had drowned the troop *en masse* after it had become bogged down in mud while trying to race across a low, wet floodplain.[16]

Drowning seems reasonable. When taken together, the information that is now available for L333 suggests that torrential waters poured across a floodplain next to a streamside that was lined with a thick woodland or gallery forest.[16] The vegetation was probably somewhat denser than it is today at Hadar and the climate moderately wetter. The fact that so many hominids were swept up by flood waters suggests they were somehow trapped and overcome by greater than normal flooding, similar perhaps to the monsoonal flooding in India that periodically claims so many lives. It may be that the hominids sought refuge in trees or that they were sleeping in trees when the storm struck, perhaps at night, when such storms are brought on by cooler temperatures. There is ample anatomical information, such as curved fingers and toes, to indicate that the hominids were ape-like and partially arboreal, like Lucy.[17]

Although there was no reported damage to the hominid fossils attributed to crocodiles or other carnivores, it is likely that the corpses were at least partly exposed to the surface as the flood waters receded. The absence of scavenging by large animals could be due to the temporary isolation of the death site by ponded waters on a floodplain until the rotting of the flood victims was well under way. Regardless, we can assume that any exposed flesh was at least scavenged by vultures, carrion-eating storks, hawks, or other birdlife, as well as multitudes of ants, beetles, flies, and other insects—all pecking, chewing, tearing, gnawing, and nibbling. The scattering of exposed body parts would have accompanied this process, along with the natural weathering of bones and the eventual trampling of the skeletons.

In *Lucy* Johanson gives Taieb credit for the curious idea that the "strongest evidence for drowning [of the hominids] . . . would have to be the absence of [other] animal fossils. Animals would not have frequented a ravine inhabited by hominids, and thus would not have been caught with them in a catastrophe."[18] In some of the more bizarre passages in the book, Johanson refers to nonhominid fossils at Hadar as "noise" or "background scatter" and then says that it is the absence of "noise" at L333 in association with the hominids—non-noise—that is so mysterious.[18]

CHAPTER 17

The Miocene

A fter our return from the Awash in late July 1975, Herb moved on to Egypt to join Wendorf, while Liz, Assefa, and I remained in Addis Ababa for the next three months writing up our field work. Liz worked on our base map of the Middle Awash and on site maps of Meadura, Wilti Dora, and Matabaietu. Assefa wrote a report on the primitive elephant molar from Wee-ee and attempted to identify the rest of the elephant teeth we had collected.[1] He tentatively called the Wee-ee specimen a mammoth, an opinion shared by Maglio. Together Liz and I worked on the geology and stratigraphy of the Middle Awash.

In late August, I recruited another Zemacha student. Sleshi Tebedge was a senior undergraduate at the university and a classmate of Assefa's in biology. Of medium height and balding, he had powerful arms and shoulders that he attributed to chopping firewood as a boy. He was born in Debre Tabor just east of Lake Tana in the Blue Nile province of Begemdir, where his father served as pastor of a small Seventh Day Adventist Church. His family later moved to Addis Ababa, where he began attending the university. He failed several university courses because he refused to take exams on Saturday, the Adventists' sabbath. Before joining the RVRME, he had worked as a research assistant at the Central Laboratory helping develop commercial strains of fish. He was a gifted musician, his proficiency acquired from years playing the organ and piano in his father's church. Sleshi was to become our specialist on fossil suids (pigs), an important faunal group for dating purposes.

By late September Sleshi had begun to identify the fossil suids as best he could, and together Assefa, Sleshi, and I had provisionally identified the remaining fossils in the RVRME collection.[2] From a total of 250 fossils, we distinguished 45 taxa representing 12 mammalian families from 15 sites from Andalee to Wee-ee. We then photographed much of the collection and sent photos to various specialists in the United States and Europe for further identification. We had no reference collections of fossils to use, because those at the National Museum were either in packing crates (those from the Omo) or locked up (the IARE collection). Nevertheless, I had built up a modest reference library over the years, and we used skeletons of modern animals kept by the Biology Department of the university for comparative purposes. In addition, several years earlier I had discovered that abandoned water cisterns built by the Italians in the 1930s were a bountiful source of skeletal remains. The cisterns were dry, open pits made of stone and cement, 3 to 4 meters deep and about 5 meters wide. They were commonly littered with the bones of animals, ranging from baboons to cobras. Although such death traps might be somewhat dangerous to crawl around in, I remain convinced that systematic excavation of these open graveyards across Ethiopia could produce an outstanding regional collection of animal skeletons. In one cistern near Awash Station, I retrieved six complete skulls of *Papio* baboons of varying ages and sexes.

Using 24 key taxa for faunal dating purposes, by the end of October we had tentatively divided the eastern Middle Awash fossil beds into ten stratigraphic and biostratigraphic units. We estimated the lower fossil layers at Wee-ee to be 4.0 to 4.25 million years old, the Oldowan levels at Matabaietu to be 2.0 to 2.5 million years, and the Acheulean layers at Bodo to be 600,000 to 800,000 years.[3] We also distinguished 18 volcanic layers that would eventually provide a chronometric framework for the region.[*]

With no diagnostic fossils from Ananu yet identified, I guessed from their regional position that the deposits there were not less than the age of Wee-ee, and probably several million years older. Shortly I received a letter from Augusto Azzaroli, an equid specialist at the University of Florence, who identified the single horse molar that I had collected at Ananu in May as *Hipparion primigenium*.[5] He said that the species is known in western Europe and North Africa from deposits of "late middle Miocene" or "Vallesian" age. I assumed the latter was a European geological time unit of some sort, but surely Ananu was not late *middle* Miocene. That would put the age at about 11 to 12 million years. After reading the letter at the post office, I rushed back to the office and looked up Vallesian in a newly published reference source.

The book, *The Late Neogene,* by W. A. Berggren of the Woods Hole Oceanographic Institution and J. A. Van Couvering of the University of Colorado Museum, described the Vallesian as a Land Mammal Age, a time period characterized in the fossil record by a particular animal or group of animals.[6] Since the introduction of laboratory dating techniques in the 1960s, such faunal time periods have been given ages by dating the strata containing the fossils. Berggren and Van Couvering described the Vallesian as one of five Land Mammal Age subdivisions for the Miocene epoch in Europe. The base or beginning of the Vallesian period is defined by the first appearance of *Hipparion* in the fossil record—called the *Hipparion* Datum—in Europe and Asia, following the migration of this three-toed

---

[*]The current radiometric date for the "mammoth tuff" at Wee-ee is 3.9 million years; the Oldowan and Acheulean beds at Matabaietu and Bodo are currently dated at 2.5 and 0.6 million years, respectively.[4]

horse into Eurasia from North America. The authors reported the timing of this first appearance to be between about 11.5 and 12.5 million years ago.

The fossil record of horses in North America begins some 55 million years ago. However, it was not until the late middle Miocene, a period of low sea levels, that the ancestors of the Old World *Hipparion* crossed the Bering landbridge into Siberia, a time of temperate climate and open grassy plains, perfectly suited for hungry immigrant grazers.[7] So well adapted were these horses to this environment that their offspring are said by Van Couvering to have spread west across Eurasia and into Africa at "geological lightning speed."[8] The leader of this stampeding horde was *H. primigenium*, the earliest known species of *Hipparion* in the Old World and the very same species that Azzaroli said we had found at Ananu.

The swiftness of the *H. primigenium* migration is what makes this horse so important as a chronological datum, because this species, or modest variations of this species, are said to appear "instantaneously" in the fossil record of Eurasia and Africa. Berggren and Van Couvering estimated that the *Hipparion* herds might have traveled from Siberia to Equatorial Africa at a rate of 10 kilometers per year. This means that in about 10,000 years these equids thundered across central Asia, over or around the Himalayan, Caucasus, Zagros, and Pyrenees mountains, through Europe, Asia Minor, and the Middle East, across the Gibraltar and Red Sea landbridges, and into God's Country—Africa. The authors suggested that the rate of migration of *H. primigenium* might have been comparable to "the invasion of rabbits into Australia or English sparrows into North America."[9]

As I reread Azzorali's letter, I recalled that *H. primigenium* was reported by Schönfeld from Chorora, where it was dated radiometrically as between 9.0 and 10.5 million years old.[10] The maximum age of the Chorora horse was close to an 11.0-million-year age for fossils identified as *H. primigenium* from a site at Ngorora, Kenya.[6] I was still not convinced that Ananu was as old as even the minimum age of Chorora.

To jump ahead momentarily, following dogged research over the last 20 years by numerous workers, much of it led by Ray Bernor of Howard University in Washington, D.C., current estimates for the age of what is now designated the *"Hippotherium"* Datum (the new genus name follows revisions in equid taxonomy) is currently placed between 10.8 and 11.0 million years.[7] This age is supported by refined radiometric dates of the Chorora Formation, which also place the Chorora horse between 10.8 and 11.0 million years old,[11] allowing our gregarious charger something less than 200,000 years—and perhaps as little as 10,000 years—to make the one-way trip from eastern Siberia to eastern Africa, via the Gibraltar or the Afar–Yemen landbridges. This means that the oldest known radiometrically dated horse in Africa apparently comes from ancient lakeside deposits on the southern margin of the Afar Depression.

I certainly knew in 1975 that it would take more than a single horse tooth to tell us the age of Ananu. My plan was to go back there and find a *Stegotetrabelodon*, and that would be the end of it. The ancestor of all elephants. Then no one would argue that this was not a Miocene animal, or close enough.

Stegotetrabelodon. In Greek, *stego* means "roof" and refers to the roof-like rows of plates on the molar surfaces; *tetra belodon* means "four tusks" in reference to two upper and two lower tusks. A prehistoric ivory hunter's dream: two for the price of one. Italian paleontologist Carlo Petrocchi was the first to name and describe this pachyderm, recovered from Libya in the 1940s.[12] He was also the first to conclude that *Stegotetrabelodon* was an ancestor of the modern two-tusked elephants. He described the genus as transitional in morphology between more primitive proboscideans and the more progressive elephants. The additional set of tusks are relic features passed down from mastodon-like ancestors and are believed to have been used for foraging herbaceous vegetation, which was widespread in earlier Miocene times. The rows of low, widely spaced plates were an adaptation useful for shredding the coarser vegetable diet that was becoming increasingly abundant as the climate grew more arid. In 1970, Vincent Maglio described the first *Stegotetrabelodon* known from sub-Saharan Africa, from Lothagam, Kenya, whose fossiliferous strata bracket the boundary (c. 5.2 million years old) between the Miocene and Pliocene epochs.[13]

By early November 1975, we were ready to go. Miocene ivory or bust. I had just received $6300 more from FORGE and used $1750 of the money to purchase a 1968, long-wheelbase Land Rover with a winch attached to the front bumper. We were to be joined later by two Italian paleontologists from the University of Florence, colleagues of Azzaroli's, with funds they obtained from the National Research Council of Italy. Others would be joining us throughout 1976 with additional funds, so I felt the RVRME was shaping up financially.

Once again we loaded up two vehicles. Our team included Liz, Assefa, Sleshi, and a writer for *The New Yorker* magazine, Bill Wertenbaker. Bill was researching a book on the search for human origins and wanted to see what exploration surveys were like. He also contributed to field costs. I emphasized to him before we left Addis Ababa that for now we would be looking for elephant ancestors, not human ancestors. With that said, I must note that I hate those accounts by the explorer–scientist who describes with hindsight the scene where he looked out over the gray mist and pronounces that he "just had a feeling" that the next morning at 8:00 A.M. he would make the discovery-of-his-life. Nevertheless, three days later, after we had crossed the Awash on the ferry at the Dobel Plantation and made camp at Ananu, we took the vehicles, rumbled up the boulder-strewn Saragata stream, crossed over to a smaller tributary called Saitune Dora, and, in a very small clearing of sedimentary exposures among a sea of basalts, found the First *Stegotetrabelodon* Family.

The locality was a jumble of molars and bones that, judging from the teeth, appeared to represent three adults and two juveniles. It is unlikely that this particular First Family had been trapped in trees by rising flood waters, but it did look as though the remains had been scooped up and concentrated in a flood wash. We realized immediately that the fossils were early elephantids, but it would be a long time before we figured out exactly what was there. Maglio's verdict two months later, based on photos we had sent him, was that the Saitune Dora specimens were indeed *Stegotetrabelodon.* Then it would be years after that before we realized that at least some of the "family" were neither *Stegotetrabelodon* nor elephants. They were *Stegodon,* a two-tusked elephantid closely related to elephants that is abundant in the Pliocene and Pleistocene fossil record of Eurasia, but rare in Africa.[14]

Our second major field objective was to visit the area that I had first seen in the central Awash the previous year on my reconnaissance flight. This is the area northwest of Gewani that is a patchwork of faults, rolling hills, and basalt-capped hills, which I would later call the Central Awash Complex. A large graben five kilometers wide cuts the area in half and is drained by the Hatowie River, which gives the graben its name. Its eastern flank is crossed by a dozen streams draining into the Awash, the two largest, the Sagantole and Aramis, creating low reliefs across a heavily dissected terrain. In the few hours we spent driving through the area we found excellent exposures, but few fossils and none that had "pre-Hadar" written all over them. I had Bill Wertenbaker with me that day, and my disappointment was palpable: "Son of a bitch, *son of a bitch!*"

It's unfortunate that he never published his book on the search for human origins, because that abbreviated trip to the central Awash was the first glimpse of an area that has since produced "the find of the century."

When we arrived back in Addis Ababa in mid-November 1975 to meet the Italian paleontologists, I was greeted by a letter from one of them saying that he and his colleague had to postpone their trip because of illness in his family. Very disappointing. While I remained in the city attending to administrative matters, toward the end of November I sent Liz, Assefa, Sleshi, and several others back to Ananu to renew surveys of the area. They had one week before I would join them in the field and Liz would return to the United States. She had one more semester to complete her undergraduate degree in geology in California.

I instructed Liz, who was to drive my Land Rover, to take the highland route through Kure Beret. I did not want her to risk using the precarious ferry crossing at the Dobel Plantation. She already had with her an Afar guide, Abraham, whom I had brought to Addis Ababa at the end of our last field trip. We agreed to meet at the ferry crossing on December 4, at 9:00 A.M.

A week later I drove to Dobel with Dejene on the low road through Awash Station. After dropping me off, he was to take Liz back to Addis Ababa. While we were visiting nearby Gewani to pick up a box of food left at the police garrison, the officer in charge told me that five days earlier—the day Liz had driven down the escarpment—the plantation had been attacked by Issa and 29 workers had been killed. When Dejene and I reached the ferry crossing, soldiers were everywhere. I was somewhat apprehensive about Liz and the others, but then I saw her sitting on the west bank of the Awash waiting for me. She was off to one side from the hub-bub of ferry traffic with her head buried in her field notebook.

"Hey you!" I yelled across the river. "Can you tell me how to get to Los Angeles from here?"

When Liz saw me, she jumped up waving her arms. "Kalb, I found the most primitive elephant in Africa! I mean *primitive*—you will not believe how primitive! I mean really *primitive!*" She was all grins, darkly suntanned, wearing shorts, sandals, and a red T-shirt. Her brown hair was sun-bleached and drawn back into a bushy ponytail tied with a handkerchief.

"What do you think it is?" I replied.

"We're not sure. Come see for yourself!"

"What does Assefa think it is?"

"He's not sure. Stop talking and get over here!"

"Liz, I hope this isn't your idea of a joke!" But I knew better, recalling her discovery at Wee-ee.

By this time soldiers, Afar, and plantation workers around the ferry were looking over, wondering what the two firenjis were yelling about.

The ferry was tied up on the opposite bank, so I emptied my pockets, took off my shirt and sandals, gave them to Dejene, jumped into the Awash, and swam the 30 or so meters to the other side in my shorts. As soon as I had climbed up the bank, Liz started telling me excitedly that she and the others "must have walked a hundred miles in five days." She had found the "*unbelievably* primitive" fossils the day before, at sundown, on her way back to camp. They came from just north of Ananu at the base of a hill called Bikirmali. When we reached the Land Rover parked under some trees, Assefa, Sleshi, and the rest were waiting.

I exchanged greetings as Liz retrieved her field bag. Everyone gathered around. She removed two small bundles wrapped in toilet paper and slowly unwrapped them, handing me two relatively large molars one by one.

I stared at them, "What the *hell?*"

I had no idea what animal I was looking at. I was not even sure what phylum the fossils belonged to. One molar was about the length and width of a pack of cigarettes; the other was a bit bigger but somewhat narrower and broken at one end. Liz said she had found them side by side, which suggests that they came from the same individual. She thought more of the animal was buried. The surfaces of the teeth were not plates as in elephant molars or *Stegotetrabelodon*. Rather, the crowns were made of two rows of vertical pairs of compressed enamel cones, except that one of the molars was well worn, revealing only the oval cross sections of the cones. The second molar was unworn and looked like rows of twin peaks separated by a deep cleft running the length of the tooth.

Liz was bouncing on her toes. "What d'ya think? What d'ya think?"

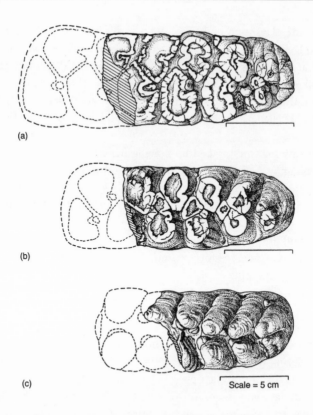

(a)

(b)

(c)   Scale = 5 cm

FIGURE 5.   RIGHT LOWER THIRD MOLARS OF ANANCUS *from (A) Wee–ee,*
(B) the Hatowie graben, and (C) Bikirmali; late Miocene to early Pliocene.

"Liz, I think you're a marvel. You have found something that looks like
pre–Everything, maybe even pre–Haile Sellassie."

Assefa thought the teeth were the first and second molars of something more
primitive than *Stegotetrabelodon.* They were not like anything figured in Maglio's
monograph, which Assefa retrieved from the Land Rover and laid out on the hood.
The fact that the fossils, or anything like them, were *not* in the monograph—a
comprehensive study of all recognized species of the family Elephantidae—was all
the deep, scientific reasoning I needed to conclude that we were dealing with a different breed of proboscidean, one likely to be very primitive.

By then Dejene had found us, after crossing on the ferry, and began exchanging
the latest political news from Addis Ababa with Assefa and Sleshi. Liz and I went
over the aerial photos of the areas she and the others had surveyed. After she gave
me a sketch map of Bikirmali, just north of Ananu, where she had found the fossils, and some information about the geology, it was time to part. I had a full day
planned for surveys to the west, and Liz and Dejene had a long drive back to Addis

Ababa. Assefa was going back with them and would photograph the molars and immediately send the prints off to Maglio for his opinion.

Liz and I hugged and kissed good-bye. We had worked together since that stormy and blazing month of July 1974 at Geraru. It was 18 months later, and she was now all of 23 years old.

She grabbed her gear from the Land Rover and took off for the ferry with Dejene and Assefa, with some last-minute instructions for me.

"Yeah, Liz," I responded. "Sure I'll write and tell you what the fossils are, as soon as we find out."

"No press conferences after I leave?"

"No press conferences."

It would be several months before we learned the identity of Liz's latest discovery, which in fact was to be her last in the Afar. Although she would later return to Ethiopia, political developments beyond our control would intervene. Somehow it was fitting that her new discovery was made on her last day in the field, during the last hour of sunlight. And yes, the fossil was definitely primitive.

Before we received Maglio's assessment in mid-January 1975, I got my first hint as to what the fossils were. I was visiting the home of an American petroleum geologist in Addis Ababa, and sitting on the mantle in his living room was a large fossil tooth. It was a molar with rows of prominent cones and a deep cleft running down the axis of the tooth.

"May I ask what this fossil is and where it's from?" I asked.

"It's a mastodon," my friend said. "I picked it up in the U.S."

"A *mastodon*?"

"That's right, a mastodon."

Mastodons (Mammutidae) are primitive, and far more so than elephants or *Stegotetrabelodon,* although all three proboscideans possessed a grossly similar body structure, including a proboscis, or trunk, from which the name for the order Proboscidea originates. The molars of mastodons are simpler than those of elephantids. The morphology on the unworn chewing surface resembles rows of female breasts, with a valley-like cleft running down the center, similar to the unworn molar that Liz found. Not surprisingly, the Greek origin of the name *mastodon* means "breast-like tooth" (*mastodont*). The term *mastodont* itself is simply a descriptive term meaning "mastodon-like." Functionally, opposing upper and lower mastodont molars crush and grind food, like a mortar and pestle, whereas we know that the opposing ridge-like molars of elephants cut and shred food, like a pair of scissors.[15] The two basic tooth morphologies of mastodons and elephants survived for millions of years and allowed these animals to live on different diets while occupying similar or adjoining habitats. In the Pleistocene, for instance, we know from coprolites and other evidence that North American mastodons browsed largely on trees, shrubs, and bushes, whereas mammoths grazed mostly on grasses and sedge.[16]

Mastodons originated in Africa in the early Miocene and migrated into Eurasia by the middle Miocene. One branch went north and east and eventually crossed the Bering landbridge into North America (near the time when hipparionines crossed into Asia). There it lived until its extinction at the end of the last Ice Age (c. 11,000 years ago).[16] Another branch of mastodons gave rise to gomphotheres, a diverse group named after the ancestral genus *Gomphotherium*. Gomphotheres radiated throughout much of the world, including North and South America, before they too disappeared at the end of the last Ice Age.[17] In the early Late Miocene in Africa, however, one of the gomphothere lineages gave rise to *Stegotetrabelodon* and another to the genus *Anancus*, which migrated into Eurasia.[18] Elephants went on to colonize much of the world, including North America in the Pleistocene, where mammoths and the more primitive mastodons alike were hunted by Paleo-Indians. In North Africa, *Anancus* lasted until the Pleistocene, but south of the Sahara, in eastern and southern Africa, the genus reached a dead end about 4 million years ago.

Anancus is the very same animal that Liz found. In his letter to us, Maglio described the Bikirmali fossils as being very similar to those from the lower Lothagam beds. Thus Liz's discovery was probably at least 5.0 million years old.

After our final good-byes to Liz, Sleshi and I headed back to the Ananu area. Over the coming days we extended our surveys north and south of Ananu, adding more fossil localities and another dozen names of streams and hills to our base map. Much of the area was too rugged for a vehicle, so we covered it on foot. More than once we walked across the coverage of a 1:60,000 aerial photo in the morning, then turned around and walked back in the afternoon by a different route, a round trip of at least 40 kilometers.

As we went into new areas, we met and conversed with Afar chiefs, arranged for local guides, recorded the names of landmarks, and on several occasions attended to medical matters. One morning, we were asked whether we could help a young girl in a nearby encampment who was blind. She was so excited by our visit that she fainted. As I was looking into her eyes, her legs crumpled, and she collapsed before I could catch her. She then went into a seizure, and her mother and sisters crowded around to help her. That was not the first time I had seen an Afar go into a seizure. On another occasion, our guide Abraham fainted and went into a seizure after hearing gunshots on the other side of a hill where his best friend was. More specifically, he fainted after he had raced around the hill and discovered that his friend was unharmed. His friend had simply fired at a gazelle. I have wondered about the origin of these seizures among the Afar, or epilepsy, if that is what it is, and whether this malady is more prevalent among these tenacious people than others.

When the girl revived, it was clear to Sleshi and me that there was nothing we could do for her. Her eyes did not appear at all infected, which might have been readily treatable. My intention was to take her later to Addis Ababa to an American friend who was an eye doctor, but he soon left the country and I knew of no other specialist in the city. The following year I did arrange for the girl to be examined by the Swiss doctor at the Red Cross clinic in Gewani, but apparently that too led nowhere.

Over two weeks, Sleshi and I documented 28 fossil localities in the greater Ananu area. Most were clustered at Bikirmali just north of Ananu, where Liz had found the molars. The rest were spread out over surrounding areas in the gouged-out and downfaulted terrain made striking by the diatomite, dark-green stratified tuffs, and basalts that I had seen six months earlier. As before, I concluded that the ancient Ananu lake had drained following tectonic activity, perhaps as a result of sea-floor spreading that originated hundreds of kilometers away.

We found two more kinds of proboscideans during these surveys: *Deinotherium* and *Primelephas*. The latter is the ancestral elephant from lower Lothagam first described by Maglio.[13] Deinotheres are distinguished by two or three blunt ridges on their molars and are even more primitive than the most primitive mastodons.[19] They are also one of the stranger-looking creatures from the Old World, with recurved lower tusks that on first appearance make them look like the Animal Kingdom's answer to the backhoe. The tusks may have been used to chop or strip vegetation, much as one would use an adze or a pickaxe.[20]

While we were camped at Ananu, we received several harassing visits from a man called Morado, who lorded over the area, imposing his personal tax on trade goods and contraband moving up and down the trail to and from the highlands. I did not like this man's telling me that "something terrible could happen" to us without his protection. The local Afar feared him and warned us that he had "killed many men." Morado came from higher up the trail and was Oromo; at least that was the language he spoke when he came riding into our camp on a mule accompanied by several of his minions, who were on foot. At the time, I had two very capable Oromo with me, Kefyalew Belachew, our antiquities representative, and a camp aide, Menda Wordofa, neither of whom was intimidated by Morado. Both refused his orders to serve him food and to water his mule. Kefyalew, who was studying law at the university, suggested that if Morado wanted to impose a tax on travelers, then he should take his case to the proper government authorities. Morado snorted, "Ankober [the name of the old Ethiopian capital] has nothing to do with what goes on at Ananu."

I have to say that Morado was a fierce and malevolent-seeming character with a face that looked like it had been hammered together with rocks. I called him Bent Finger because one of his fingers obviously had been broken and had grown back together at an odd angle, making it look like the spur on a fighting cock. He wore jodhpurs, boots, a shamma around his shoulders, and around his head a handkerchief that was tied in the back. Pulled squarely over his head was a wide-brimmed hat in a fashion made famous by Emperor Menelik in his later years.

On several occasions Morado's henchmen approached our camp at night and lobbed stones into our midst, but by then I had sought the assistance of the local Afar chief to provide guards for our camp, and a few rocks hardly disturbed our sleep. The showdown, if that's what it was, came at the end of two weeks in the late afternoon the day before we planned to leave the area. Morado rode into camp and insisted that we leave unless we paid him protection money, *or else.*

What would Henry Stanley have done in a situation like this, I thought? That was easy. He would have had 20 of his Zanzibari riflemen armed with Remingtons poised to reduce Morado to fertilizer unless he headed into the sunset. But this was not the Congo, so as Morado sat on his mule a few yards from me, I began eyeing his rifle in its saddle holster—a look not unnoticed by its owner—and to calculate how fast I could reach over and grab the rifle, and shoot his goddamn mule. Could I do it in, say, three seconds, four? But then, what if Morado grabbed the rifle first, in say, *two* seconds . . . ?

"Okay, everybody," I hollered out, "let's break camp and get the deuce out of here."

In record time we packed up and left Morado and his men standing there. That was the last I saw of him. When we returned to the area a few weeks later, he was gone. Perhaps the Afar chiefs had something to do with his disappearance.

———

During our stay in the western Awash, Sleshi and I spent a day surveying a sedimentary ridge that forms the extended western flank of the Hatowie graben in the central Awash. The ridge bounds Lake Yardi to the north and projects toward Gewani, causing the Awash to bow to the east several kilometers before turning north again. The broader floodplain is the area of Beadu, and the elevated peninsula is known as Buri. Both are dotted with stonework, and one or the other is heavily settled by Afar year-around, depending on the level of the river and the prevalence of mosquitoes. Well-worn trails follow the ridge crest to a river crossing at the Awash. Judging from artifacts we found throughout the area, it was apparent that the immediate Buri area had been inhabited on and off for tens of thousands of years, at least. As we mapped the names of the various Afar encampments and landmarks along the peninsula, we found more artifact localities, mostly Middle Stone Age or late Acheulean, to add to those Herb and I had found the previous May. At one locality just above the Awash floodplain were large, fresh, crudely manufactured handaxes and some cleavers made from flakes, unlike the more refined bifaces we found across the river. The area was called Dakanihyalo, which apparently means "elephant boundary," perhaps in reference to the natural barrier the ridge forms between the Awash wetlands to the east and the eroded terrain to the west.

While walking along the ridge, I guessed the bottoms of the nearby river and swamp were paved with *dakani* remains. I also imagined earlier days when slavers passed through the area, going from the highlands to the coast with freshly captured "goods." Places like Yardi, Buri, Beadu, and Dakanihyalo were ideal habitats for elephants, and surely excellent sources of ivory to be hauled to the coast by unpaid porters. The use of slaves ("black ivory") to transport elephant tusks ("white gold") was fundamental to commerce across the continent from the beginnings of African history.[21] Ironically, the Arab traders, who for centuries profited from live humans and dead elephants, were among the first to introduce Islam to coastal and inland tribes such as the Afar and Issa.[22] By the time David Livingstone and his missionary followers had begun their life's work in southern Africa in the 1840s, promoting religion and civilization in exchange for commerce, they were following a well-trodden path. There was a major difference between the Arab traders and the Europeans, however: The latter also became colonizers and walked off with most of the continent.

The widespread introduction of firearms—repeating rifles and particularly breech-loading cannons like the Holland & Holland that fired .557 Express slugs—turned the killing of *Loxodonta africana* into big business.[23] Considering the hundreds of thousands of African elephants slaughtered during the last century, it is about time that these compliant beasts are given their proper place in history—for making the slave trade even more lucrative than it already was, for promoting the cause of Islam and Christianity in Africa, and for ensuring that pubs and clubs the world over had an adequate supply of billiard balls. This probably is not what the Bible had in mind when, in Job 40:19, the elephant is referred to as "the masterpiece of God."

In mid-1973 when I was traveling around the southern Afar, I came across an old man wearing an ivory bracelet, something rarely seen among the Afar.[24] He told me that his father had made it for him from the tusk of an elephant he had shot. I asked if he knew the spot where the elephant had been killed. He said he distinctly remembered the place, near the Kesem River in the extreme southwestern Afar, because he saw the bones of the elephant there many times as a boy. I asked if he would be willing to show me the locality and he agreed. The chances were certainly remote that any remains would be found, but at the very least I could pay homage to one of the last elephant kill sites in the Awash Valley. I guessed the old man was in his middle sixties, and it sounded as though he was about 12 years old at the time the elephant was killed, which would have put the date in the early 1920s.

The story fit with accounts of early travelers passing through the southern Afar. Several sportsmen at the turn of the century reported an elephant herd of at least 40 individuals in the area of Bilen, 20 kilometers north of the Kesem River.[25] Bilen is the site of some warm springs that drain into a large swamp that probably attracted the elephants. When Nesbitt traversed the Afar in 1928, he mentioned no elephants anywhere, but in 1934 Thesiger reported a single animal in the area of Bilen, probably one of the last of the herd.[26]

With the old man as my guide, and accompanied by Kelati and Joseph (my linguist friend), I forded the Kesem in the Land Rover in midafternoon. After several hours of searching, we found no elephant remains. However, we did find a lush riverine setting of grassy wetlands enclosed by the Kesem River to the south, the Awash to the east, and the Kebena River just to the north—a perfect setting for the largest terrestrial herbivores.

In the late afternoon, we returned to recross the Kesem, when we discovered—to no great surprise, because it was the rainy season—that over the last few hours the river had risen at least a meter from unseen rains in the highlands. The Kebena had also risen, so we were trapped and had no choice but to wait it out. Throughout the night I waded into the surging Kesem to check its level, ready to cross at a moment's notice.

By daybreak the water had subsided, but not enough, so I waded and swam to the other side and walked to a nearby plantation. I soon found the manager, a muscular young Italian with hair and beard the color of lava. I asked him if he could help us cross the river by pulling my Land Rover with a tractor. Instead, he suggested we wait a few hours and see if the water did not drop further. He then took me to the kitchen and gave me a mug of strong, steaming tea, warm bread, and a handful of bananas grown on the plantation. After I had finished the last banana, I lay down on a cot in an adjoining room, intending to rest for a while. I instantly fell asleep amid dreams of swirling waters and bleached elephant bones.

I awoke two hours later, aghast at how long I had slept. I jumped up and raced back to the river to find Kelati and Joseph impatiently waiting for me with the old man sitting under a tree, probably deep in thoughts of his youth. The water had indeed dropped, but just barely enough, so we took the belt off the radiator fan to prevent water from blowing onto the engine and slowly crossed the river with water seeping in over the floorboards.

After driving to the plantation to tell my Italian host that we had crossed the Kesem safely, we drove to Awash Station Buffet, where we had a big meal. I then took the old man back to his encampment and thanked him generously. With DNA fingerprinting and isotope analysis, there is perhaps much that his ivory bracelet could have told us about the origins and migrations of one of the last herds of Afar elephants. I was sure I could have offered our friend enough to entice him to part with his bracelet, but even science has its limits.

CHAPTER 18

Crossing the Talalak

By December, 1975, it was a little more than a year since the overthrow of Haile Sellassie and the mass execution of the nobles. The month also marked the anniversary of the Derg's declared path of socialism, which led to substantial changes, ranging from rural and urban land reform to the establishment of a national trade union. Most recently, the government had proclaimed the formation of peasant associations, urban leagues, and agricultural cooperatives.[1] Neighborhood defense squads were also organized to combat the growing number of dissidents. An immediate casualty of these armed groups were the hyenas that prowled the city at night looking for garbage. Shots would ring out in the late-night hours, and you knew that another carnivore had been sacrificed for the revolution.

Late December was also the beginning of the extensive holiday period, made more lengthy because the Ethiopian Orthodox Christmas is celebrated on January 7. Turkey was hardly served on our Christmas Day, but Judy did a terrific job of roasting hens and baking pies out of squash, yams, or even canned cherries that American friends picked up for her at the U.S. commissary. Once we had an Egyptian goose that I had bought from a street vendor, but it was so thin that by the time we removed the feathers there was scarcely anything left. For presents, Justine and Spring received toys or clothes that our families had sent them from the United States, and Judy made them holiday dresses. That Christmas was made particularly memorable for Justine, then five years old, by a dollhouse Judy had built for her. At night after the girls were in bed, she made all the holiday decorations out of soap and match boxes and bits of cloth.

On New Year's Day of 1976 I returned to the field with my crew. We were joined by Craig Wood, a doctoral student at Harvard, whose interests were paleoecology and microvertebrates, such as rodents and insectivores. Craig was to help us reconstruct past environments and ecosystems of selected sites by analyzing where and how animal remains originated, accumulated, and were buried. Taphonomy, another of Craig's interests, attempts to answer such questions by using quantitative and statistical methods of fossil collection and by determining the mode and degree of damage to animal remains before, during, and after burial. Such investigations can make it possible, for instance, to reconstruct more confidently the cause of death and original composition of an assemblage, or "family," of hominid fossils.

Taphonomy as a scientific discipline originated with the Russian paleontologist I. A. Efremov, who coined the name in 1940 to mean "the laws of burial."[2] His intent was to learn more about the pathway of animal remains "from the biosphere to the lithosphere"—to appreciate how, for instance, a group of animals together in life can be mixed with entirely different animals at the time they are buried, by actions such as flooding, scavenging, and/or predation. As one taphonomist has put it, "Fossil animals do not live in social groups; they have no home range or preferred habitat; and they do not move, feed, play, learn, reproduce, fight, or engage in any other behaviors."[3] Fossils are inert; it is the challenge of the taphonomist to make prehistoric life vital and interactive again.

Over the next two months, we moved camp 20 times while traversing the Middle Awash along both sides of the river, from Gewani to Andalee, and from Ananu to the Busidima River west of Hadar (see Map VII on page 140). We added more pre-Hadar pieces to the regional section in areas south of Wee-ee and north of Ananu, finding more *Anancus* in both places. At Amba, in the Hatowie graben, we found abundant fossils, including yet more *Anancus* and early elephantids. Another traverse through the Sagantoli and Aramis areas revealed enough fossils to make the area one of high priority for future exploration.

Surveys south of Lake Yardi revealed another Late Stone Age site with abundant microliths amid a bone bed of fish and turtle, similar to the occurrence in the Nemay Koma basin. North of Dakanihyalo, around the margins of Dalu Ali, we found small late Acheulean handaxes similar to those discovered previously at Buri. Further north we located more artifact sites, including one east of Kada Meshellu stream and northeast of Meadura that is pre-Acheulean, and another to the west, at Issie, that may be Sangoan. One night in late February at Alkanasa, just north of Meadura, we made camp after dark, and when the moon came out after dinner, we discovered that we had camped in the midst of hundreds of Acheulean bifaces. Craig and I had the novel experience of conducting an archeological survey in the moonlight.

To Craig's satisfaction, we found a few small mammals at Ado Kaho north of Gewani and at Matabaietu. When we reached Andalee, however, Craig pointed out that the site was *buried* under rodent fossils. He estimated that there were tens of thousands, perhaps hundreds of thousands, of microvertebrates. Herb and I had walked back and forth across the site a dozen times nine months earlier and had seen only monkey fossils and artifacts. As Craig said, it is just a matter of thinking small and knowing what to look for. Nine months later I could bring another specialist to Andalee and he would politely point out that the site was *really* buried under millions of microscopic lizard turds.

Even sediments themselves require careful perusal. There is a renowned sed-
imentologist at the University of Texas at Austin, Robert Folk, who discovered
a mysterious, thin, translucent layer of opalized glass in Israel, while excavating a twelfth-century B.C. archeological site.[4] At first he thought the layer was
slag, possibly from an ancient iron smeltery, but careful study revealed something else.

Folk described the layer—eschewing academic jargon—as "ice-blue to pale
robin's egg blue," its surface "frothed into large bubbles . . . more fragile than the
thinnest Christmas-ball glass." The layer was studded with siliceous fibers like
"coarse whiskers" and it "crunched" when walked on. Beneath the glass was
another mysterious layer, "a strangely brilliant yellow powder . . . like ripe grape-
fruit rind," containing small, "wiener-shaped" ceramicized pellets. Folk and an
associate concluded from electron microscopy, X-ray analysis, and laboratory
experiments that the glass probably resulted from the fusion of silica-rich plant
remains, perhaps from rye or wheat, caused by accidental burning. The underlying
baked pellets appeared to be rat feces that possessed characteristic "radial sphinc-
ter marks at the blunt end (like the little nub at the end of a tied sausage)." Presumably the rats fed on the grain.

In several hours at Andalee, Craig and the rest of us recovered nearly 500 fossils. The order of size of the animals ranged from rodents to monkeys, medium-
sized bovids and pigs, large bovids, and a few very large animals such as *Equus*.
Craig readily distinguished five species of rodents that included a porcupine and
the large cane rat, which is eaten in parts of Africa. The size distribution of mammals was roughly the reverse of that at sites like Bodo, Meadura, and Matabaietu,
where large herbivores were dominant. Craig quickly observed that one reason for
the plentiful microfauna at Andalee was abundant, fossilized bird pellets, the undi-
gested diet of birds that is regurgitated. Pellets of terrestrial birds of prey typically
consist of bundles of bones, teeth, skin, and hair from small animals, particularly
rodents, that when discarded from roosts can accumulate into thick middens on
the ground below. One of the most common raptors worldwide is the barn owl,
which is present in the Awash Valley today.[5] It feeds in open areas and roosts in tall
acacias and giant fig trees, like those that line the banks of the Awash River just a
few hundred meters from Andalee.

Again the evidence suggested that Andalee was once forested, but until more
fossils were identified we still did not know the age of the site, other than that it
was broadly later Pleistocene, nor did we have more information about its archeology, other than Herb's suggestion that the tools were Sangoan and may have
involved woodworking. Nevertheless, Craig helped us put more pieces of the site
together: rodents, bird pellets, and predators that roosted in trees.

Soon I came to appreciate Craig's inquiring mind, even though, as a Texan, I
shared Lyndon Johnson's deep conviction that Harvard boys were not to be
trusted. Of medium build, Craig was bookish and hippie-like with longish blond
hair and a beard. He shuffled when he walked and wore a yachting hat in the field

and a perpetual grin on his face as though there was always something funny going on. He smirked a lot. The guy reeked of Harvard. All my suspicions came true after only a week in the field.

We were camped at Wee-ee, a place without water during the dry season, where as a matter of convenience we drew water from a nearby swamp. Because the swamp water was odoriferous, brown, and mysteriously organic, I doubled our purification procedures for the camp water supply. One morning Craig propped a small mirror on the hood of the Land Rover, filled a tin cup with water, and started to trim his beard. When I happened by, I found him staring into the cup.

"What's the *problem,* Craig?" I asked.

"There are creatures swimming in the water," he smirked.

"So? It's rich in marine life."

"It's not the marine life I'm worried about." A typical Harvard response.

"And what's *that* supposed to mean?"

"Schistosomes, amoebas, and giardia."

"So, throw some Halozone tablets in the water."

That he did. As an experiment he put two tablets in the cup, an amount that should purify two liters of water, and left it sitting all day. Halozone is a chlorine-based chemical that is normally effective against biological agents harmful to humans. It was widely used by U.S. soldiers during World War II and the Vietnam War. We usually carried it with us in tablet form when on foot to use for treating local water if we ran out of water purified in camp. There we used a special chlorine-based chemical in powder form, as well as iodine crystals, on the theory that if one did not kill a particular germ or parasites, the other would. Apparently the water Craig was using that morning had come from untreated water in jerricans used for bathing and for the Land Rover's radiator.

That afternoon when we returned to camp, the "creatures" in Craig's cup were still alive, perhaps even mutating. Later he asked the doctor in the Red Cross clinic in Gewani to examine a sample of the water under his microscope for schistosomes, the fluke that parasitizes blood vessels of humans and causes schistosomiasis. Also known as bilharzia, the disease can cause severe damage to the kidneys, and even death. The doctor found no worms, but he did find a thriving microfauna. Coincidentally, in the mid-1970s, epidemiological studies conducted on Afar plantation workers in the Awash revealed no infection from schistosomiasis, but it was found among agricultural workers brought down from the highlands.[6]

I have since learned that Halozone becomes unstable after a certain period of time and loses its toxicity to parasites and other organisms, such that today iodine is widely used in its place. Soon after Craig's observations, we began purifying all the water kept in camp, even that used for vehicles or for washing, lest it be used unintentionally for other purposes. We also began filtering the water and used alum when needed to settle out organic material. In 1978 we camped for several days beside the same swamp that we had drawn from when Craig was with us in early 1976, again drinking its water—fully treated—with no ill effects.

In mid-January 1976 we began surveys in the upper Millé River basin in north-central Wallo Province, an area that formed the landward extension of the greater Red Sea rift. I chose this area as another of the three RVRME permit areas on the periphery of the original IARE concession (see Map VIII on page 143), because I believed it to be one of the oldest sub-basins in the Afar. Hence it promised to contain fossiliferous deposits older than those in the Middle Awash. I reasoned that with the counterclockwise rotation of the Danakil Alps, or miniplate, during the Miocene, the opening of the northern Afar should have progressed from west to east and from south to north, in tandem with the opening of the Red Sea.[7] Evidence of this early rifting includes a 3000-square-kilometer graben in the upper Millé area, bounded to the west by the Ethiopian escarpment and to the east by a basalt plateau (see Map III). Basalts 15 to 25 million years old from the western margin of the graben, granites 22 to 26 million years old lie to the north and east, and the Millé plateau is capped by basalts 6 to 8 million years old.[8] Tazieff's team described the younger rocks as being interbedded with terrestrial sediments,[9] and German geologists tentatively identified Mesozoic limestones and possible Neogene sediments at the northeast end of the area.[10] I thought these sediments just might correlate with the Pliocene-Miocene Red Beds in the northern Depression. Finally, the same Afar guide who had taken me to the primate-rich site of Amado south of the area in late 1973 told me that more fossils were present upstream from Gura Ali volcano.

We devoted only one week to surveying the area, in part because we were repeatedly warned by our guides that the "Liberation Front" (Eritrean, Afar, Tigray?) was camped "on the other side of the hill." The area to the east was remote and hidden by hills, which made it easy for rebels to attack trucks and other vehicular pickings traveling the nearby Bati-Assab road. We found no Miocene fossils, but we did spend enough time there to dispel the notion that limestone outcropped in the area or that the Red Beds were present. Of potential interest, we did find thick exposed Pleistocene sediments, and some Pliocene tuff-filled channels containing fossils that may have been part of the same rift-river system that fed lake Hadar.

For the first half of February, we returned to the western Middle Awash and established 20 new fossil localities north of Ananu. One of our most interesting finds, from Adu Dora, north of the Hatowie River, was the mandible of yet another early elephantid, *Stegodibelodon,* first described from Chad by Coppens as a form intermediate between *Stegotetrabelodon* and elephants.[11] The molars and elongated chin of the Chad elephantid most resemble *Stegotetrabelodon;* however, as in elephants, the lower jaw lacks incisors. The Adu Dora mandible was missing its molars, but in other respects the jaw was nearly identical to the Chad specimen. At the spot where we found the jaw, we also found a large foot, probably of the same animal, with the tarsals, metatarsals, and phalanges perfectly articulated, an extremely rare discovery for any elephantid and invaluable for locomotion studies of large proboscideans.[12]

At the time, we were camped on the upper Hatowie River, a white-water stream cascading through the foothills, at one of the most picturesque spots in the entire Awash Valley. The stream is protected on either side by low hills and flows beside a spacious, shaded grove of high umbrella acacias with a ground cover of leaves and fronds. The water was clear, cold, and full of fish. One afternoon, while Craig and Assefa were off at a microfauna locality, Sleshi and I returned to camp to catalog fossils. While I was taking a break for a swim in the stream, it occurred to me that if there were a way to catch the abundant tilapia (similar to perch) nibbling on my toes, we could have a feast that night. Because we had no fishing hooks, the only real alternative for efficiently catching fish was a net of some sort.

After careful thought, I decided that the only net really adequate for the job was the new nylon mosquito net Craig had brought with him. I was certain he would share my enthusiasm. I disassembled the net from Craig's bedding and, with Sleshi's help, spent the next hour or so dragging it through the stream, having a wonderfully productive time hauling in load after load of tilapia. When we had finished, the net had a few rips in it and was dripping with algal slime and fish scales, and of course there was the fishy smell, but I was sure this was nothing Craig would notice.

To round out our pending banquet, we had a quantity of "Afar punch" that Sleshi had brewed over the previous days from barley, yeast, and sugar. By the time Craig and Assefa returned to camp that afternoon, Sleshi and I were in a holiday mood with the smell of fresh grilled fish filling the air.

Wouldn't you know that the first thing Craig asked was "How did you catch the fish?"

I spent a sleepless night inside his mosquito net while he slept comfortably beneath my own.

In mid-February, we headed north to get a sense of the geology and logistics between the Jara and Busidima Rivers (see Map VII on page 139). As we traveled through the area, we passed over Pleistocene sediments, met more Afar chiefs, skirted around an alleged stronghold of Afar "rebels," and recorded which rivers were flowing (the Borkana and Talalak) and which were dry but contained hand-dug water wells (the Jara, Dowie, Kada Halibee, and Busidima). North of the Dowie River, we observed a spectacular wall of basalt, which, judging from my topo map, rises 550 meters above the western floor of the valley—a rock climber's dream. The cliff is bounded by steep faults probably as great as any between there and the Gulf of Zula 500 kilometers to the north, or even perhaps the Gulf of Suez.

Just before reaching the Kada Halibee stream, at Issie, we came across another microfauna site mixed with artifacts, somewhat higher stratigraphically than Andalee across the river. Almost immediately we were surrounded by children from a nearby encampment whom Craig enlisted to help collect small fossils. As he described it in his report of the field trip, the youngsters "were running around picking them up like grass seeds from a new-sown lawn."[13] The site contained concentrations of taxa similar to those at Andalee, including fossil rodents, monkeys, small bovids, and a few large animals. There was also a rich assemblage of flaked tools made from quartzitic pebbles.

After picking up a local guide to help steer us through the badlands that I knew lay ahead, we barreled north. We headed up the dry Kada Halibee stream as far as we could go, cut overland through a morass of upper Pleistocene silts, dropped down into another stream, plowed overland through more Pleistocene silts, dropped into another stream, climbed onto a low plateau, and by late afternoon reached the high bluffs overlooking the Talalak River, which was flowing, but not deep. I knew of no one who had crossed through the river gorge in a vehicle.

Our task was to get to the other side and then find a route to the Busidima River that we could later use to return to the area from the north for detailed surveys. Taieb and I had traveled from Hadar as far as the Busidima, and I knew that he had tried to drive to the Talalak River from the Busidima but had failed. Nesbitt crossed both rivers with camels in 1928 from the direction of Gewani, but he did so by climbing up the Borkana and Dowie drainages to the plateau, by-passing the badlands altogether. He described the climb to the highlands: "The oppressive miasma of the Danakil lowlands along the Awash had been left behind, and here the air was light and pure."[14]

I judged the Talalak gorge to be about 75 meters deep and the south bank to have a slope of about 50°. I pulled the Land Rover to the edge of the plateau and walked down to the river with the others. We filled in low spots along the path and cleared away boulders, but we left bushes and small trees intact to slow my descent. The path down would be a straight line, because any turns would certainly roll the vehicle—there would be no turning back.

With everyone lined up along my route to the river to mark the way, I jumped into the Land Rover, clamped on my seat belt, checked to make sure I was in the lowest gear of four-wheel drive, and popped off the emergency brake. With pebbles and dust flying, bushes and acacias snapping and crunching, the engine whining, the wheels smoking, everyone yelling and whooping, and crocodiles plunging into the Awash 10 kilometers away, I rumbled down the slope pumping the brakes with my hands locked on the steering wheel . . . *yahooo!*

It seemed like a heartbeat before I reached the bottom of the gorge, hit the narrow floodplain, and slammed on the brakes at the water's edge.

Getting down to the river was the easy part.

Before the descent, our guide had shown us a wide camel trail leading up the opposite side of the valley. It was not so steep as the path on the south side, but still it was steep. Frankly, I didn't care. I could think of nothing better than to be marooned in the area for two weeks surveying, while also taking the Land Rover apart piece by piece and carrying it up the north bank. Granted, the engine might be a problem, but there was plenty of game in the area, and in a real emergency there was always sport fishing with Craig's patched up mosquito net.

Three times we tried to climb out of the gorge. Each time we removed more and more gear from the vehicle to make it lighter. The first time: jerricans, tools, spare parts, tow chain, food boxes, butane canisters, spare tires, tarps. The second time: cooking gear, canteens, personal gear, bedding, medical supplies, Craig's $300 high-tech, Alpine hiking boots. The third: ball of string, my Swiss knife, marker pen, film, bar of soap, Craig's Frisbee, and Maglio's monograph. Our equipment was scattered up and down the slope; it looked like Mount Everest after the ice melts.

On the last run, I backed the Land Rover as far down the river's bank as I could. By the time I roared up the trail, I was going fast enough to be in third gear, and then I downshifted to second and first as the trail got steeper and steeper. As I neared the summit, everyone was waiting and leaped behind the vehicle, pushing for all they were worth—legs pumping, pebbles churning, rubber burning, and lungs sucking exhaust.

We made it.

After hauling the gear up the hill and repacking the vehicle, we continued north. That night I planned to camp next to the Busidima River, a map distance of only 10 kilometers. Our guide told us the trail was good all the way: "three camels abreast." The sun was low in the sky; we had one hour before it got dark.

As it turned out, we reached the Busidima well into the night. When we were about halfway there, the trail led into badlands and narrowed along precipitous slopes, forcing us to dig our way through tight spots and reinforce embankments a dozen times. Worse, we ran out of water, and by the time we reached the Busidima and found the Afar wells in the dry river bed, the moon was high in the sky. We drank until we could drink no more. When we awoke the next morning, we found that the wells were full of baboon shit.

The trip from the Kada Halibee to the Busidima was a boondoggle. In my haste to simply traverse the region, we missed the chance to explore the area. Of course I planned to return; but when opportunities come, you should take them, because you never know what will happen down the road, as I would learn to my regret. Of particular interest were the areas along the Awash between the Kada Halibee and the Talalak, and the area south of the Talalak nearest the Awash. There aerial photos showed superb sedimentary exposures—surely a treasure of Pleistocene archeological sites, fossil beds, and hominid skulls waiting to be discovered.

Our plans to go then to our Lake Abhé permit area turned out to be a bust. Perhaps it was because of Ali Mirah's Afar Liberation Front, or because of tribal maneuvering connected with the pending independence of the French Territory, or both. Whatever the reason, the Ethiopian Army had closed off the area. When we arrived at the Tendaho Plantation, the officer in charge, Colonel Zewdie Tafara, Brigade Commander of the Hararghe Ground Forces, advised us to avoid the lake area. If we felt it was imperative to go, he would provide us an armed escort, but he would not allow his men to stay in the area overnight.

"Thanks, Colonel, but I guess we'll try another time."

CHAPTER 19

The Central
Awash

At the end of February 1976, Craig returned to the United States, via a trip down the Nile from Khartoum, and I remained in Addis Ababa for the next four months, analyzing data and expanding the research base of the RVRME. News of arrests, purges, and executions, the escalating war in Eritrea, and louder war drums from Somalia periodically filled the city. I made greater and greater efforts to keep the RVRME in tune with the more positive aspects of the revolution.

My days were filled with meetings with government officials as I sought to promote our work, student training, and the expansion of storage and laboratory facilities at the National Museum. The Derg was calling for a new era of openness and public debate to thrash out political differences, so I frequently invoked the call for openness in my expanding list of needs for the RVRME. In all cases I argued for greater support of *"Ethiopia*-based research."

Because I felt Johanson's CIA allegations were never far below the surface, I also made special efforts to keep the Ministry of Culture and other government offices fully informed of the nature and progress of our work, often inviting officials to visit our offices or see our collections at the National Museum. In April, at the time the RVRME research permit was to be renewed, I submitted a 182-page annual report to the minister that spelled out all aspects of our work, our funding, our activities, and my particular brand of RVRME nationalism.[1] I sent summaries of the report to other government offices, whether they wanted them or not.

My efforts at openness had one adverse effect: It gave Berhanu Abebe the opportunity to concern himself increasingly with RVRME affairs. With the zeal that characterized his days as director of the Haile Sellassie I Prize Trust, he continued to believe in his own imperial omniscience. Disturbed because I had recently recruited two geology students, one through the Zemacha and one from Yale University, Berhanu wrote me to say that the RVRME was too concerned with geology "in quality and quantity" and not enough with paleontology, "geological research being only *complementary* . . . to paleontological research." Even the name of the new office he created for himself at the Ministry reflected his thinking: Department for Research, Restoration and Preservation of National Heritage. Aside from the purely geological question of whether elephants, rodents, and humans millions of years old constitute national heritage, my problem was that I never knew from one day to the next what Berhanu's identity might be, and this could be crucial to whatever administrative assistance I needed. On Tuesdays he might be a paleoanthropologist, on Thursdays an ethnologist, a "legal historian" every third Friday, and so on. I remember going to his office one day wondering what he would be this time. *Now, which day of the month is he a government administrator? . . . a legal historian every "third" Friday. . . .* That's when it hit me. Of course: the phases of the moon! Suddenly it all made sense: Berhanu ran his office on the basis of lunar cycles.

On another occasion, we were meeting by appointment in his office when there was a knock at the door and a Japanese man stuck his head in. He identified himself as representing some cultural exchange program and politely reminded Berhanu of their appointment, which had apparently been scheduled at the same time as mine. Berhanu asked the man in and the two shook hands, meeting for the first time. That being the case, Berhanu immediately launched into his story.

". . . and yes, when I was in France studying for my doctorate under Claude Lévi-Strauss. . . ."

There was a blank look on the Japanese man's face, so Berhanu repeated, "When I was in *France* studying for my *doctorate* under *Claude Lévi-Strauss*. . . ."

As word spread about the RVRME discoveries, more and more people were writing me who were interested in joining the RVRME research, either in the field or in Addis Ababa, studying one fossil group or another. Also, by mid-1976 we had five Zemacha students working with us, including two new recruits: Tsrha Adefris, a biology student, who was to become our resident anthropologist, and Emanuel Tola, one of the geology students Berhanu complained about. Tsrha (Amharic for Sarah) was from Addis Ababa but had grown up in Asmara; Emanuel was from southwestern Ethiopia. Tsrha had previously worked for the Zemacha organizing women's associations and literacy programs, and Emanuel had worked for the Geological Survey. Like Dejene, Assefa, and Sleshi, both were university seniors and came recommended by their professors. I hoped eventually to see at least some of these students enter graduate schools in the United States or elsewhere pursuing RVRME research.

We still had major gaps in our senior personnel. We needed more geologists—specialists in radiometric dating, depositional systems, and structural geology—and I wanted a senior paleontologist to oversee the fossil studies, just as Wendorf would direct the archeological studies. Jolly and Conroy were in charge of hominids and other primates.

By the end of June, some of our funding had increased and approval for other funds was pending, all in small grants to one RVRME member or another. Most important, Wood had convinced Bryan Patterson of Harvard to join us and to apply to NSF for major funding. Patterson was interested in renewing work in Africa following the highly successful field work he had carried out at Lothagam and elsewhere in Kenya in the 1960s. The partial mandible his team found at Lothagam, a gracile australopithecine recently redated to between 4.2 and 5.0 million years, was still the oldest hominid known at the time.[2] Still vigorous at the age of 67, Patterson was a self-made man who became a Harvard professor of paleontology without earning a single college degree; in fact, he never finished secondary school.[3] Born in London, he was shipped off to the United States at the age of 17 when his family ran low on money. His father, a colonel in the Indian army and a building engineer, had written four books about his life experiences. His most famous, *The Man-Eaters of Tsavo,* is about a pair of lions that killed 128 of his work crew during the building of the Kenyan railroad in the late 1890s.*

It's curious: Neither Patterson, nor Mary Leakey, nor Richard Leakey received advanced degrees in anything, yet all made names for themselves in the hominid business, and none turned to a life of crime. The lack of a Ph.D. is something I have had to live with, and neither my father nor my spouse was an Africanist. The only camping trip I recall my father going on was as one of several chaperones accompanying my brother's Boy Scout troop outside of Houston. That night a huge storm blew in from the Gulf of Mexico, leveling the camp. Before things got out of hand, however, my father jumped into his car, drove down the road, and checked into a motel. The next morning be returned to the camp bright and cheery. The scoutmaster was not impressed, and my father was not invited to attend future campouts, which suited him just fine.

Wendorf also planned to submit a major proposal to NSF for the Middle Awash, as did Conroy and Jolly, but all wanted a "big find" during the pending fall field season to bolster their applications. I anticipated more funds from FORGE, especially because its director, Alfred Kelleher, hoped to visit Addis Ababa in the coming months. The evidence increasingly indicated that the Middle Awash was one of the most continuous paleontological and archeological records in Africa. What perplexed me, though, was the apparent absence so far of Hadar-age sediments anywhere in the Middle Awash. We had found pre-Hadar fossils on either side of the Awash, and post-Hadar fossils in Matabaietu-age beds in the eastern Awash and possibly to the west, but nothing obvious in between. Even if the paleolakes had migrated to the north as was apparent, it seemed that we should still find some Hadar-age beds to the south, at least fluvial sediments deposited by rivers feeding

---

*In 1997 Paramount released a movie about the colonel's experiences at Tsavo. It was called *The Ghost and the Darkness,* and starred Val Kilmer.

into the Hadar paleolake. I felt the key must be in the Central Awash Complex, which was still largely unsurveyed and, therefore, was our next objective.

On July 7, 1976, I drove to Bole Airport to meet Doug Cramer's 7:15 A.M. flight from Europe. He was a colleague of Conroy and Jolly at New York University, where he taught human anatomy at the medical school. We had corresponded for several months, and Doug had picked up several small grants to contribute to field expenses. He was a big guy, weighing something like 150 kilograms. His hobby was power lifting. Forty years old, he had long, prematurely gray hair and a long beard to match. He strode through the customs line and we introduced ourselves; I told him I thought he looked like Santa Claus. "Yeah," he said, "that's what the little girl sitting next to me said before I threw her off the plane." He wore faded blue jeans, a jacket to match, and a navy cap, and he had a backpack slung over one shoulder. He had spent three years in the U.S. Navy before going to the University of Chicago, where he received his Ph.D. under F. Clark Howell. Prior to that he had earned a Master of Fine Arts degree in medical illustration from Johns Hopkins University. As I soon appreciated, Doug possessed a contradiction of talents: he could bench-press 200 kilograms, but he was also an expert artist and craftsman who could render fossils into superb drawings or casts. He later told me that he could not only give me a lobotomy but also turn it into an illustrated lecture. He was married with two children; his wife was an opera singer.

One of Doug's classmates at the University of Chicago had been Johanson, with whom he had worked in the Omo. For his dissertation, Doug studied the craniofacial morphology of chimpanzees, and Johanson studied chimpanzee dentition. Doug had little good to say about Johanson and thought he had freely used other people's data. We had instant rapport.

Five days later we left for the central Awash, accompanied by Alemayehu Asfaw, again our antiquities representative, and with our field crew but no student assistants. They had just completed their Zemacha service and were back in the university for a brief summer semester before they would rejoin us (at increased salaries). The Derg had relented, allowing seniors to take accelerated courses and graduate in September.

After picking up our guides in Gewani and crossing the ferry at the Dobel Plantation, we drove to the south side of the Hatowie River, which was full of water from bank to bank (see Map VII on page 140). The spot I had chosen for our camp, however, was on the *north* bank of the river on the eastern side of the Hatowie graben. The only way to get our equipment and supplies across the 30-meter-wide river was to carry everything by hand.

I pulled the Land Rover up to the river and turned off the motor. We sat there for a moment watching the muddy waters churn by. When we had camped on the upper Hatowie in February, during the dry season, the river was white water and dry on its lower reaches; now, during the rainy season, the lower Hatowie was full and looked like thick, brown gravy.

"Doug, there's something I forgot to ask you."

"Yeah?"

"Can you swim?"

"Yeah."

"That's fortunate."

"So, what's your plan, wise guy?"

There were two reasons not to use the vehicle: I wanted to survey the entire area east of the Hatowie in the central Awash by foot; and in order to cut down on weight in the Land Rover, we had brought no extra fuel.

After disconnecting the battery cables, we crossed the river, chopped out a clearing in the gallery forest, and then carried everything across. Doug had bought a pith helmet in Addis Ababa, and while wading back and forth across the waist-deep waters carrying equipment, he looked like a late-nineteenth century newsreel that might be entitled:

PORTAGING THE LUALABA WITH WHITE MANPOWER

It rained almost every night, as it had the previous July across the river when Conroy was with us, but there were no lions to harmonize with the other night sounds. On several occasions, however, our Afar guides leaped up in the night, brandishing their rifles after hearing terrifying growling sounds, only to realize it was Doug snoring. At other times, we would hear him snarling in the early morning hours as he struggled with his undersized tent.

For the next two weeks we walked the outcrops, mapping fossil localities over the entire 350-square-kilometer eastern flank of the Central Awash Complex. Doug claimed he lost 35 pounds during that time, which, given his size, was like scraping just enough ice off an iceberg to make a snowcone. I think he lost that weight because he refused to sit on the grass mat with the rest of us when eating meals in camp. Instead, he ate his food while circling around us, like a predator, all the while carrying on a furious commentary about whatever subject was on his mind. It was unnerving.

"Take a load off, Doug! You're burning calories as fast as you consume them."

"As piss-poor as this food is, who the hell wants to share it with half the insects in Africa?"

The first important find we made was near camp. There we discovered that the fossil beds Craig Wood and I had found a few months earlier beneath a basalt at the eastern margin of the Hatowie graben extended its entire length, a distance of 12 kilometers. The sediments immediately beneath the basalt were baked, which meant that the lava flow had cooked the underlying layer and, therefore, was younger than the fossils contained in it. Nearly everywhere we looked below the basalt there were fossils, including those friends of antiquity that we have come to love and trust: *Anancus, Stegotetrabelodon,* and *Primelephas.*

The 80- to 100-meter basalt plateau overlooking the Hatowie graben was heavily faulted into smaller, sediment-filled grabens, but it contained fewer fossils. Further east, between a prominent basalt hill called Dalu Ali and the Aramis stream, were the rolling, variegated strata that had stood out so clearly during those 1974 and 1975 flights. Here we found more early proboscideans and an assortment of fossil pigs, carnivores, horses, rhinos, bovids, hippos, and monkeys. Most fossils were scattered and isolated, but at Aramis concentrations were greater and would eventually yield "sensational" fossils.

All together, in the two weeks during which we surveyed the area, we located dozens of fossil localities, and sampled 28, throughout six stratigraphic levels within perhaps 200 meters of sediments and basalts. We estimated that the entire sequence was lower Pliocene, ranging from about 4.0 to 5.0 million years old.[4]

One of the most intriguing fossils we found, high in the section in the Sagantoli area was an *Anancus* molar unlike any I had seen in the Central Awash Complex (see the figure on page 200). It did compare, however, with molars from Ado Kaho, Wee-ee, and Matabaietu in the eastern Awash. All were noticeably larger than molars found lower in the section, and the enamel on the crown surface seemed thicker and more complex, suggesting a more evolved species than those found elsewhere. When later studying the entire collection of *Anancus* fossils we had recovered, Assefa and I thought we recognized an evolutionary trend in the lineage: an "early stage" of small, simple molars represented by those that Liz found at Bikirmali near an intermediate "middle stage" from the Hatowie graben and an "advanced stage" of larger, more complex molars in the upper layers of the regional section of the central and eastern Awash.[5] We also found marker beds at the top of the section in the Central Awash Complex that seemed to match those in the eastern Awash. I thought the position of the layers, and of the "advanced" *Anancus* on both sides of the river, tied the two areas together into a continuous, unbroken sequence—all pre-Hadar in age.[4] The next-highest strata in the regional section still appeared to be the upper Pliocene beds at Matabaietu and elsewhere, which meant that Hadar-age strata in the Middle Awash were missing.

In March 1976, Taieb, Johanson, Coppens, and geologist James Aronson published a description of the Hadar geology in *Nature*, formally naming the sedimentary sequence the Hadar Formation, which they divided into four stratigraphic units 140 meters in thickness.[6] Given radiometric dates of 3.0 million years for the basalt near the middle of the formation, and faunal evidence, the beds seemed to range from about 3.5 million years at the base of the formation to about 3.0 million years at its top (Appendix III).* The evidence the RVRME had gathered sug-

---

*This formation does not, or should not, include the younger artifact-bearing layers at Gona and elsewhere that unconformably overlie the older Hadar strata.

gested that the Middle Awash strata were both older and younger than the Hadar beds. If this was correct, then where the hell were there *any* Hadar-age strata in the Middle Awash?

Near the end of July, I climbed Dalu Ali, the highest hill in the centered Awash, some 150 meters in elevation. It was situated nearest the Awash River and offered a commanding view of the strata on the east bank. In the distance along the river's marshy edge was a herd of waterbuck, my favorite of the large antelopes—magnificent animals with swept-back horns and powerful shoulders that allowed them to hurtle full tilt through knee-deep water. With my eyes focused alternately on the herd and the strata rising from the across the river, I went over the names of the various streams and landmarks on the opposite bank with my guide, Abraham: Sibabi, Asalala, Ado Kaho, Wee-ee, Belodolie, Maka. . . . Unless I was seeing things, the strata had a distinct northward dip. This appeared to be a regional pattern. With increasing subsidence in the central Afar during plate separation, sediments washed down from the highlands into lakes and rivers occupying lower and lower elevations, but at higher and higher positions stratigraphically. Like a deck of cards lowered at one end, the highest cards fall forward to lower levels and remain the "youngest." The Law of Superposition applies: Each layer is younger than the one beneath it. Whereas the lowest layer at Wee-ee might be 4.0 million years old and 650 meters above sea level, the "top card" at Hadar—80 kilometers downriver—might be 2.5 million years old and 550 meters above sea level. We can assume that older strata (the bottom cards) were highest in elevation and most subjected to erosion, whereas younger strata farther down the valley were buried by Pleistocene sediments. Thus, to the south, Hadar-age strata that were once present had been heavily eroded, whereas those to the north are at least partially covered by younger sediments. That the beds at Hadar itself are exposed at all is due to the eastward swing of the Awash River that cuts through the area.

It is also true that Hadar-age strata, which at one time had been present in the Middle Awash, would have been downfaulted *into* the Awash rift and buried by younger sediments. By analogy, a deck of cards lifted up on its side would slip *laterally* to lower levels, with the same pattern of "younger is lower and older is higher." And if we lifted the deck up on one *corner* only, the cards would slip both sideways and forward. In this case, if we took two decks of cards, placed them side by side, and lifted up the opposing outside corners at the end of each deck while slowly moving the decks apart, the cards in the intervening space would slip into and down the "valley" between the two decks, becoming lower in elevation and remaining stratigraphically younger in both case. This model should approximate how—with extension and subsidence across and down the Middle Awash rift—Hadar-age strata could be found beneath the Awash floodplain and could occupy lower and lower positions to the north.

By the end of July, Doug was at his fighting best. He would yell out when the sun was highest in the sky, *"Pain, I love it!"* He thought we could make a fortune by running a "fat farm" in the Middle Awash.

"No shit. Look, I live in New York City, and I know there are people there who will pay *anything* to come out here and suffer to lose weight. Endure the fucking heat, eat lentils and rice for a month, drink mud, and run up and down these hills all day banging rocks together. Man, we're sitting on a gold mine!"

The time of day when the heat got to us both was during our long treks back to camp. Near the end of our stay, in late July, we were often walking straight into the setting sun. At the time, we were averaging 18 to 25 kilometers a day just to get to and from our survey areas. By late afternoon we were worn out and scorched, but the shadows from the basalt hills would grow longer and longer, and soon the sun would set behind the distant escarpment, and peace and harmony would flow over the land. Back at camp, Doug and I, and Alemayehu and our Afar companions, would bathe in the Hatowie River, then sit around drinking hot tea while discussing the events of the day and waiting for dinner.

We broke camp on July 29 and portaged everything back across the still-waist-deep Hatowie. After reconnecting the battery cables of the Land Rover, we headed for the ferry at Dobel and then on to Gewani. When we reached the village, it was a seismic event. Doug crashed around from souk to souk, laughing and talking to people—in a language no one understood—while buying cigarettes, beer, and things to take back home, as though he had just been released from a labor camp. Within ten minutes everyone in the village knew he was there. Crowds of kids followed him around as he dispensed candy, thinking that he was some kind of a legendary giant, while some of the older people wondered whether perhaps he was the reincarnation of Lij Iyasu, the renegade crown prince of years past who had periodically invaded their land.

CHAPTER 20

Baboons and Pigs

W hen we returned to Addis Ababa, the city was still buzzing over a coup attempt in mid-July. Major Sisay Habte, a moderate Derg member, was accused of leading a CIA-style "destabilization" campaign. Trained in electrical engineering in the United States, he, along with 18 others accused of crimes against the state, were arrested by the Security Command and executed. Two alleged accomplices, military officers, fled to Gojam Province, where they were hunted down and killed by the People's Militia; another, head of the Zemacha, was arrested and died in prison.[1]

For months a schism had deepened in the Derg ranks between the hardliners and the moderates. One favored continued military rule, increased ties with Moscow, and a military solution to the ever-worsening war in Eritrea. The other faction favored civilian rule sooner rather than later, a more neutral position toward the Soviet Union, and a negotiated settlement in the north.[2] The catalyst for the coup attempt was a disastrous "Red March" on Eritrea in May, made up of government troops and tens of thousands of armed but pitifully trained peasants who had been promised land in Eritrea in exchange for victory. The "human wave" was butchered before it even reached the border.[3]

There followed more dissension, more executions, more territory lost to the Eritrean rebels, and increasing war hysteria in Mogadishu: A friend, Charlie Harrell, who was a military attaché to the U.S. embassy, predicted that in a year Ethiopia would be at war with Somalia. "If you have any need to go to Dire Dawa," he told me, "you'd better go now, or start taking lessons in Somali." Dire Dawa is the largest town on the Hararghe escarpment, near the contested Ogaden region.

Following my return from the field with Cramer, I spent the remaining weeks of the summer of 1976 exchanging letters with Patterson, Wendorf, and Conroy over the details of the NSF proposals, while also preparing for the fall field season. This one would involve 15 people, with our time divided between the archeological sites in the eastern Middle Awash and the older beds in the western Middle Awash. Until then there was ample work to do at the museum. Cramer remained in Addis Ababa for most of August, preparing and casting fossils. He was joined for part of the time by Clifford Jolly, who refined the identifications of our primate collection. In addition, H. B. S. Cooke from Canada studied our suid fossils.

Assefa, Sleshi, and Tsrha worked at the museum part-time while finishing up course work at the university. Assefa was able to use his reports on the Middle Awash elephants for his senior thesis, and Sleshi and Tsrha used their reports on the suids and primates similarly. All three, along with Dejene, graduated in mid-September, but it would be another year of revolution before they actually received their diplomas. Meanwhile, all four resumed working full-time with the RVRME at an increase from the Zemacha salary of $25 per month to $300 per month, which I paid out of my own pocket. I was more determined than ever to see the four of them begin graduate studies, at least partly because I could not afford their salaries!

I first met Jolly in mid-1973 in Awash National Park when he was running a field school sponsored by New York University, at the same time that he was studying the *Papio* baboons in the park. After my break with the IARE in late 1974, Clifford was one of the first to join the RVRME, and he soon recommended Conroy and Cramer. Clifford, who is English and resembles the actor Michael Caine, received his Ph.D. in physical anthropology from the University of London in 1965. He wrote his dissertation on the evolution of baboons, using a large collection of undescribed *Theropithecus* and *Papio* fossils lent to the British Museum by Louis Leakey on behalf of the National Museums of Kenya. Following study of more collections in Nairobi, Clifford's efforts resulted in two major accomplishments: a monograph on the classification and natural history of *Theropithecus;* and an ability to mimic Louis's accent perfectly.

Jolly's monograph, published in 1972, substantially revised the systematics of the genus.[4] It also convincingly demonstrated that the present-day gelada baboon, which is confined to the Ethiopian highlands, is the living relic of a *Theropithecus* ancestor that roamed much of the continent in the Pliocene. By 1976 Clifford had published a series of related papers on baboons, as well as four books, and other works on topics ranging from human origins, primate behavior, and functional anatomy to the molecular genetics of monkeys. He also co-authored a paper entitled "The Riddle of the Sphinx-Monkey," which traces descriptions of the monkey from classical writings to those of sixteenth- and seventeenth-century travelers and concludes that the inscrutable creature is the gelada baboon.[5] The article suggests that the gelada's lion-like mane, quadrupedal body, round human-like face, conspicuous mammae, and chest decoration, which resembles a string of beads (*sphingion* in Greek), may explain the origin of the name sphinx-monkey—and possibly even that of the myth of the sphinx itself.

Of Clifford's publications, his most widely recognized is an essay published by the Royal Anthropological Institute in 1970, wherein he drew on his eclectic research to propose a new causal model of human origins.[6] The paper begins modestly enough by rejecting the model Charles Darwin proposed a hundred years earlier in *The Descent of Man.* Darwin proposed that human origins began with upright walking, which freed the hands to make tools and weapons. This adaptation was to have led to a reduction of the canines (used by apes for defensive purposes) and incisors (for tearing and cutting food) and an accompanying increase in relative brain size. If this scenario were true, Jolly asked, then why do the tool-using chimps that Jane Goodall studied have such large anterior teeth, and why are stone tools not found in association with the remains of some fossil hominids that have reduced anterior teeth? Jolly also questioned the Darwinian view that climatic changes in the Neogene, which resulted in diminished forests and increased savanna grassland, caused apes to become bipedal carnivores by forcing them out of trees in search of new food resources. If this were true, he asked, why have baboons and some chimps adapted so well to savanna environments without becoming bipedal or major meat-eaters? And finally, if tool use was supposed to coincide with the first hominids, why have no stone tools been found in deposits as old as the earliest hominids (such as Lothagam), which antedate the first known tools by several million years?

Instead of subscribing to models of human origins that cannot be tested, Jolly proposed studying a group of present-day primates that live in a savanna-like environment to see how these animals have survived without becoming tool users or dependent on meat. Living on the high plateau of central Ethiopia are the animals Jolly selected for this comparison: the gelada baboon, the most terrestrial of all nonhuman primates.

Although the gelada does not live in a savanna lowland, it does inhabit open country in the highlands that is treeless and arid for much of the year. There it has adapted to a foraged diet of mostly grass, seeds, and roots. To accommodate these food resources, the gelada has developed a suite of physical characteristics analogous to those of early hominids: opposable thumbs that allow it to be an efficient small-object feeder; an upright trunk adapted to maintaining a sitting and feeding posture for long periods at a time; and especially powerful molars that enable it to chew hard seeds and coarse foods, features analogous to the molars and presumed diet of australopithecines. Jolly also reasoned that the extinct relatives of the gelada and early hominids alike ground their food between their molars in a sideways milling manner that favored the reduction of canines, just as the reduction of incisors accompanied refined hand-to-mouth eating of small food parcels.

Although gelada molars mince (and do not grind) food, and their incisors (but not their canines) are reduced only relative to other living species of baboons, the core of Jolly's hypothesis linking diet, instead of culture, to human origins, caused a major shift in the ideas of human evolution.[7] Anthropologists began to think more of "preadaptive" traits in prehominids, such as the flexible arm and the prehensile hand used for manipulative skills, the upright trunk with its predilection to bipedality, and powerful jaws and strong cheek teeth that could be used for crushing more demanding types of food resources.[8] These features gave focus to the notion of an early hominid that did not have to kill other animals and make stone tools to survive. That is, Jolly suggested a bipedal ape that was precultural, or non-cultural—one that occupied an intermediate evolutionary position between more primitive hominoids of the late Miocene/early Pliocene and more progressive hominids of later times.[9] Jolly's insights were all the more noteworthy because when he published his acclaimed essay in 1970, he had not observed a single gelada baboon in the wild.[10]

By late August 1976, assisted by Tsrha Adefris, who helped catalog and organize the fossils, Clifford had examined the several hundred specimens in the RVRME primate collection. He concluded that there were at least 12 taxa, representing 5 major groups of monkeys and baboons in the Middle Awash, most with relatives living in the Awash Valley. The two largest collections—*Theropithecus* from Mat-abaietu and Wilti Dora and *Cercopithecus* and *Colobus* monkeys from Andalee and Issie—were post-Hadar in age; six species of monkeys or baboons from as many levels were pre-Hadar in age. It was evident that an expanded collection from the Middle Awash would surely offer a scientific feast of species variability, demographic patterns, and evolutionary changes in primates that preceded and succeeded those at Hadar. An additional prospect was the use of primates still living in the Awash Valley as analogs to the extinct primate populations, as Clifford had done for the geladas on the plateau.

From 1971 to 1973, Clifford and his colleagues and students at New York University had begun to study the interaction and microevolution of *Papio* baboons in Awash National Park.[11] For that purpose the team trapped and tranquilized hundreds of baboons, which they weighed, measured, photographed, fingerprinted, and sampled for blood, saliva, and hair; they also made casts of their teeth. After a hiatus during the revolution, in 1982 Clifford and anthropologist Jane Phillips (by then Glenn Conroy's wife) resumed the project, which continues to the present.[12] Unexpectedly, early in the study the demography of the baboons served as an analog to human populations, and vice versa: The drought of 1973 had an adverse effect on the baboons, as it did on the Afar people in Wallo Province. In both cases, the highest mortality figures were among the very young, leaving disproportionate numbers of adults to survive.[13] One of Clifford's students, Pat Shipman, even published a paper in *Nature* on the broader implications of the drought on *fossil* vertebrate assemblages.[14]

Whereas Jolly was of the new school of African paleoanthropology, H. B. S. Cooke, who evaluated the RVRME suid collection that same August, was of an earlier generation of African vertebrate paleontology. He was ten years old when he read the 1925 article in the *Illustrated London News* about Raymond Dart's "man-like ape" discovery at the Taung limeworks. Born in Johannesburg, Cooke was one of many young people influenced by Dart, just as later generations were influenced by Louis Leakey. After studying anatomy and paleontology at Cambridge and serving five years in the South African Air Force during World War II, Cooke received a doctorate in geology at the University of Witwatersrand in 1947. Over the next two decades, he developed a reputation as a meticulous researcher. He became particularly well versed in the use of fossils for dating the hominid cave sites of South Africa, where volcanic materials for radiometric dating are lacking.[15]

In the early 1970s, Cooke's expertise in biostratigraphy drew him into the controversy of his career after he was asked by Richard Leakey to describe the pig fossils at Lake Turkana. Although the elements of this story are well known, some of it is worth repeating because of its bearing on Afar research.[16]

After studying the Turkana suids, Cooke concluded that the critical layer called the KBS Tuff, which runs through the Turkana fossil beds, was incorrectly dated. Instead of being 2.6 million years old, as was determined radiometrically, Cooke estimated the tuff to be 600,000 years younger.[17] Richard was aghast, because the tuff had been used to date the centerpieces of the Turkana discoveries. Cooke's estimate based on the pigs called into question the age of the so-called "oldest stone tools in the world," which had been recovered from the tuff itself, and of the presumed "oldest toolmaker," the large-brained 1470 *Homo* skull, found just below it. If Cooke's 2.0-million-year age was correct, then the tools and the skull were close in age to both the artifacts and the *Homo habilis* fossils found years earlier in the lower levels of Olduvai. Such a conclusion would significantly lessen the importance of the Turkana discoveries. Not to mention that if the original date was incorrect, it would also give Richard's competitors something to crow about.[18]

Richard had relied on a date for the tuff provided by two respected British geochronologists.[19] Cooke, however, relied on a superior stratigraphic and radiometric framework used by the American team working in the Omo Valley to date the Turkana pigs indirectly.[20] Interestingly, one of the best-kept secrets of African paleontology is that at the time, Cooke was by no means a suid "specialist," as was widely believed. When he began his studies of the Turkana suids in 1970, he had written only two papers that dealt exclusively with African suids, both published in the 1940s.[21] He was, however, fully capable of following state-of-the-art procedures for relative dating.

Cooke first measured and identified a number of well-dated suid molars collected by the Americans throughout the Omo sequence, and then matched the molars to those comparable ones above and below the KBS Tuff collected by the Kenyan team. Putting all the evidence together, including supporting faunal data from Maglio, Howell, and others,[22] Cooke made a convincing case for the younger date, which he presented at conferences in Austria in 1971, Nairobi in 1973, and New York in 1974.[17] Richard, however, feared that Cooke's argument was somehow rigged. After all, Cooke had had his differences with Louis in the old days, and it was no secret that the Americans working in the Omo would jump at any opportunity to bring about Richard's downfall.[23] By the time Cooke once again made his case for a younger KBS Tuff at a meeting held in London in early 1975,[24] the Americans were already dancing on Richard's grave. *Finally,* superior American technical know-how—and nearly $1,000,000 in NSF money—had triumphed over the Kenyan "tent boy."

Richard was still unwilling to accept either Cooke's conclusion or a new series of radiometric dates obtained by the American side that supported a younger date.[25] In 1974, without first informing Cooke, Richard recommissioned the study of the Turkana suids.[26] To do the job, he recruited two members of his own team. One was John Harris, a British paleontologist and trusted lieutenant; the other was Tim White, an American graduate student from the University of Michigan, who was simultaneously studying early hominid mandibles for his Ph.D. dissertation. To carry out Richard's mission, Harris and White collected over 500 suid fossils during the 1975 Turkana field season.[27] At breakneck speed they also studied African suids housed in a dozen museums and institutions from South Africa to Berkeley.[28] In early September 1976, Harris and White completed their study, which they gave to Richard in Nairobi in the form of a manuscript they intended to submit to *Nature.*[29]

Among their many interesting findings was their absolute conviction that Cooke had been right all along.

There is no evidence that Richard exploded on the spot, although several accounts agree that as he read the paper line by line in his office at the National Museum, he went through a severe mood change.[30] He insisted that key portions of the manuscript be changed, particularly those written by White concerning the age and description of the Turkana hominids. White, who had a reputation for being hotheaded and self-righteous, was incensed at what he viewed as scientific censorship. Apparently he was also under great mental and physical strain after working nonstop for months on both the suid study and his dissertation. According to Richard, an altercation followed in his office. White referred to him as a "dictator" and called his scientific leadership a "disgrace." He then stalked out of the office and slammed the door behind him.[31]

White immediately left Nairobi for a previously scheduled trip to Nice to attend a conference, telling Harris in essence that he couldn't get out of Kenya fast enough.[32] In his own account in *Lucy,* Johanson picks up the story: "Badly shaken up, Tim left for the airport. If the *Nature* paper was to be emasculated by removing all reference to the implications for the hominids, then Tim wanted no part of it."[33] Probably if there was ever a time in his life when the distraught young man (he was barely 26), at the beginning of his career, needed reassurance from someone with compassion and professional understanding, this was it. As fortune would have it, such a person was also attending the Nice conference. Professor D. Carl himself.[34] When the two met on the shores of the Riviera, we can only guess that Johanson opened his arms to the troubled confidant of Richard Leakey. In any case, within days Richard received a letter from White in which he resigned from the Turkana team.[35]*

———

Unaware of this high drama unfolding in Kenya, in May 1976 I had tried to recruit White to study the RVRME suids. In the course of doing so, I sent him photographs of some key specimens from the base and top of the regional section in the Middle Awash, hoping for his evaluation. But in June he replied that he was overcommitted to other projects.[27] In 1975 I had made similar overtures to Harris, but that relationship was waylaid by some caring person who whispered in his ear that the RVRME was operating in the country illegally.[38] As for the suid photographs I had sent White, he never mentioned or returned them, but four years later I found identifications and references to the specimens in a monograph that he and Harris published on the African suids. Molars from the greater Ananu area were identified as *Nyanzachoerus tulotos,* known from the lower levels of Lothagam and the late Miocene site of Lukeino in Kenya; those from Andalee were identified as *Mesochoerus majus,* present in the upper levels of Olduvai and sites of later ages.[39]

Cooke's main reason for being in Addis Ababa in August 1976 was to study the Hadar, Geraru, and Amado suids, an arrangement agreed upon between Cooke and the IARE in 1973. For Johanson, Cooke was a double bonus: one, his expertise at the expense of Richard Leakey; and two, his expertise. I had mixed feelings about Cooke's studying the RVRME suids, because I figured that anyone associated with Johanson meant trouble, which of course was irrational, but that was my frame of mind in those days. Bryan Patterson and Cooke, however, were close associates from years past, and that changed the equation; also, Cooke was willing to work with Sleshi.

Cooke's study of both the IARE and the RVRME suids was in fact the best of all arrangements; it would allow him to examine the evolution of the group throughout the entire Hadar/Middle Awash sequence. One of the early problems in the Omo/Turkana research was that all three teams—the American, French, and Kenyan—had their respective paleontologists studying the same faunal group. The Americans had D. A. Hooijer from Leiden study the fossil rhinos from one part of the Omo, the French had Claude Guérin study the rhinos from another, and John Harris studied the rhinos from just south of the Omo (Turkana), all from similar time ranges. A similar personnel overlap was created for the study of the equids,

———

*The White and Harris manuscript was rejected by *Nature,* probably for political reasons, but was published by *Science* on October 7, 1977; a monograph of the suid study, with reversed authorship, was published by the American Philosophical Society in June 1979.[36] During the late 1970s and 1980s, the KBS Tuff was exhaustively dated and redated by several radiometric methods. There is now uniform agreement among experts that its age is 1.9 million years.[37] Remarkably, Cooke was off by only 100,000 years.

bovids, carnivores, monkeys, and (of course) hominids.[40] This was a potential source of trouble between the RVRME and the IARE, except that the geological time periods that our teams studied were not the same, and I was confident that Patterson, Jolly, and Conroy could reach collaborative arrangements with the IARE paleontologists.

From a preliminary examination of the RVRME collections at the museum, Cooke confirmed, as Sleshi had correctly determined, that five lineages of suids were present in the Middle Awash: *Nyanzachoerus, Notochoerus, Mesochoerus, Metridiochoerus,* and *Phacochoerus.* Among these, Cooke distinguished at least 11 species, 5 of which belonged to *Mesochoerus,* the genus that precipitated the Turkana "pig war." At least 2 of the 5 species were new.* Overall, the evolutionary record of *Mesochoerus* in the combined Awash sites held the promise of being the most complete succession for the genus in Africa. It was my hope that such a claim could be made for other faunal groups as well, including our own.

**Mesochoerus was later replaced by the genus name *Kolpochoerus.* One of the new species, which Cooke first identified at Hadar, he later named *Kolpochoerus afarensis.*[41]

CHAPTER 21

The Artifact Trail

We left Addis Ababa for the field in mid-September 1976, just a few days after I had learned that Alfred Kelleher, the director of FORGE, had died of a heart attack in South America while visiting grant recipients. He had hoped to come to Ethiopia in June but had postponed his trip. The acting director of FORGE said he would try to continue to support my work. Because of the politics surrounding the IARE, FORGE never received the credit it deserved for supporting the IARE's work during its formative days, although I later credited FORGE's contributions to my own research in a dozen publications. I was always grateful to Kelleher for sticking by me when things got tough. I had never told him of Johanson's story that FORGE was a CIA front.[1]

Our objective over the next six weeks was to raise the level of archeological investigations in the eastern Middle Awash one more notch. It was also understood by all that, this time around, finding a hominid had high priority. I considered the fossil arrays at Matabaietu and Bodo to be our best shot, as did Craig Wood, who was rejoining us. In addition, we had with us two more archeology graduate students of Fred Wendorf's from Southern Methodist University: Bill Singleton and Paul Larson, both experienced excavators. Others with us included Paul Whitehead, a graduate student in anthropology and geology at Yale University; and Sleshi, Assefa, Tsrha, and Alemayehu Asfaw.

The first stop was Matabaietu (Map VII on page 140). No hominid fossils; however, I was soon more certain than ever that humans had lived there in the late Pliocene. On our second day we staked out a 6600-square-meter grid at 5-meter intervals over the area where we had found Oldowan tools in July 1975. By the fourth day, Singleton and Larson had mapped and collected 150 artifacts. They had also dug three test trenches but had failed to find a source layer for the artifacts. On the sixth day, Whitehead and I used a level and a measuring rod to record a number of elevations across the grid map area. By comparing the distribution of the artifacts with measured geological sections of the site, we concluded that the artifacts originated from the middle or upper part of the sequence. The test trenches had been dug too low. A small but significant number of artifacts were found upslope from the main artifact cluster and from the highest test trench. Even more compelling, Larson found a bifacial basalt chopper in mint condition eroding from a tuff at the top of the sequence. The stone tool retained its natural gray color, was crudely flaked at one end, and was impregnated with tuff.[2]

Many of the artifacts throughout the grid showed moderate patination, or none, on their unexposed sides and only minor edge damage. Singleton and Larson classified 20 artifacts as tools and 130 as debris, primarily cores, flakes, and chips. The tools included 6 choppers, 7 heavy-duty scrapers, 2 light-duty scrapers, 2 polyhedral hammerstones, and 3 chisel-like tools. Also present in the grid area were the remains of bovids, elephants, and other animals; on its south margin were two Afar graves partially built with hippo and elephant fossils.

The evidence suggested that the stone tools were manufactured and utilized by early humans along a streamside. The artifact site had been inundated and sealed by volcanic ash and had remained so until recently, when all, except the in situ chopper and any artifacts still buried, were progressively lowered from higher levels by erosion (deflation). Some of the artifacts may have come from the middle layers of the sequence; or the uppermost "chopper tuff" may have been the source of all the artifacts, as well as some of the more complete fossils in the area.

On the basis of more identifications of Matabaietu fossils, we still estimated the site's age to be 2.0 to 2.5 million years. I also continued to believe the Matabaietu beds were related to similar sediments at Wilti Dora and Bodo. To help confirm the correlation, we explored the Gemeda stream area, midway between Wilti Dora and Bodo, and found sediments and fossils similar to those to the south and north.

———

For the next 17 days we worked our way down the eastern Awash and back up again: Bodo, Hargufia, Meadura, Buyelle, Alkanasa, Andalee; then Harata, Subalealo, Kada Meshellu, Meadura, Gaenelea, Hargufia, Bodo. Sites that we missed when traveling downriver were visited on the return trip upriver.

At Bodo and Meadura, Singleton and Larson once again excavated test trenches to determine artifact and fossil source layers, although one could argue whether this was necessary because fossils and stone tools were nearly oozing from the ground. Nevertheless, digging holes in the ground is what archeologists do, and important determinations were made that confirmed associations of animal remains and artifacts. Numerous other localities revealed isolated stone tools and skeletons or body parts of animals that had resulted from single-purpose, short-duration butchery of animals. Large-bladed basalt flakes were recorded next to hippo skeletons, small scrapers next to medium-sized bovids. Apparent butchery marks on bone were observed, as was artificial breakage of skulls or postcrania. Such purported observations were a recurring source of debate between Singleton and myself. I was convinced we were seeing human-derived breakage of bones and cut marks, but Bill—an obsessive devil's advocate—always had alternative explanations: trampling by animals, stream damage, gravity waves. One day I came across a partial suid cranium with a perfectly formed concentric depression fracture in its center. I took Bill to see it.

"Okay, how does this happen other than by a blow from a rock?"

He pondered it for a second. "Meteor impact crater?"

A more careful look also revealed distinct deep tooth marks from the canines of one, perhaps two, large carnivores. The association suggested that the pig was first killed by a blow from a rock, or some type of blunt instrument, and later drew the attention of one or more carnivores.

Discrete piles of rock debris were observed throughout the Acheulean sites, apparently where stone tools had been made on the spot from rock cores. Other tools we know were made at the quarried factory site on the basalt flow at Meadura and from the core boulders at Bodo. We again collected artifacts and fossils from measured quadrants, this time tied into a 22,880-square-meter grid. A tabulation of artifacts revealed a modest number of heavy tools—bifaces, large retouched flakes, ovate handaxes, choppers, and cleavers—and a high percentage of rock debris reflecting on-site tool manufacturing.

Both Singleton and Larson considered the tool assemblages at Bodo, Hargufia, and Meadura to be middle rather than early Acheulean, largely on the basis of the toolmaker's innovative technique of knocking off sharp flakes from preshaped rock cores, as Wendorf had observed at Meadura. I still believed the Bodo Acheulean levels to be older than Meadura or Hargufia for geological reasons, but the ample fresh tuffs running throughout the sites would answer that question in due course.

Our search for a hominid during the week we were camped at Bodo was a flop. Several days before we moved north, I returned to camp in midafternoon with Whitehead and found everyone else sitting under the work tent. Larson was absorbed in drawing obscene pictures on the roof of the tent with a marker pen; Singleton and Wood were locked in an arcane debate about the core curricula of SMU versus Harvard; Tsrha, Assefa, and Sleshi were gossiping about their university chums; and Alemayehu was reading a paperback romance novel. Of course I gave everyone a severe tongue-lashing about why-this-is-not-the-way-to-find-a-hominid, after which my audience immediately resumed their activities. Stanley would never have stood for such impudence ("Okay men, on the count of three, I'm going to open fire . . ."). When we decamped a few days later and headed north empty-handed, in a rare rainfall for that time of the year, I knew we had missed our opportunity.

—————

For three days we camped at Andalee, beneath a fig tree nearly 6 meters wide at its base. The gurgling Awash flowed nearby. Leaning against the tree was a ladder lashed together with strips of bark, presumably left there by the Afar to retrieve figs, sap, bird eggs, or honey.

Right away the paleontologists among us began adding to our sample of fossil rodents and monkeys that were prevalent throughout the site. Altogether, identifications revealed at least 26 species of animals, including snake, lizard, frog, and the talon of a raptor. Our collecting procedures were more pick-and-choose than random, but some patterns of abundance were striking enough to suggest relative numbers of animals; we were also able to make inferences about Andalee's past by comparing the habitats of paleofaunas with those of their modern counterparts.[3]

Over half the identified animals (monkeys, bushbuck, bushpig) lived in wooded areas, whereas only a third of that number lived in open country (gazelle, bovini, alcelaphine antelope, giraffe, the giant *Theropithecus* baboon, *Equus*). Most of the remaining animals lived along the waterside (crocodile, otter, cane rat, frog), in water (fish), or on the fringes of these zones (reedbuck, impala). A third of the total individuals—it would surely have been more in a truly random sample—were small mice; however, because many of the rodents had been brought to the site by raptors (as indicated by their pellets) that roost in woodlands, we could only suggest that the abundance and preservation of the microvertebrates was consistent with a wooded environment.

The archeologists once again collected and tabulated artifacts from gridded squares. Most of the tools were heavy-duty picks, choppers, and core axes; many were light-duty scrapers and flakes; some were large cutting tools.

The diversity of both fauna and tools at Andalee made it apparent that the Andaleeans exploited a broader range of food resources than the Acheulean big-game hunters. Had my colleagues and I been swept up into some sort of a Pleistocene Outward Bound time warp, I am confident that we would have quickly appreciated being able to use all of the stone tools. Aside from rendering large animals with large cutting tools, we might have used picks for digging up roots or tubers or for digging into termite mounds or riverside burrows housing cane rats. We might also have used picks for digging fire pits, water holes, or animal pitfalls; for making shelters and protective enclosures; and as weapons to use on paleopaleoanthropologists. And we might have used the axes for chopping up meat or edible plants and for clearing vegetation, cutting bark, or stripping branches to make spears.

After identifying the Andalee fossil collection as best we could, the closest age we could come up with was still very approximate. For a stab in the prehistoric dark, I guessed 100,000 to 150,000 years, a time that seemed consistent with the regional archeology. If the Sangoan was a cultural adaptation that had followed the mainstream Acheulean tradition, and had preceded or coincided with the beginning of the Middle Stone Age, then Andalee fit this order stratigraphically. As previously concluded, the Acheulean levels lay beneath the plateau gravels, and Sangoan and Middle Stone Age industries lay above the gravels.

Whatever the age of Andalee, the prolonged dry period that had produced the underlying gravels in the later Pleistocene severely altered the landscape; in the process, the massive gravels carried down from the highlands had "buried" the Acheulean way of life. No more lakeside resorts and fertile river bottoms, no plentiful herds of large herbivores, and no populations of archaic humans living off the fat of the land. Whatever conditions had helped perpetuate the Acheulean tradition for over a million years in Africa were over, at least in the Middle Awash.

Widespread aridity is indicated by cores drilled in the Indian Ocean in the early 1970s, which provided evidence of a dramatic increase in global cooling between 125,000 and 175,000 years ago.[4] Confirmation of this cooling period was part of the convincing evidence in support of the Milankovich astronomical theory of the ice ages, which correlates solar cycles with periods of climatic change on Earth. Such cooling in eastern Africa may have turned the Middle Awash into a gravel wasteland. Did people remain in the Afar during a winter that lasted for 50,000 years? Corvinus's work at Hadar hints that this may have been the case. Recall that she described not only crude stone tools lying on the surface of the gravels at Hadar but those that were in situ as well.[5]

The abrupt global *warming* that followed in the late Pleistocene, between 100,000 and 125,000 years ago, was also recorded in the Indian Ocean cores.[4] This period of warming may have witnessed the first appearance of anatomically modern humans in the Middle Awash—bands of Andaleeans that followed linear oases (gallery forests along rivers) into the Afar lowlands. As the stone tools and paleofauna at Andalee indicated, these would have been resourceful people who used an increasingly diverse toolkit to pursue an increasingly diversified diet.

On our last day at Andalee I walked north of the site by myself. I followed an animal trail along the Awash that soon became very narrow, with a thin margin of vegetation on my left next to the river and a steep wall of boulder gravels on my right. The gravels reached well above my head and ran parallel to the riverbank for some distance. Almost immediately I found large, broad flakes and long, bladed knife-like tools made of quartzitic rock raining down loose gravel slopes. Whoever had made them had done so quickly, not with the leisure that allowed for refinements; they were made by people who were on the move and purposeful. I had a curious, uneasy feeling about being there, perhaps brought on by the narrowness of the path, but more, I believe, by the massive, chaotic, and violent nature of the deposits and the boldness of the stone tools.

From Andalee we made our way back upriver to the south branch of Hargufia stream, in the midst of the Acheulean sites, where we camped. The next morning Wood and I drove to Gewani with Larson, who was returning to the United States. In the village we met Dejene. He had just driven down from Addis Ababa, bringing with him Charles Smart, who was studying fossil African antelopes for his dissertation at Princeton University and would study the RVRME bovids. Dejene would also join us in camp. As the RVRME's administrative officer, I wanted him to see what field conditions were like and learn firsthand something of our logistics. His presence delighted Sleshi, Tsrha, and Assefa, who teased him relentlessly with stories about wild animals and what Afar "cannibals" do with city boys. That night, however, we all stayed at the spartan AGIP motel and had dinner outdoors, seated on rickety metal chairs arranged around tables on a cement slab. The outline of Aleyu volcano was barely visible in the distance.

Once we had been served beers and were waiting for our plates of spaghetti in the semidarkness, I began going through the mail Dejene had brought me. A floodlight attached to a high pole with a cloud of insects buzzing around it made it just possible to read. Dejene, seated across the table, was intently watching me open the first letter, from Judy, with a pall cast over his face. What was *that* all about? I wondered, returning his gaze. Dejene had a way of holding back bad news. What I read in Judy's letter was that two days previously Kelati Abraham, my former field assistant and now close friend, had been dragged from a bar in Addis Ababa by government thugs, forced to kneel down in a parking lot, and then shot in the head. Apparently several others had been similarly executed. Dejene had been with him and apparently had come close to being killed himself.

The trouble that led to Kelati's death had actually begun shortly before we left for the field in mid-September, during festivities staged to celebrate the second anniversary of the revolution.[6] The antimilitary Ethiopian People's Revolutionary Party called for a boycott of the celebration by its supporters, but the effort met with only limited success. A week later, the EPRP called for a general strike protesting the government; it too again met with limited success. The following evening, on September 23, the motorcade carrying Mengistu, still a top strongman in the Derg, was ambushed in an assassination attempt, injuring him and several bodyguards. A wave of arrests followed, as suspected supporters of the EPRP were indiscriminately picked up in the streets and in bars or were executed on the spot. One of these bars was just a few blocks from the Hilton in the Casa Incis part of the city, where Dejene and Kelati had met. Apparently, whoever had tried to kill Mengistu had escaped in a green Land Rover, like the one Kelati was driving.

Some months earlier I had helped Kelati get a job as a driver for an American engineer with a road construction company. Soon after that, Judy and I had them both over for dinner, and I sensed that my American friend was a little taken aback to find us so close to Kelati, since he was practically a member of the family, often joining us on birthdays and holidays.

It made no sense to kill Kelati. Practically every Land Rover in Ethiopia was that same factory-painted green color.

CHAPTER 22

Human Origins

I t was midmorning, October 15, 1976, and I was in camp at Hargufia (see Map VII on page 140), waiting to meet an Afar chief. I had arranged the meeting the previous day to plan future explorations in the northwest sector of the Middle Awash. He was coming some distance from across the river, so I knew I would be stuck in camp for a while. Just as well. There was always more catching up to do with field notes, and I needed time to ponder a curious discovery made the day before.

East of camp I had found a recent human skull eroding from the base of a hill. Very recent by geological standards. From the preservation of the skull, I judged it to be less than a hundred years old. Its teeth were not carved to a point according to Afar custom, and I had never seen Afar burials anywhere but in open areas beneath mounds of rocks. Strange. I left the skull where it lay but planned to return to the locality later. My immediate thought was that it was somehow connected to the Gortani and Bianchi expedition in 1938. The Italians had camped in the area, and I assumed they were responsible for a small building, the ruins of which I found not far from the skull's location. The building had probably been built for storage purposes. Only its foundation, made of slabs of gastropod limestone found along the valley margin, remained. Situated slightly north on the plateau at a place called Terena was a similar ruin of another building, next to a long-abandoned airstrip, evidently also built by the Italians.

Earlier that morning I had driven back to the skull locality in one of our Toyota pickups (newly acquired from Tenneco) with Bill Singleton. I wanted to show him the unusual burial. When he probed the area around the skull, his knife struck something hard. Brushing away the surrounding clay, he found a *second* skull. Unlike the first, the front teeth were missing. As Bill exhumed the skull a hollowed-out area was revealed. It was a small cave! An animal den? I stood up and backed away from the hill, looking again at the position of the skull. It appeared that the overlying clay had collapsed, covering the cave opening.

We removed more sediment from around the skull, exposing more of the cave, as well as the upper torsos of two skeletons. Bones of other animals were not immediately apparent. Was it then not an animal den? If it had been a hyena den, there would be chewed and half-eaten bones all over the place. Perhaps a grave had been hastily dug into the base of the hill. Basalt boulders had been used to help cover the entrance. However it was made, the grave seemed undisturbed, except for the erosion that had exposed the first skull. We assumed that the two individuals had died at or very near the same time. The question was how they died.

I wanted to do a proper excavation of the cave from top to bottom, but there was too much overburden to justify the time such an effort would take. Reluctantly, at this point I had to return to camp to meet the Afar chief. I left Bill there to pursue the investigation and walked back to camp, leaving the Toyota with him.

I had been back in camp about an hour, keeping an eye on the direction of the Awash for the chief, when in the distance I saw a plume of dust raised by a vehicle speeding toward me. That's odd, I thought; it was my Land Rover, and it was coming from the direction of Bodo, where I had sent Charles Smart and several others early that morning. I had not expected them back so soon; in fact, I planned to go to Bodo after my meeting. I was eager to learn Smart's identifications of the many bovid fossils scattered around the Acheulean site.

It was when the Land Rover was about a hundred yards from camp that I saw Singleton barreling toward us in the Toyota. Smart arrived first. He jumped out of the vehicle.

"We found a hominid at Bodo! A lower face, with a nearly complete maxilla. *Definitely* a hominid. Whitehead agrees."

"Great! What is it?"

"I think it's australopithecine."

"At the Acheulean site? That would be surprising." Australopithecines became extinct about 1 million years ago; no one believed the Bodo fossils were that old or that Bodo's Acheulean industry was that early.

"Who found it?"

"Alemayehu." Not surprising. Alemayehu Asfaw, the same antiquities representative who found the first jaws at Hadar—three individuals in two days. That was mid-October 1974: two years before, almost to the day. Born in the province of Ilubabor in the southwestern part of the country, within the drainage of the White Nile, Alemayehu was then 28 years old and was the only person on earth with field experience at Hadar, in the Middle Awash, and in the Omo. He was also a former librarian and had obtained a diploma in personnel management from a college correspondence course.

Just then Bill Singleton rolled up in the pickup, climbed out, and strode over to us with a big grin on his face. Rounding a hill in the distance was the Afar chief, Ismail, with two companions. Bad timing.

"Bill! Guess what?" I said. "They found a hominid at Bodo!"

"Yeah, but wait 'til you hear what I found!"

"*Bill,* did you hear what I said? They found a *fossil* hominid at Bodo!"

"Yeah, I heard you, but you gotta hear what I found."

"Okay, what did you find?"

"Those two skeletons? One was buried on top of the other."

"On top of the other?"

"That's right, *face to face.*"

"*Face to face?*"

"That's what I said: *face to face.*"

"So you're thinking that one of the skeletons has to be female?"

"You bet'cha. Cave-sex." He guessed the male was in his early twenties and the female was in her late teens.

"Bill, this could be our big break. We could get onto a TV talk show in the U.S. with this one."

"Exactly."

The scenario was obvious. Two people in the heat of the day had crawled into a shallow cave, perhaps some kind of an animal den; during their love-making there was a cave-in.

"There's something more here, Bill. Depending on the age of those skeletons—if they are Pliocene, for instance—we may have the earliest known evidence of human reproduction. Fossil intercourse."

"Absolutely: human origins."

Wow, things were popping! Three hominids in two days, just like Alemayehu at Hadar. Now I had to decide which discovery to go see first. Okay, that would be Bodo, but to conclude about the cave find—

Bill collected both skeletons, which we took back to Addis Ababa and gave to the university. An assistant, Saladin Abubaker, later carefully cleaned and labeled the skeletons and coated them with a preservative. It was my intention to have him study the skeletons and write a report on their age and sex, but the project was not completed. In 1977 Clifford Jolly examined the skeletons and judged they were both males, greatly complicating our claim of "human origins." I later wrote several people in Italy who I thought might have knowledge of Gortani and Bianchi's work, in hopes that there might exist a diary or field notes somewhere that expanded on their publications—and that might also explain the deaths of the two individuals in the cave. My inquiries were unsuccessful.

The Bodo hominid consisted of most of a maxilla and palate. The dental arcade was thick and broadly parabolic in shape, a distinctive feature of *Homo*. Also preserved were the nasal area, much of the cheek bones, and part of the eye orbits. Unfortunately, the teeth were sheared off, leaving only tooth sockets and roots.

After we moved camp from Hargufia back to Bodo and plotted out a refined 2-meter grid around the hominid locality, I left for Addis Ababa to call Wendorf and Jolly with the news. Wood went with me to attend to other matters. I hoped that Fred or one of our New York University colleagues would jump on a plane and join us, but none could get away from their teaching loads. I also sent a telegram to one of FORGE's supporters, telling him of the "hominid find," which he later told me was mysteriously relayed to him as "homorid fird." While in Addis, Wood ran into Johanson, Taieb, and Tom Gray, who were preparing to go to Hadar, and arranged a meeting for the five of us in the bar of the Ethiopia Hotel. Wood had the notion that we could all work together for a common goal—brotherly love, that sort of thing—but there was no chance. I had written Taieb several letters after forming the RVRME proposing that we collaborate on regional stratigraphy and on the study of particular fossil faunas. I had received no replies. He had hitched his wagon to Johanson's star, and that was the end of it.

When we returned to Bodo, I took Justine, nearly seven years old, out of school to join us. When Selati, our chief guide, saw this, he insisted that his ten-year-old daughter, Maleeka, also join us. The two girls ran around the site together, asking everyone questions about this or that, one in English and Amharic, the other in Afar. By the time we left, Justine had picked up enough Afar words to start asking questions in Afar as well. She asked so many questions that Singleton told her he would pay her 25 cents an hour not to ask questions. Justine's reply was "Why?"

I had always anticipated that the first hominid evidence we found would be a single tooth or part of a limb bone and that people would consequently argue whether it was hominid or a zebra. But there was no disputing the fact or magnitude of our discovery: Eight days later, while mapping bone and artifacts within the grid, Whitehead and Wood discovered the upper face that perfectly matched the lower face found by Alemayehu.[1] The new find was 11 meters from the first and consisted of much of the eye orbits, the complete brow ridges and forehead, and part of the basicranium. Smaller pieces of bone, mostly from the skull vault, were scattered in the immediate area and were easy to distinguish from those of other animals, most visibly hippo remains, probably from a single individual.

The overall aspect of the cranium was that of a very large, ruggedly built individual, probably a male. The face was robust and broad, with widely set eye orbits and a low, greatly enlarged nasal opening. Flaring and heavily buttressed cheek bones wrapped around the outer eye orbits and joined massive brow ridges that protruded and met above the nose area in yoke-like fashion. Viewed from the side, the lower face, nasal bridge, and maxilla were prognathic, whereas the frontal braincase was smooth, low, and receding. Judging from the thickness of pieces of the skull vault (up to 13 millimeters) that we recovered, a concentrated head butt from this individual would have been a terminal experience. Likewise, it would have taken more than a glancing blow across the head with a rock or a tree limb to subdue this creature.

Debate raged about the identity of our discovery, and I was certain my colleagues were plotting their individual press conferences. During breakfast, following sleepless nights, everyone was bleary-eyed, sullen, and shifty-looking. Wood judged the hominid to be a large Neanderthal. Singleton favored *Homo erectus*, mainly because *erectus* fossils were associated with Acheulean tools at Olduvai. Whitehead and Smart argued both ways. Taken together, the arguments suggested that "Headcracker Man" was something in between an *erectus* and a "proto-Neanderthal." Nevertheless, for the first time in Africa, the skull of an Acheulean toolmaker had been found directly associated with tools.

Over the next week we recovered much of the rest of the cranium by carefully searching and probing (but not yet screening) the top 10 centimeters of loose sediment within our refined grid, being careful not to disturb any underlying archeological or animal remains. A test trench Singleton dug around the spot of the lower face resulted in the recovery of cranial fragments to a depth of 30 centimeters, which suggested that excavation of the entire grid area to at least that depth would be fruitful. Throughout the remainder of the area we found eight basalt artifacts—five crude flakes, one possible scraper, and two apparent core fragments—interspersed with pieces of the skull remains. Systematic cut marks on the braincase and face were later identified by Tim White; he attributed them to intentional defleshing, perhaps even cannibalism, potentially giving the artifacts a sinister significance.[2] Using a scanning electron microscope, he counted at least 25 individual cut marks on the cranium, surely made with a stone flake.

Judging from the sand and fine gravel that had impregnated part of the brain cavity, it is possible that the cranium was at least partially broken at the time of burial, or that it was detached intact from the rest of the skeleton, allowing sediment seepage into the basal opening of the skull. Some breakage could have occurred prior to burial; perhaps it was caused by trampling or was associated with defleshing. If the cranium was "de-brained" by cannibals, for instance, this could be revealed by careful study of bone breakage, particularly around the basal opening of the skull—and perhaps even by very fine scratches or scrape marks *inside* the brain cavity.

The presence of cranial fragments found at some depth during excavation of the test pit, as well as their recovery from higher levels, indicated that burial had probably taken place in stages, caused by recurrent low-energy flooding and streambank deposition. This was consistent with our interpretation of the geology of the site as a sandbar inside a stream meander. It was apparent that surface exposure, including animal traffic, had contributed to the continued fragmentation of the skull and that deflation of the surrounding sediment had helped concentrate and sort bone fragments and artifacts on the surface. It is likely that the fossils and artifacts were left at the immediate hominid locality by natural, not human, agents but that whatever transport took place was not very great.

While most of our team worked at the hominid locality, I concentrated on the geology, particularly trying to determine the relationship of tuff layers in the vicinity of the sand unit bearing the hominid, which we later called the Upper Bodo Sand Unit (UBSU).[3] The relationship of the tuffs to the 5- to 7-meter-thick UBSU was not at all clear-cut and was complicated by lateral thinning of the unit by erosion, intervening stream drainages, unconformities, and faulting. Nevertheless, I concluded that the UBSU was probably bracketed stratigraphically by a "lower tuff" and an "upper tuff," which thin out or are faulted out of the section to the west, as the UBSU thins out to the east. I guessed that after all the tuff outcrops in the area had been carefully mapped, "Bodo Man" would be datable radiometrically. For the time being, we guessed that the hominid and associated fossils were between 500,000 and 700,000 years old.[4] Radiometric dates have placed the UBSU at about 600,000 years, although much geological work on the site remains to confirm this age.[5]

On our return to Addis Ababa from Bodo, our field cook suffered a stroke. One of his relatives came to the office to tell us. I took Smart with me to the hospital and found Makonnen, probably in his middle forties, in one of the wards, lying on a mattress with a stained plastic cover, partly covered by a linen cloth. His lower body was exposed, revealing a massive hernia, apparently of some duration. He never should have gone with us to the field, but he had five children and had said nothing about health problems. His breathing was heavy and rapid. Smart, from his experience as a U.S. Marine medic in Vietnam, said he was hyperventilating and guessed he would die very soon because of depressed blood pressure. He died the next day. I took up a collection of $1000, equivalent to two years' salary, and took it to his widow. She said she would use the money to open a souk.

For the next nine days, we were all occupied with report writing. Smart described and identified the bovids from Bodo and all the other sites, and he wrote a preliminary appraisal of the paleoecology of the Bodo site based on 21 identified taxa. He reasoned that the modern counterparts of the paleofauna suggested a permanent waterside environment next to an open tree savanna. He also wrote that "the conspicuous absence of gazelles and elands, which are migratory forms present in open tree savanna areas during local wet seasons, suggests to me that we are dealing with high depositional rates from the first rains at the end of the dry season, before the arrival of migratory forms."[6]

Wood curated and cataloged the Andalee rodents, assisted by Mesfin Asnake, a new student working with us; Whitehead described the primates, assisted by Tsrha; and Sleshi and Assefa curated the rest of the newly collected fossils.[7] Before returning to the United States, Singleton wrote reports on the archeology, concentrating on the Acheulean, and Larson later sent me a report on the Matabaietu Oldowan tools.[8] Singleton also cataloged an assemblage of artifacts that Smart and I had collected at Dakanihyalo on the west side of the Awash the day after we left Bodo.[9] Accompanied by four Afar, we swam across the river at Buri and walked to the site, retrieving enough stone tools for Singleton to evaluate. The artifacts, dominated by large, crude handaxes, were made out of a dark gray basalt, similar to nearby basalt outcrops, and were in excellent condition. On their age Singleton and I had a difference of opinion. He believed the stone tools were middle to late Acheulean, and I believed they were early Acheulean, pre-Bodo. Assefa and I later argued back and forth whether elephant molars from Dakanihyalo were older or younger than Bodo. I argued on the side of Singleton, and Assefa argued, correctly, on the side of me. They were early Acheulean.[10]

During November, we divided our time between surveys in the western Middle Awash and the Blue Nile basin. Over the years I had taken numerous trips into the Blue Nile region for one reason or another, and I felt confident that vertebrate fossils could be found there in terrestrial deposits of Mesozoic age (65 to 230 million years), which much intrigued Wood, Patterson, and others at Harvard, who were particularly interested in finding early mammals. The major radiation of mammals followed the extinction of the dinosaurs at the end of the Cretaceous (65 million years ago), but mammal-like reptiles and the earliest true mammals date back to the early Mesozoic. Major blank spaces lie in between. I also believed fossil mammals could be found in lignitic sediments interbedded in the thick flood basalts overlying the Mesozoic rocks, which antedated or coincided with the early history of the Afar. Not entirely convinced it was not being bullshitted, Harvard nevertheless put up the money for the surveys and sent an experienced field person, Chuck Schaff, to join us.

Over the next ten days, Wood, Schaff, and I surveyed five areas in the southeast Blue Nile basin, along the Guder, Muger, and Jema tributaries. We found no fossil mammals, but each locality we visited was probably worth a paper in *Nature*, or so I believed. At our first locality, in the first 15 minutes, we found the first dinosaur fossil found in the Horn of Africa, a single tooth of a carnosaur recovered from Cretaceous sediments. In upper Jurassic sediments we found fish, shark, turtle, masses of paleoflora, and perhaps some of the earliest flowering plants (angiosperms) in the world. I later returned to one area we surveyed, in the Muger Valley, and found fossil resin (amber), commonly a rich source of fossil pollen and insects. Interbedded between flood basalts we found fossil leaves in lignites and volcanic tuffs containing carbonized fossil tree trunks, limbs, and twigs. The Ogaden was not completely overrun by rebels, so we surveyed an area on its northern margin, in the uppermost Shebelli River basin, finding more plant remains of Jurassic age. At the same time we visited Dire Dawa and the ancient town of Harar, but when we sought permission to survey in areas near the outpost of Jijiga near the Somali border, the local military commander suggested we come back when "tourism is more convenient."

We spent only 11 days in the western Middle Awash, laying the groundwork and planning the logistics for a more lengthy stay in 1977. I wanted Wood, Smart, Schaff, and Whitehead to see the Central Awash Complex fossil localities that Cramer, Alemayehu, and I had located the previous July. After surveying the

Hatowie graben for five days, we moved camp to the east side of the complex next to the Awash at Aramis. While in the area, we collected a number of fossils and mapped new localities for future collecting. Smart added significant numbers of bovids to our collection, including primitive boselaphines and bovines; Whitehead collected molars from lignitic deposits that he thought were similar to living mangabey monkeys, which live in swamp forests; and Wood and I concentrated on stratigraphy. Alemayehu, Schaff, and the rest of our Ethiopian colleagues surveyed new fossil areas.

While we were at Aramis, bellowing hippos kept us awake throughout the night. Because hippos come ashore in the dark, our guides kept a fire burning, fearing that one of us would be flattened in our sleep. This caution was not to be taken casually. A friend camping on the bank of the Omo River awakened one morning to find fresh hippo tracks 15 centimeters deep next to his bedroll. Nevertheless, at Aramis as elsewhere, we still bathed in the Awash—all except Whitehead, who feared crocodiles of any size and was teased accordingly. (His fears were justified: Several years later, while doing field work in Kenya, he was grabbed by a crocodile while walking at night along a riverbank. His left arm was nearly ripped from his body, and he narrowly escaped with his life.)

After moving camp to the fish-filled waters of the upper Hatowie River, we visited more fossil localities. Although fossils were rare in the area, we all felt that one locality in particular (Asa Koma, "Red Hill") would produce a rich fauna accompanying a full-scale excavation. The base of the hill was littered with fossils that had been reduced to rubble by the trampling of animals, but clearly whole specimens were eroding to the surface. After spending two days in the area, we headed back to the ferry crossing at Dobel. En route with Smart, the right front door of my Land Rover kept popping open and soon would not close at all. The same thing happened to the back door, spilling out equipment and people. An inspection of the undercarriage revealed that the chassis was buckling just behind the engine block.

Back in Addis Ababa, the Greek chief mechanic at the Mitchel Cotts House Garage told me that he had worked on Land Rovers for 25 years and that mine was the first he had ever seen "broken in half." The next day he gave me a typed estimate for repairs, which I still have. At the end of an impressive list of parts to be replaced or repaired—the chassis and most of the remainder of the vehicle—the total cost was given as E$5109.00, or U.S.$2468.11. After pondering this surprisingly modest figure, I read the fine print at the bottom of the second page:

> N.B. Please note that the indicated amount of E$5109.00 is for labour charges only. Spare parts will be charged according to actual cost of parts plus expendable material, in addition to the labour charges listed above. Approximate cost of Spare Parts will be from E$11,000 to $12,000.

I sold the Land Rover for scrap.

By the end of the field season, in early December 1976, Major Mengistu was promoted to lieutenant colonel, and 50 more "dissidents" and "anarchists" were executed for crimes against the state.[11] By the end of the year, however, the Derg was dealing with much more potent antigovernment forces who were not to be dispatched so easily. Two liberation fronts in Eritrea merged into a united command; major battles were being fought in parts of the Ogaden with the Western Somalia Liberation Front; there were still active liberation movements in Tigray and among the Oromo; Ali Mirah's Afar loyalists were still waylaying trucks on the roads to the coast; and the Sudanese were threatening Ethiopia's western frontier.

Given these developments, and the graphic reporting of Ethiopia's problems, it was hard for people outside the country to believe that it was safe to travel anywhere. However, most of the fighting was in the far north, the far south, and the far east, and it did not take an intelligence network to know generally what was safe where. Much of this information was reported daily by the media in a half-dozen languages, including English broadcasts by the BBC, Reuters, Voice of America, and the Voice of the Gospel, all of which were routinely listened to and supplemented by word of mouth according to each person's knowledge and experience. It mattered little whether some of the reporting was sanitized, propagandized, or sensationalized, because the truth of the matter always seemed to come out, whether it was during the next broadcast, the next day, or the next week. People always seemed to know what was going on. There is nothing subtle about a revolution when there are two sides vying for the attention of we the proletariat.

CHAPTER 23

Espionage Science

O n November 27, 1976, over my objections, Wendorf held a press conference at SMU announcing discovery of the Bodo skull. One of the articles that appeared in the Dallas papers paid special tribute to the president of the Lone Star Cadillac dealership, an important contributor to our surveys.[1] Another article was headlined "For SMU Students It Was the Find of a Lifetime."[2] Johanson, always quick to promote goodwill, sent me copies of the articles. At least I assumed it was he, because the envelope was postmarked Cleveland. Someone also informed the Ministry of Culture of the SMU press release: I was soon reprimanded for "gross violation" of their regulation that required that discoveries first be announced through the ministry.[3] Fortunately, Berhanu Abebe, still head of antiquities, was out of the country at the time, or I surely would have had a fight to retain my research permit.

By mid-December all our visitors had gone, but not before the Canadian ambassador and his vivacious daughter threw a memorable party that assembled a unique blend of the diplomatic corps and her friends. On December 16 I also left Ethiopia with my family for a working vacation in the United States. As Judy and I were packing last-minute items, I noticed Spring, now two and a half years old, running around collecting things in her purse.

"Spring, what *are* you doing?" I said, grabbing her untied dress strings as she came flying by.

"Gettin' ready to go to Amirrorcah," she replied, tugging to get away. Tug. tug.

"Now *wait* a minute! I want to ask you a question." She stopped tugging for a second; questions from adults were still tolerated.

"Are you collecting some valuables to take to America?"

"Uh huh, valubls. Wanta see?"

"Sure."

She opened her purse and there were three pieces of broken colored glass, four olives, and a crust of bread.

"What's the matter Spring—don't you think there's any *food* in America?"
"Maybe."

Just like an Ethiopian, always planning for contingencies.

While in the United States, I divided my time between family and friends on the Gulf Coast and the West Coast, and colleagues and foundations on the East Coast. In Seattle, where Judy's family lived, I also visited Vincent Maglio, who was then in medical school at the University of Washington. Brooklyn-born, he had received his Ph.D. from Harvard in vertebrate paleontology in 1970 and had obtained a chaired position at Princeton the next year. Three years later, however, he decided to change his career to medicine. One of his reasons for leaving paleontology was his distaste for the politics surrounding research in East Africa. He had been a member of Bryan Patterson's team at Lothagam in 1967, when the famous hominid mandible was discovered, yet Patterson was refused follow-up funding by NSF for continued work. Because of differences with Richard Leakey over the Harvard field permit, Patterson gave up working in Kenya altogether.[4] Maglio then joined the Turkana research effort for two years, but he was drawn into the "pig war" over the age of the KBS Tuff, on the side of Cooke, and dropped out of East African studies altogether.[5]

Not only did Maglio change professions from paleontology to medicine; he also changed his identity. He is now Jonathan Dutton, and he has served this field well, too; he is a full professor, chief surgeon, and director of a medical center at Duke University. His résumé lists a unique blend of scientific papers on fossil elephants and ophthalmology, his specialty. He was also very generous during my visit in Seattle, giving me several boxes of books and rare reprints on African paleontology, and a valuable research collection of photographs of type specimens of fossil elephant molars. When we parted, he agreed in principle to join the RVRME field work, if funding and time permitted. I was determined to lure him back into the African fray.

In New York, I met the acting director of FORGE, who had paid for my ticket from Ethiopia to attend a fund-raising luncheon at the University Club to be attended by FORGE's board of directors. On the morning of the event, however, a blizzard blew in, causing most of the city to shut down and those living in the suburbs to flee before the snow clogged the roadways. The fund-raiser was a bust.

In late December Wendorf submitted a proposal to the NSF Anthropology Program, asking for $168,405 for archeology studies in the Middle Awash, and Conroy and Jolly submitted a proposal to the same program, asking for $125,000 to pursue the search for and study of more Middle Awash primates. The proposals were interrelated and were packed with the specifics of our discoveries to justify the funding requests. The combined list of research objectives to be pursued if the requests were awarded ranged from investigating butchery practices of Acheulean hunters to seeking more fragments of the Bodo skull and the ancestors of the Hadar hominids. Implicit in both proposals was the possibility that the Middle Awash could prove to be one of the longest single records of human fossil and cultural remains known. To establish a chronometric framework for the area, the Conroy/Jolly proposal included funds for specialists to take charge of dating. Both proposals contained the required budgets for travel, per diems, field expenses, modest RVRME overhead expenses, and salaries, including those for me and our Ethiopian colleagues.

A third proposal would be submitted in the spring by Patterson to cover non-primate paleontology studies, paleoenvironmental investigations, and more geologists, including two graduate students in geology from Harvard and two more students from Addis Ababa University. Although Patterson was in South America at the time of my visit to Cambridge, Craig Wood and Charles Smart had helped map out the proposal. I agreed to write the first draft after returning to Ethiopia. Otherwise, my visit to Harvard was a nightmare.

The day I arrived in early January 1977, another blizzard slammed into the northeast United States, driving the temperature in Cambridge down to 1°F. It was said to be the worst winter in Massachusetts in 100 years. It was not much warmer in the Museum of Comparative Zoology, because I had arrived during the Christmas break and there was little or no heat in the buildings. Worse, Spring had been right: There was no real *food* in Cambridge. Everywhere I went, people were eating tofu, large bowls of seaweed, or quiche. After a week of this I felt like I was being quiched to death. Craig told me that if students were caught eating hamburgers at Harvard, they could be expelled, and, for eating cheeseburgers, dismembered.

The highlight of my visit was near the end of my stay, when Craig took me to a party. Actually it was a "wake" for a friend of his, Bob Trivers, a sociobiologist who had just been denied tenure at Harvard and was leaving for The Other World (another university). As I understood Trivers's research, he drew parallels between the social behavior of insects and that of, say, humans. For example, he might compare the roles of the queen bee and her thousands of worker bees to the hierarchy of the Third Reich. "Sounds good enough to me, Bob," I told him after we had chatted a bit. Otherwise, the party was full of Harvard luminaries, such as E. O. Wilson, who enjoyed an international reputation for spending much of his life studying ants. Everyone got drunk and danced around to the beat of Jamaican music. At one point I yelled out, "Hey! You people want to study bugs? Come to Ethiopia—I'll show you REAL bugs!"

A few days after I arrived back in Addis Ababa at the end of January, the U.S. and British cultural centers were firebombed.[6] Two days after that, while we were hosting a party for Justine's seventh birthday, thousands of demonstrators marched down Churchill Road in a show of support against the Eritreans and Somalis.[7] The chants of marchers blended menacingly with the sounds of children dancing to country music tunes.

Six days later sounds of a gun battle burst from the Derg's downtown military headquarters during a daytime shoot-out that left eight people dead, including the nominal head of state, Lieutenant Colonel Tafari Banti.[8] This left Lieutenant Colonel Mengistu as the chairman of the Derg and leader of the country. According to street talk and published accounts, the gunfight erupted during a meeting when Mengistu's supporters pulled out weapons and blazed away at Banti and his men, who had apparently been set up for the killings.

While this power struggle was taking place in the capital, the leadership of the IARE was having its own problems. Johanson's field work that season was limited to the last few weeks of January, because he had not finished cleaning and casting the "First Family" hominids recovered in 1975.[9] The Ministry of Culture had recalled all fossils exported for study. Two other American expeditions were also affected by the directive, both based at the University of California at Berkeley. One was headed by F. Clark Howell (who had moved there from the University of Chicago), hoping to renew work in the Omo, the other by the archeologist J. Desmond Clark, who was working on the Hararghe plateau. The recall of both fossils and artifacts was part of a larger policy shift by the ministry to promote the development of the National Museum and the training of Ethiopians. This was partly my doing, because I continued to preach to anyone who would listen that Ethiopia would never raise money for modern laboratories for prehistory studies, or train Ethiopians to use them, as long as collections were shipped abroad for study, and as long as foreign researchers were not required to include Ethiopians in their research. In mid-February of 1977, the minister of culture began to require all research groups recovering prehistoric materials from Ethiopian sites to construct facilities to house their collections.

Another reason for Johanson's delay in getting to the field was the gala opening in mid-December of a lavishly outfitted laboratory at the Cleveland Museum of Natural History for the study of Ethiopian fossils.[9]

When the program director for anthropology at NSF, Nancie Gonzalez, learned by chance in mid-November that Johanson was not participating in field work that NSF was paying for, he belatedly told her that he was busy casting hominid fossils because the loan of fossils to his museum had been revoked.[10] However, he wrote, "I prefer not to give you the details of the reasons for the Ethiopians' change in policy."[11] What mysterious "details" Johanson had in mind must have been a bell-ringer.

During a brief visit to Hadar in January, Johanson took with him Jack Harris, a doctoral student in archeology from the University of California at Berkeley, to pursue a discovery made in November by a French archeologist who had returned to France. Following up on the work of Corvinus in the upper Gona area, Hélène Roche had discovered an in situ stone tool locality of the greatest antiquity, which she planned to excavate the next field season. Johanson immediately set Harris to work on the site.[12]

In an article later published in *Science,* the discovery was heralded as the "world's oldest" artifacts, approximately 2.6 million years old, for which Johanson took partial credit, because he had helped Harris excavate the site.[12] The episode created a bitter feud between the French and the Americans—and eventually among the Americans themselves—that continues to the present.[13]

Meanwhile, I was ruthlessly advancing my own career. Following Wendorf's announcement of the Bodo skull, the RVRME had an announcement of its own, authorized and published in the *Ethiopian Herald* on January 30, 1976, under the title "Human Skull, Tools Discovered in Awash." The story was later carried internationally.[14] As part of our media blitz, the Ministry of Culture scheduled a press conference for the RVRME in early February, but the day before it was to be held, the permanent secretary of the ministry—a position answering directly to the Derg—was assassinated. Then an interview I gave on Ethiopian National Television turned into an embarrassment when I referred to our "stone tool" discoveries in the Middle Awash as "stone *stools.*"

The January announcement came two weeks after Jolly, Conroy, and Cramer—using a wooden crate for a work table on the lawn of the National Museum—reconstructed and studied the skull, which they identified as most like *Homo erectus* but exhibiting features shared with pre-Neanderthal *Homo sapiens*. Much follow-up study was needed, but first we hoped to recover more skull fragments accompanying excavations to be carried out by our SMU colleagues. However, some problems developed.

At the end of January, according to NSF procedure, Nancie Gonzalez had sent Wendorf's grant application to five specialists to judge as anonymous peer reviewers. Two of the reviewers were also members of a panel that would meet at NSF in March to evaluate the reviews as a whole, along with those of other proposals. Following the panel meetings, Wendorf and other grant applicants would be given copies of the formal reviews and a summary of the panel's evaluation.[15]

As I later learned through the Freedom of Information Act, all three outside reviewers were from the University of California at Berkeley. One was archeologist Glynn Isaac, a regular recipient of NSF funds for his work in East Africa. He was also the Ph.D. supervisor of Jack Harris, who had just finished working at Gona with Johanson.

Although Isaac gave the SMU proposal an "excellent" rating, as did three other reviewers, he accompanied his review with a confidential letter to Gonzalez, which "after some hesitation" he thought should be brought to the attention of the panel and higher NSF officers. Accompanying a distorted history of the formation of the RVRME, he took me to task for claim-jumping the original IARE permit and ended his letter by raising the subject of my status, and indirectly my funding, in Ethiopia. There were "various seculations [sic] and rumors . . . of a kind, which must be of concern to the international community of anthropologists." I learned that in a follow-up telephone call to Gonzalez, Isaac was more specific about my reported CIA connections.[16]

During the panel deliberation of the SMU proposal, Gonzalez read Isaac's letter to the panelists and then launched into the CIA issue. She emphasized that anyone engaged in CIA activities would blacken NSF's name and bring discredit to the scientific community. There was no significant discussion of the scientific merit of the

RVRME discoveries, of the proposed research, or (surprisingly) even of the political situation in Ethiopia. Rather, Gonzalez suspended discussion of the proposal indefinitely, telling the panel that she would investigate the CIA matter further. There was no mention of the spying charges in the minutes or in the official summary of the meeting.[17]

Over the coming weeks, Gonzalez conducted a secret inquiry, approved at the highest levels of NSF, into my alleged covert activities. Questions central to her investigation concerned my source of funds, my "ability to move freely around the country," and my "seemingly close relations with all sorts of Ethiopians."[18] She later described to me one of her unnamed informants as "a person of vast integrity with long experience in Africa" who had "*direct* knowledge" of my activities in Ethiopia. It was apparently this same source who made a convincing case to her that I had "infiltrated" the Ethiopian government.[19]

That was the end of the SMU proposal. The NYU proposal was also declined, because it "was closely tied to the one by Wendorf."[20]

Unaware of my rising status in the underworld, I returned to the United States and Cambridge in mid-March 1977, carrying with me a 53-page draft of the Patterson proposal. Patterson was still in South America, but following lengthy correspondence with him and others at Harvard, Wood, Smart, and I refined and completed the proposal. The final points were approved by Patterson in phone calls to Argentina. Everything not in the NYU and SMU proposals was included in the Harvard proposal: phylogenetic, paleoenvironment, taphonomy, paleoclimate, and basin development studies. The projected budget was $241,860.

Deciding that I would not endure the blizzard conditions that once again turned the Harvard campus into a "white hell," I holed up in an apartment during the time I was in Cambridge. I made forays to the Museum of Comparative Zoology only for urgent business. Day in and day out I toiled over revisions of the proposal, lists of fossils, and geological sections, while howling winds blew across the tundra and sleet clattered against my windows. Periodically, I could hear the distant trumpeting of woolly undergraduates as they roamed the frozen campus seeking relief from the unrelenting winter. I attribute my own survival to a Greek submarine sandwich takeout two blocks from my apartment, where I felt safe to venture only by leaving scribbled pages of the draft proposal scattered along the hoary landscape to lead me back to my shelter. Finally, at the end of two tormented weeks, the proposal was finished.

My final task at Harvard was to go through the chain of approvals for the proposal, an ordeal I found even more convoluted than the renewal of my Ethiopian driver's license. That done, in early April I flew to Washington, D.C., and hand-delivered the required 20 copies of the proposal to NSF's Office of Central Processing, after which it eventually landed in the Ecology Program. I next retired to the home of friends I'd known since childhood who were then living in the suburb of Kensington, Maryland. There I was to prepare for a lecture I was to give the next day at the National Academy of Sciences. Instead, I stayed up with my friends retelling stories of our youth in Houston, drinking pitcher after pitcher of banana daiquiris, until I fell asleep in a stupor at 3:00 A.M.

The next morning when I awoke, it was all I could do to suppress a scream. I found myself standing behind a podium at the National Academy with the audience looking at me as though they were staring into the sun. My God! I had no idea how long I had been there, or what I had said! I looked at my watch. I guessed I had been there 30 minutes. I was terrified.

With no other choice, I began speaking.

". . . and so . . . and so . . . and so," I stammered, "because of all these reasons . . . because of *all* of these reasons . . . the RVRME is a progressive institution and the expedition style of research is a thing of the past. African countries do not need or even want 'expeditions.' If the U.S. National Science Foundation does not realize this, then it is decades out of date and is promoting neocolonialism. Ethiopians want their people trained and educated; they want modern institutions in which to conduct their *own* research; they want to discover their *own* country; they want to exercise their pride and gain international respect."

Sobered by my rhetoric, I continued. "It is no longer relevant for NSF to say that development in the Third World is not their business. Development is their business if they want American scientists to continue working in Africa. Africans are no longer content to follow the white *b'wana* around as second-class citizens. It is not a question of 'science' on the one hand or 'development' on the other. The issue is that in the developing world the two must be interchangeable. *The two must be interchangeable.*"

The lecture over, there was a pattering of applause. As the 30 or so people silently filed out of the auditorium, I noticed Mary Greene, the associate director of the anthropology program at NSF, whom I had met the previous day. I had hoped to see Nancie Gonzalez, but she was out of town.

In mid-May, back in Ethiopia for several weeks, I received back-to-back letters from Wendorf, who had just returned to the United States from Egypt.[21] He had visited Gonzalez in Washington, D.C., and learned that his proposal had been turned down. There were two major reasons for the decline, he wrote (1) the question of whether he could handle both the Middle Awash project and his work in Egypt and (2) concern over "gossip . . . which stems from Johanson." Fred urged me to send him a statement about the CIA allegations and specifically an accounting of my funding over the years. In the second letter, written to Gonzalez with a copy sent to me, he complained to her of "McCarthyism" and asked that the CIA issue be taken to the director of NSF for further investigation. If it were determined that I was not a spy, Wendorf said, then NSF would be obligated to have the proposal reconsidered by the panel and by "any other reviewers or others who participated in the initial handling of this matter." He told her that he was personally "convinced that Kalb is *not* a CIA agent."

In his letter to me, Fred then said that with "20–20 hindsight" he believed the planning of the Middle Awash work was in fact fundamentally flawed because the NYU and Harvard research interests conflicted with his own. The work outlined in the Patterson and Conroy/Jolly proposals should have been limited to precultural time periods. Fred then said that in order to salvage our efforts, he had asked Berkeley archeologist J. Desmond Clark to "help out" with the Middle Awash archeology. Clark would assume principal responsibility for the sites, and if all went according to plan, Fred and his wife, also an archeologist, and Bill Singleton would join Clark's expedition at Bodo in February. All that was required was my agreement and diversion of Clark's NSF grant money from his work in Bale Province in southern Ethiopia to the Middle Awash. Apparently Gonzalez had already agreed in principle to the redirection of grant funds. Clark's Bale project was important—an Early Stone Age site on the Hararghe plateau—but it contained only fragmentary fossils and no hominids.[22]

After digesting the intricacies of Fred's letters, I telephoned him to hear more about his exciting plan firsthand, particularly the chronology of how it had evolved. I learned that Fred had first heard of the decline of his proposal, and of the role the CIA matter played in its rejection, from Glynn Isaac, even before hearing of it from Gonzalez.

Before I had responded to Fred's proposal, Clark had already written Gonzalez telling her of his intention to take over the "Central Afar sites at Bodo . . . provided Jon Kalb's position is clarified to general satisfaction." Clark copied his letter to Wendorf, Howell, and Johanson.[23]

John Desmond Clark was born in London in 1916 and early developed keen interests in history and archeology. He pursued both subjects at Cambridge University, where he graduated with honors in 1937. The following year he took a position in Northern Rhodesia (now Zambia) as curator of the Livingstone Memorial Museum, which is located 5 kilometers from the Zambezi River and 10 kilometers from Victoria Falls. The posting allowed Clark to pursue his two passions, archeology and rowing. He joined the Zambezi Boat Club and later asserted that the sport had taught him the importance of teamwork. Over the next few years Clark investigated a series of Iron Age and Stone Age sites along the Zambezi River, until his work was interrupted by the outbreak of World War II. From 1941 to 1946 he served in the British Army in Somaliland and Ethiopia, first with an ambulance unit and then as a civil affairs officer. During his spare time he collected enough

Stone Age artifacts to use for partial fulfillment of a Ph.D. dissertation, which he completed in 1951 following a year of residence at Cambridge.[24]

After resuming his position at the Livingstone Museum at the end of the war, Clark discovered his most celebrated site in 1953 at Kalambo Falls, at the south end of Lake Tanganyika. The site contained a lengthy sequence of Stone Age to Iron Age cultural levels and valuable paleofloral remains, but no hominids or other vertebrate fossils. Over the next eight years Clark produced a flood of writings and several books covering the breadth of sub-Saharan archeology, developing an impeccable reputation as a researcher. Then in 1961, just before the decolonization of British East Africa, he accepted a professorship with the Department of Anthropology at Berkeley.[24]

In 1966 Clark was joined at Berkeley by Glynn Isaac, then 29 years old. A South African who had also studied archeology at Cambridge, Isaac established his reputation working with Richard Leakey on the australopithecine site of Penini, in Tanzania, and independently at Olorgesailie, an Acheulean site with archeology and fauna similar to that of Bodo, but no hominid fossils.[25] Clark became Isaac's mentor, and together they formed a strong partnership based on their shared interests in early hominids. Beginning in 1970, however, Isaac's work at Turkana with Richard Leakey—uncovering "the oldest artifacts" made by "the oldest toolmakers"—soon stole the limelight. Whether Clark was affected by this is unknown, but we can guess that by 1977, after spending four decades reading about "sensational" early hominid discoveries by others in South Africa, Olduvai, the Omo, Turkana, Hadar, and now the Middle Awash, he was ready for some changes in his advancing career. Kalambo Falls was certainly important, and Clark wrote two magnificent volumes on the site, but they had not placed him among the glitterati.[26]

I first met Clark in late 1972, when he came by my office at the Ethiopian Geological Survey, seeking maps and reference materials on the southern Afar and the Hararghe escarpment for surveys he was proposing for early 1974. Concerned that his intended research area might overlap the IARE permit, I asked Clark to draw the boundaries of his proposed field area on an aeronautical chart pinned on the wall of my office. The area that he drew on the map, which I still have, covered an area west and south of Awash Station, well removed from the IARE concession. Several months later, however, I learned that the actual boundaries of Clark's field permit engulfed most of the southern half of the IARE area (see MAP VI on page 91).[27] I alerted Taieb and the Antiquities Administration to the problem, which was finally straightened out in mid-1973. It was agreed that Clark's group would investigate younger sites in the area where the two permits overlapped, whereas the IARE would pursue older sites.[28] Although I doubt that Clark was happy with this arrangement, we remained on friendly terms, and in March 1974, following the end of his field season—mostly on Late and Middle Stone Age sites—I let him use my office and compound to process his collections and field data and to store his Land Rover. A year later we met again, at the end of his second field season. This was also the time when the RVRME was authorized by the Ministry of Culture and

Wendorf was visiting Addis Ababa. I invited Clark to a small party at my house, attended by officials from the Ministry of Culture and by Wendorf and others, to celebrate the authorization of the RVRME. I had just taken Fred on a survey flight to see the exposures in the Middle Awash, and he bubbled over to Clark about the potential of the area. I sensed that Clark did not share Fred's elation.

Over the next two years I saw Clark or his students from time to time when they were in and out of Addis Ababa between stints of field work, and I regularly told them of the progress of the RVRME surveys. One of Clark's students, Ken Williamson, often dropped by the office or my home and, over bottles of Meta beer, routinely beat me at games of checkers (as did everyone else). Ken was working on a Middle Stone Age cave site near Dire Dawa that in the 1930s had produced a 70,000-year-old neanderthaloid mandible. We routinely swapped stories about our work and the latest bizarre directives of Dr. Berhanu Lévi-Strauss. . . . (For example, fossils impacted in cave rock matrix that were to be shipped abroad for removal from the rock matrix had to be individually labeled and cataloged before leaving the museum.) Regrettably, after the rejection of the Wendorf proposal by NSF, I no longer trusted anything connected with Berkeley, and my friendship with Ken came to an end.

At the time that the SMU proposal was declined in the spring of 1977, Clark had spent four field seasons and $232,000 dollars of NSF money in Ethiopia, most of it in an entirely new permit area, in northern Bale Province (see Map II), that had been annexed to his original permit area in Hararghe Province.[29] Although his team documented important Early Stone Age industries at Gadeb and early controlled fire use in a unique highland setting, Clark must have known that he was stretching it to ask NSF for more money on a site that lacked fossil hominids or even significant faunal remains. But he did, and his proposal was turned down for lack of adequate scientific merit.[30]

In mid-May 1977, Conroy and Jolly officially heard from NSF that their proposal was declined because of its ties to Wendorf's, "plus the political situation in Ethiopia."[31] Rightly so. By then politics was very much an issue. On April 23 Mengistu had solidified his courtship of the Soviet Union and expelled what remained of the long-standing U.S. military missions in Addis Ababa and Eritrea. He had also closed the U.S. consulate in Asmara, the U.S. Navy medical research unit in Addis Ababa, and the U.S. cultural affairs office. At the same time, Ethiopia began taking delivery on 500 million dollars' worth of Soviet military hardware to fight the Eritreans, and the Somali rebels moving north through the Ogaden and Bale Province.[32] Night after night, in the early morning hours, Judy and I lay in bed listening to the dull roar of tanks, armored personnel carriers, and military trucks rumbling along Churchill Road.

Although the United States had been a strong ally of Ethiopia ever since the end of World War II, its prolonged support of the Haile Sellassie regime had caused lasting bitterness among the country's reformers and intelligentsia. This resentment became particularly strong when the United States sided with the Emperor during the 1960 coup attempt, and in the years that followed, when he increasingly relied on his U.S.-backed military to retain power. Those who would later vent their anger against the United States included young military officers, who railed against the old guard of aristocrats in the military for resisting reform while using their positions to amass personal wealth. One such officer was Mengistu Haile Mariam, who reportedly carried an extra grudge against the United States over a racial incident he experienced in the 1960s while at Fort Leavenworth, Kansas, for special military training.[33]

Anti-American hostility in the Ethiopian military increased just before and after the overthrow of the Emperor in 1974, when the United States proved unwilling to offer Ethiopia increased military assistance to counter the gains made by the Eritreans and the growing Somali threat. With the rise in power of Mengistu, the Derg made no secret of its contempt for "western imperialists." During a speech in Revolutionary Square, Mengistu reportedly threw a Coca-Cola bottle filled with blood on the pavement and declared that the "blood of American imperialism will be spilled."[34] In turn, U.S. criticism of the Derg increased with each new wave of executions, human-rights violations, and denunciations of the United States. American discontent with the Derg carried into the presidency of Jimmy Carter, who took office in January 1977, just days before the moderate leader Tefari Bante and his followers were gunned down.

By April 23, 1977—the date on which 300 American officials and their families were expelled from the country—the Derg was desperate for a massive infusion of arms to meet the combined assaults of its external and internal enemies. The Soviets agreed to provide Ethiopia with the needed arms only on the condition that it sever its military ties with the United States.[34]

Within days after the Americans were ordered expelled, several hundred students, many caught demonstrating against the government, were massacred by security forces.[35] Within weeks the estimated number of youths killed—during a period known as the "red terror"—rose to a thousand, as death squads roamed the city in search of dissidents and "terrorists." In the middle of this I was walking across the university campus one afternoon with a student, Solomon, when one of these squads approached from the opposite direction. One of the thugs, wearing a faded blue-jean jacket, a chrome-plated pistol stuck in his belt, and a sailor cap pulled out over his forehead, slammed into my shoulder as he passed. I instinctively swung around, just as Solomon lurched over and grabbed me, leaving marks on my arm that lasted for a week. Shortly before that incident his best friend, a medical student, had been arrested and killed for helping to make cyanide capsules to be used by dissidents in case they were arrested and faced torture.

On May 1, 1977, Mengistu gave a speech during the May Day celebration, telling cheering crowds that with the expulsion of the Americans, the era of "slavery" was over. In late May the U.S. embassy staff was reduced to a skeleton crew.[36]

As Wendorf had urged, in early June I sent Gonzalez a statement describing the origins of the CIA story that had arisen following my dispute with Johanson at Hadar in 1973. I also described my finances, while defending FORGE as a reputable organization. I used the opportunity to promote the construction of a storage/laboratory facility at the National Museum to house and properly maintain the collections recovered by NSF-financed expeditions, most of which were still heaped around the museum in packing crates or dumped wherever space allowed. At the end of the statement, I said that the RVRME did not need Clark and his team to do the archeology for us in the Middle Awash—that Wendorf was capable of doing the job.[37]

I received a note from Gonzalez saying that her tenure at NSF was just then ending, so she would give my statement to her successor, John Yellen, to answer.[38] Yellen never replied, nor did Wendorf receive a reply from Gonzalez or Yellen to his request that his proposal be reconsidered.

In late July, Clark reviewed the Harvard proposal for NSF, warning that my "possible involvement with the CIA must, until this has been cleared up, remain a matter of concern for any scientists."[39] Two weeks later Clark submitted a $94,000 budget (from his previous grant award) to Yellen for a 3-month field season in Ethiopia in early 1978, half in Bale Province and half in the Middle Awash. He asked for funds to take to Bodo a 23-member team that included Wendorf and Wendorf's wife (an archeologist) and son (a student in archeology at Berkeley).[40] Clark also submitted a permit request to Berhanu Abebe for field work in the Middle Awash, listing the Wendorf family as new members of his team.[41]

Unaware of these developments, my colleagues and I were working around the clock in Addis Ababa. In addition to my Ethiopian staff, Liz Oswald rejoined us, and I was able to enlist several other part-time assistants, both American and Ethiopian. Everyone had a job to do—compiling or analyzing data, making illustrations, drawing maps, or labeling and describing fossils at the museum. At one point, following the expulsions of the Americans, I hired three Ethiopian secretaries who had lost their jobs at the U.S. embassy to help us complete a 432-page annual report for the Ministry of Culture.[42] To do the job, the embassy let me borrow three Remington electric typewriters. This was not without risk, because neighborhood militias were still conducting house-to-house searches for duplicating machines or typewriters used in preparing antigovernment literature. On one occasion when I was away, our home was searched by armed soldiers with Judy and the girls present. Apparently everything went smoothly until Spring, then three years old, had a fit when the soldiers started going through her belongings. On another occasion, a dozen soldiers entered our compound at lunchtime. I walked outside to meet them, thinking it would be another house search, when suddenly their leader, standing 10 meters away, asked me over a bullhorn whether we had any empty bottles to donate to a fund-raiser.

Much of my time during these summer months—during the heavy rains, when cloudbursts or hail periodically pounded the capital—was spent writing papers for publication and letters to foundations or universities, seeking money or scholarships. I did receive a $7000 grant from the Wenner-Gren Foundation for Anthropological Research to buy another secondhand Land Rover, and the Explorer's Club gave me modest funds to help with field expenses. Then Doug Cramer picked up some money from several sources, so we were set financially for a return to the field, which we planned for early 1978.

September 11, 1977 was a banner day for the RVRME: Its research staff finally graduated from college. Even though Sleshi, Assefa, Tsrha, and Dejene had completed their degree requirements the previous summer, the graduation ceremony was delayed a year. Like their fellow graduates, they had weathered the periodic closings of the university during the tumultuous final years of the imperial regime and three years of military leadership and revolution. An unknown number of students who should have graduated died trying—during protests and riots and following arrests in the middle of the night. All were part of the conscience of a nation, and all, in fact, had simply graduated to an institution of higher learning before their time.

What was later named Addis Ababa University began as University College in 1954. Its enlarged successor, Haile Sellassie I University, was chartered in 1961 and gained its imperial name in 1974.[43] In 1977 the university was divided into five colleges (agriculture, business administration, public health, technology, and theology) and five faculties (arts, education, law, medicine, and science). On the oft-delayed graduation day that September, some 500 students marched up to the stage to receive their diplomas, in an amazingly efficient ceremony as such affairs go.

The event was held under a giant tent on the grounds of the People's Palace. Mengistu, the keynote speaker, extolled the achievements of the revolution and later handed out the diplomas. He was a man of medium height, 36 years old, with regular features made unforgettable by posters plastered all over the country. I was seated near the back of the audience and recognized a number of friends, foreign and Ethiopian; faculty members and graduates; and some diplomats. At the end of the ceremony, I was one of the first out of the tent. So was Mengistu. I was standing off to one side waiting for friends when he and his bodyguards strode down a stone walkway toward some steps. A dozen or so graduates and their friends stood at the bottom of the steps, hugging and kissing one another in animated conversation. Suddenly, they saw Mengistu and immediately fell silent, while bowing their heads and stepping back to clear a path down the walkway. Instead of proceeding, however, Mengistu seized the moment and began exhorting the graduates with calls to protect the motherland, while shaking his fist above his head. The graduates, knowing their roles, responded in kind.

"Down with feudalism!"

"*Down with feudalism!*"

"Down with imperialism!"

"*Down with imperialism!*"

"Down with counterrevolutionaries!"

"*Down with counterrevolutionaries!*"

This was no mere impromptu pep rally. With steeled oratory, Mengistu was telling them what they *must* believe and what they *must* embrace for the revolution to succeed.

"Down with feudalism!"

"*Down with feudalism!*"

"Down with all enemies of the revolution!"

"*Down with . . .* "

Suddenly, I caught Mengistu's eye and he turned and looked directly at me. What he saw was a tall, blond foreigner, a *firengi*—probably an American—watching him, spying. What I saw was a brutal reformer, who knew that the revolution's only chance of survival, and his own, was to win over this tiny group of graduates. And he wanted them to know that he was prepared to do so *by any means necessary.* This is what he was telling them and what he was having them repeat over and over.

"DOWN WITH CAPITALISM!"

"*DOWN WITH CAPITALISM!*"

"DOWN WITH REACTIONARIES!"

"*DOWN WITH REACTIONARIES!*"

In case any of the graduates misunderstood what Mengistu was saying, a few weeks later the number-two man in the country, Lieutenant Colonel Atnafu Abate, regarded as the last "moderate" in the Derg, was executed. He was accused of being an obstructionist and consorting with CIA agents.[44]

By mid-September word drifted back to me from the just-ended Pan-African Congress for Prehistory Studies held in Nairobi that my own alleged CIA connections were more alive and well than ever. Certainly, the leaks that earlier had flowed out of NSF greatly increased the story's audience and its apparent credibility.

After a week spent attending the Congress in Nairobi, and back in Washington, D.C., Yellen was eager to resume his new duties as director of the anthropology program. He had met with both Clark and Berhanu Abebe at the Congress and was himself now fully programmed to push for the approval of Clark's proposed field work in Bale Province and the Middle Awash. He wrote Clark that he supported his plans on the basis of three arguments that addressed the security of his proposed field areas:

> The very fluid political situation in Ethiopia precludes any simple blanket geographically applicable statement about the advisability of research. Secondly, I reiterated Dr. Abbebe's [Berhanu Abebe's] assurances that when he issued a research permit for a particular region that he would take the question of safety into account. Finally, I stated that given your own wide experience in Africa that we should defer to your judgment of the situation.[45]

Had Yellen's careful analysis applied to just the Middle Awash, he would have had it right, because the area remained reasonably secure; however, a war was raging in Bale Province, an area that covered the western third of the fiercely contested Ogaden. In addition to the ethnic Somali population, which included the Western Somalia Liberation Front, Mogadishu had recently mobilized a 22,000-man army that had swept the entire region. By late August, intense fighting reached as far north as the highland towns of Jijiga, Dire Dawa, Harar, and most recently Goba, where some 1000 Ethiopian soldiers had been reported killed.[46] Goba was less than an hour's drive from Clark's archeological site at Gadeb in northern Bale (see MAP II).

On his return to Addis Ababa from Nairobi, Berhanu once again tried to invalidate the RVRME field permit. He argued that we lacked adequate money to carry on our work, but he was overruled by a newly appointed minister of culture. Berhanu had learned that the SMU and NYU proposals had been turned down. He had also been told that the Patterson proposal had been declined as well, although at the time it had not. Because the Middle Awash was no longer a dish on the table for Clark, and because he was not about to take his team into a combat zone in Bale Province, he informed Yellen that on reconsideration, he would remain in Berkeley and write reports.[47] The IARE also skipped their planned late-1977 field season.

In mid-September the RVRME became affiliated with the science faculty of Addis Ababa University, and in late October we were granted use of a spacious laboratory in the Department of Biology that became the Paleobiology Research Laboratory, the first of its kind in the country.[48] The arrangement required the radical use of the university for training, education, and research, and the loan of fossils to the university from the National Museum, so of course Berhanu vigorously opposed the idea. Once again, however, he was overruled by the minister, with strong endorsements from the commissioner for higher education, the commissioner for science and technology, and the dean of the science faculty.[49]

At the beginning of December I learned that, not surprisingly, the Patterson proposal had in fact been declined. Two weeks later I wrote the director of NSF, listing all the reasons why the RVRME-related proposals should be reconsidered.[50] I also again argued that NSF should put up the money for a storage facility to house and preserve the more than 50,000 fossils and some million artifacts that had been collected by NSF-financed expeditions over the previous decade, along with those that would be collected in the future. Finally, I requested the opportunity to give a presentation at NSF on the work of the RVRME. The next day my family and I left Ethiopia for the United States; Judy's mother was gravely ill in Seattle.

On Saturday afternoon, January 21, 1978, two days before my presentation at NSF, I checked into a sleazy hotel in downtown Washington, D.C., on the edge of one of the great monuments of government ineptitude, the sprawling black ghetto of the nation's capital. The burned-out and boarded-up buildings still stood as a reminder of the riots that followed the assassination of Martin Luther King a decade earlier.

I spent two days into the night and throughout the night preparing my lecture, while listening to the sirens of police cars racing from one murder or drug bust to the next. At two in the morning of the second night, my presentation, filled with a catalog of all our marvelous discoveries in the Middle Awash, was ready to go. I set my alarm and lay down to sleep. An hour later I was still awake. I could not escape the conviction that we stood exactly no chance of getting the decisions on the proposals reversed, something I had learned virtually never occurred. To prevent my efforts from being wasted, I decided to use the opportunity to try to convince NSF to do something about the need for a storage facility for collections in Addis Ababa. I climbed out of bed and began preparing a *new* lecture, which I worked on until dawn.

At 9:45 A.M. I walked through the glass doors of the NSF building at 1800 G Street NW, past the blind cashier selling newspapers, snacks, and coffee, and entered the elevator to John Yellen's office. I had first met him, a bearded, thirtyish, Harvard-trained archeologist, some years before over lunch at the Smithsonian. In mid-1976 I had tried to get Wendorf to recruit him to work on our younger sites in the Middle Awash but had received a strong rebuke from Fred, who said he had his own plans for those sites. I assumed Yellen had learned of my inquiry through a mutual contact and had figured out its results.

Before an audience of 25 or so NSF staff, I gave the best presentation I could, which was pretty bad, considering I had not slept in two days. I had dark circles under my eyes, I was nervous, and I rambled. I began by showing slides of letters written to me by Ethiopian ministers and other officials who authorized or endorsed the work of the RVRME. Next I presented a brief explanation of our organization and purpose. I then spent the rest of the presentation showing views of a number of Ethiopian prehistoric sites where large collections of fossils and artifacts had been recovered by NSF-financed teams—the Omo, the Lake District, the Hararghe escarpment, the central Afar, and multiple shots of Hadar—followed by views of sites in the Middle Awash that promised to expand these collections greatly. I showed slides of exquisite fossil skulls, skeletons, and arrays of bones and stone tools. Finally, I showed views of the Ethiopian National Museum and the deteriorating collections, whose "intrinsic value far exceeds the million dollars they cost the American taxpayer."

The only question I recall at the end of the presentation came from someone in NSF's International Office, who asked how NSF could in good conscience fund a laboratory in a country with a brutal dictator who slaughters students. "It would be like investing in a similar project in Vietnam or Cambodia." I replied that even in those countries cultural treasures require preservation.

Afterwards, Yellen led me to the office of the assistant director of NSF for the Directorate of Biological, Behavioral and Social Sciences, Eloise Clark, to discuss the declined proposals. Even before I sat down, she pointed to a thick stack of file folders on her desk and said that there was "no record in the files" that the CIA question played any role in the evaluation of the proposals, nor was there "any record" that the matter was discussed in the panel meetings. A tall, trim woman wearing gray slacks, she ended her monolog by saying, "There was no basis to reconsider any of the proposals."

Obviously she had better things to do with her time, and that was that.

I later learned that in a memorandum to the assistant director written before our meeting, Yellen had reviewed the correspondence "in the files" between Isaac, Gonzalez, and Wendorf, all relating to the CIA issue, as well as Desmond Clark's review of the Patterson proposal, but had recommended that NSF "skirt around what must be a very thorny issue."[51] Yellen said that in any case it would be unwise to fund field work in the Afar at this time because of the political situation in Ethiopia; however, he could "envision specific projects which could be workable." He elaborated:

> Basically, the further one goes from Addis the more unsafe it becomes; the nearer one gets to Somalia or Eriteria [sic] the more dangerous it becomes. We gave Desmond Clark permission to undertake his last field season in Ethiopia this coming Spring. Several weeks ago he decided it would be unwise to carry out further field work. Based on the above criteria Clark's area is significantly safer than Kalb's. I implicitly trust Clark's judgment and think it unwise to fund work . . . in the Awash triangle.

Once again Yellen's faith in Clark was moving, perhaps even spiritual; however, over the previous six weeks newspapers in the United States had been full of stories describing Ethiopia's preparations for a massive, Soviet-backed counteroffensive against the Somalis across Hararghe and Bale Provinces (see Map II), fueled by the largest airlift of military hardware in Africa's history. In an about-face, the Soviets had turned on its former ally Somalia deciding that Ethiopia, with its position along the Red Sea, was a bigger prize to be won.

As we walked back to the elevators after our meeting with the assistant director, I told Yellen that perhaps I should pursue my inquiry about the CIA matter through the Freedom of Information Act. His response was that I would probably be wasting my time, that records of declined proposals are routinely destroyed by "the great shredder in the sky."

About the storage facility, the assistant director had told me the State Department was firmly against the idea. I was skeptical. I knew that those with the embassy in Addis Ababa entertained no such notion. To pursue this inquiry, I was able to meet with Congressman Paul Tsongas of Massachusetts, who had served in the Peace Corps in Ethiopia's Sidamo Province in the early 1960s and spoke Amharic. It was late afternoon, almost dark outside, and freezing, with a low cloud cover visible through the windows of Tsongas's office. On his walls were pictures of Ethiopia, including one of Haile Sellassie. The congressman had just flown in from Boston, where it was snowing, and wore a rumpled dark-blue business suit and oiled, leather hiking boots, which he propped up on his desk as we talked.

I briefly summarized the status of prehistory research in Ethiopia, some of which he knew about, and then described the deplorable state of the collections "paid for by the American taxpayer." He called in a secretary to take notes. Unmoved by my rhetoric, he responded that it did not matter who had paid for the collections: Something should be done about them and he would look into it.

We then talked about the situation in Ethiopia, and he told me that he and another congressman had just returned from a fact-finding trip to Addis Ababa a few weeks earlier. He said they had prepared a report on the trip that they were giving to President Carter the next day, and would I read it and give him my opinions? Sure. His secretary fetched the report and showed me to a desk in a corner of the office, where I sat for the next hour, filling the margins of the report with comments.

I found the report packed with insight and amazingly moderate in tone, despite current Ethiopia–U.S. relations.[52] Yes, the congressmen had seen atrocities during their brief visit—bodies in the street with bound hands, others displayed in a public square—but the report urged patience with Ethiopia's revolutionary course and respect for its nationalism and its right to defend itself against invasion. It predicted that once the Somalis were driven out of the Ogaden, Ethiopia's alliance with the Soviets would soon end, and friendly ties with the United States would resume; the report also predicted that Ethiopian victory in Eritrea was "unlikely." A rare interview with Colonel Mengistu, whom the congressman described as "thoughtful and determined, but unstable," produced these lines by the leader:

> Ethiopia is endowed with great natural resources, but the rivers and lakes [under Haile Sellassie] were left to wild animals and did not serve the people. "Ethiopia," he said, "is rich enough to have three harvests a year and enough minerals for her needs and a surplus to export. "What is in short supply," he said, "is trained manpower."[53]

The next day I met with the desk officer for Ethiopia at the State Department, who said that NSF had never spoken to him about housing scientific collections in

Addis Ababa, although Congressman Tsongas's office had done so that morning. The desk officer said he saw no reason not to support such a proposal, which in fact he thought was a symbolic way of maintaining ties with Ethiopia. At his suggestion, I then met with people at the Bureau of Educational and Cultural Affairs to discuss the training of Ethiopian students. The bureau had recently approved Tsrha Adefris a Fulbright scholarship to study physical anthropology at NYU under Clifford Jolly. I assured the bureau staff that even though the U.S. Information Service, which normally handles such matters, had been expelled from Ethiopia, the Ethiopians would approve more scholarships.

CHAPTER 24

Big-Game Feeders

In early February 1978, the Ethiopian counteroffensive against the Somalis opened with a two-pronged armored and infantry attack along the Djibouti railroad in the eastern highlands. Aided by Soviet advisors and four battalions of front-line Cuban troops, the "red army" ground its way into Somali-held territory throughout the month. The war became increasingly internationalized as Egypt, Iran, and Saudi Arabia reportedly gave arms to the Somalis, and—in the enigmatic politics of the Middle East—Israel and Muammar al-Qaddafi helped arm the Ethiopians.[1]

On March 2, 1978, the war opened on another front as Doug Cramer and I, armed with a bottle of arake and five Ethiopians, headed for the Middle Awash. By the time we reached Gewani in the late afternoon, we had driven through three military roadblocks, passed two troop convoys, and skirted around the former headquarters of the Trapp Construction Co. at Camp Arba, which was then filled with row on row of neatly lined tents and soldiers speaking a dozen African languages, as well as Russian and Spanish.

We had two scientific goals on this field trip, which was to last only two weeks. One goal was to survey carefully the area northwest of Gewani for fossils, and the second was to recover more fragments of the Bodo skull. The first goal, a geological objective, was to find fossil evidence to support my preconceived notion that sediments beneath the massive basalts at Sibabi correlated with those beneath the thick basalts in the Hatowie graben (see Map VII on page 140).

The bank of the Awash north of Gewani was flooded all the way to Wee-ee owing to the "small rains," so we camped at a higher elevation near Sibabi. We found no diagnostic fossils beneath the basalt, but we did find clouds of mosquitoes rising from the marsh at sundown, every one of which swarmed to the largest living object they could find, which was Cramer. Even though we slept under mosquito nets within mosquito nets for added protection, Doug still made his growling sounds in the night, periodically yelling out, "Pain, I love it!" There were lions in the area, but none dared venture into our camp.

On the fourth day we moved to Bodo. There we shared camp with a UN team led by a friend who was studying breeding colonies of quelea, grain-eating birds that were destroying sorghum crops on plantations in the Awash Valley. Resembling the common sparrow and called "Africa's feathered locusts," these migratory birds nest along the river and are a major source of cereal loss in sub-Saharan regions. Many African farmers are said to be more successful raising these birds than their agricultural crops.[2]

We began our work at Bodo by clearing all the grass, brush, and snakes that had reclaimed the hominid locality since our last visit in October 1976. On that occasion we had simply probed the top 10 centimeters of loose sediment for fragments quadrant by quadrant in our grid, but this time we systematically screened the sediment. Over the next week we successfully recovered 200 more bone fragments, many belonging to the skull. By the time we had finished screening the site, I was convinced that recovery of more of the skull, perhaps even of postcranial bones, was possible, but it would require a large, full-scale excavation of the locality to a deeper level.

By the time we packed up camp in midmonth, I had added to our faunal catalog by finding a clutch of fossil crocodile eggs and a beautifully preserved giant *Theropithecus* baboon cranium, missing only the canines, which came from the same layer as the *Homo* skull. When living, the baboon probably weighed on the order of 70 to 75 kilograms (about 170 pounds). Two giant terrestrial primates at the same location, one bipedal, must have made a curious dance card, particularly in that one probably would have eaten the other if given the chance. In 1977 Glynn Isaac suggested that the remains of scores of giant theropiths recovered from Ologresailie, Kenya, resulted from "periodic baboon hunts" by Acheulean men using "missile stones."[3] He guessed that the baboons were then butchered and presumably eaten. Anthropologist Pat Shipman and colleagues later studied the breakage patterns of the bones and agreed, adding that "stray individuals [mostly juveniles] were probably killed at irregular intervals and brought back to the campsite for butchering."[4] Archeologist Rick Potts of the Smithsonian, however, has argued that there is not a single "convincing" butchery or percussion mark on the bones and therefore no evidence of systematic butchery.[5] This suggests that either (1) the baboons were killed by natural causes or (2) they were killed in a manner that has not yet been detected. It could also be that the baboons were killed not only for their food value but simply because they posed a threat or were serious pests. This I can relate to. I recall an uncomfortable night camping in the Menegesha forest in the Shoa highlands with friends, surrounded by extremely aggressive *Papio anubis* baboons looking for food. These animals weigh only about one-third what male theropiths weighed, so I can imagine how intimidating these giants would be.[6]

For several reasons, it seems likely that Acheulean hunters would have added baboon protein supplements to their diet if given the opportunity. After all, humans have a long, colorful history of eating each other,[7] so what's the fuss about eating an oversized monkey? Baboons and chimpanzees are known to supplement their vegetable diet with insects and small vertebrates, and chimps will also eat the offspring of other chimps and of other monkeys; they would probably eat human infants if given the chance. We can assume that giant theropiths probably would have done the same.[8]

On March 16 we packed up at Bodo and were on the road by 6:30 A.M. I wanted to reach Addis by sundown and not be passing, after dark, roadblocks manned by soldiers looking to be promoted. The cool morning air was still hanging over the valley when we reached the marshlands south of Wee-ee. Sky-blue bee-eaters darted in front of us as we slowly convoyed along the water's edge; warthogs scattered into the underbrush; and oryx dashed into the hills. We startled a large herd of the magnificent waterbucks that surged away from us in long leaps through the marsh grass like a pod of dolphins, their thick, ringed horns curving backwards and upwards like pitchforks. Stocky, with a shaggy, reddish-brown coat and wiry mane, black lower legs, and white markings, the males can weigh over 230 kilograms.[9] They feed alternately on grassland and acacia pods, or on water grasses, reeds, and rushes. Waterbuck are never far from water and will seek refuge from their enemies in bush, high grasses, or water when threatened. An account from Kenya describes wild dogs chasing down a male while the cow and calf immersed themselves in water with only their noses exposed.[10]

Watching the herd, I concluded that were I to be reincarnated, this would be the animal of my choice: one that lives on the edge of two worlds. After we emerged from the solitude of the bush and headed back to the capital past more troop convoys, I realized this *was* my reincarnation.

By the time we returned to Addis Ababa, the decisive battles of the Ogaden counteroffensive were over. The Somalis were outmanned on the ground and outmaneuvered in the air by Ethiopia's superior air force, said to have been bolstered by Cuban pilots flying MIG fighters. The biggest battle, at Jijiga, was directed by a Soviet general; it resulted in the annihilation of four Somali brigades trapped between an Ethiopian–Cuban frontal assault and armored vehicles dropped behind the Somali position by giant helicopters.[11] Somali guerrillas continued hit-and-run raids in the months (and years) to come, but their strength was severely eroded.

By April 1978 there were 17,000 or more Cuban troops in the country.[12] Despite their large numbers, they were rarely seen in Addis Ababa. One night, however, I lost my way on the outskirts of the city when Judy and I were going to a party at the home of a friend with one of the embassies. When I found what I thought might be the right house with a number of vehicles parked out front and loud music coming from within, I left Judy in the car and went to check. After I had knocked on the metal gate, an Ethiopian guard armed with a machine gun appeared—not unusual for embassy functions—soon followed by several more armed guards. I told them I was looking for the party of such-and-such a person, and they simply waved me toward the house, where the beat of the music was decidedly Latin. Through the windows I could see the house was full of young men in colorful shirts that hung loose at the waist, who were dancing and carrying on with a dozen or so Ethiopian girls. Bottles of whiskey and beer sat on the tables, and cigarette smoke filled the room. It was too late to turn back, so I knocked on the door. A man came over and motioned me inside. Playing it safe, I addressed him in Amharic, then in English, but he understood neither, so I spoke to him in Spanish.

I told him I was lost and asked if I could use his telephone, if possible . . .

"¿Es usted Norte Americano?"

"Sí." So much for my accent.

"¿Los Estados Unidos?"

"Sí."

"Venga, venga, por favor," he said, waving me into the house. He was friendly, and my appearance drew the attention of several others standing nearby. They were all clean-shaven and athletic and looked to be in their thirties. I guessed they were junior military officers or functionaries of similar status. They also seemed a bit tipsy, at the least, and asked me where I was from in the United States.

"Tejas."

"¿Tejas?"

"Sí."

"¿Galveston, Tejas?"

"No, Houston."

I doubted that they were aware that many of the arms smuggled to Fidel Castro in the late 1950s, when he was fighting the Batista regime, had originated from the Texas Gulf Coast. I was one of many young people at the time who had romanticized the Cuban revolution. You can imagine my pique when in March 1977 Fidel was the first head of state to visit Ethiopia after the revolution and I was not invited to a banquet, or even a game of darts.

They offered me a drink, but I explained that my wife was waiting in the car. Another time. I used the telephone and got the right directions, and then thanked my hosts and shook hands. As I walked out of the door, I turned and said, "Viva la revolución!"

"¿Como no?" Why not? one of them replied.

Cramer remained in Addis Ababa for ten days, reconstructing more of the Bodo cranium, while giving casting lessons to museum staff members and making a new cast of the cranium to take back to New York. All together, we recovered some 96 smaller skull fragments in 1976 and 1978 to accompany the two large pieces of the face. Twenty-six fragments had been restored to the cranium in 1976. After laying out the remaining fragments on a well-lighted Formica worktable in the Paleobiology Research Laboratory, we were able to restore 15 more fragments to the cranium. Much of the skull vault was now present, revealing a full, rounded cranium that flared sharply outward at the face and was thickest at the top of the skull. Many of the remaining fragments came from the rear and basal cranium and represented a significant portion of the braincase, but critical pieces were missing, and further reconstruction was not possible. The fossil was nevertheless complete enough for a preliminary description, which Conroy, Jolly, Cramer, and I submitted to *Nature* two months later (and which was published later in the year).[13]

In 1978 there were five recognized groups of crania of fossil *Homo*, ranging from 1.9 million years ago to the present, which for convenience I refer to as Groups A–E (see Appendix I on page 319).[14] The most primitive and earliest, which make up Group A, are crania assigned to *Homo habilis*, closely related forms, the 1470 cranium, and other fossils from Turkana now referred to as the species *H. rudolfensis*. Group B, referred to as *H. erectus*, comprises a diverse collection of crania found in Asia (including "Java Man" and "Peking Man") and at Olduvai. More primitive forms that share features with *erectus*, known from Turkana and South Africa, are referred to as *H. ergaster*. Skipping to Group D, we have the Neanderthals, first discovered in Europe in the mid-nineteenth century and named variously over the years as an independent species of *Homo* or as a subspecies of *H. sapiens*. The last group, Group E, consists of modern humans.

Group C includes crania that are considered to be "transitional" in morphology between *erectus* and the Neanderthals. The most famous of these fossils is "Rhodesian Man," found in a cave in 1921 at Broken Hill, Rhodesia (Zambia). Other crania similar to this fossil are known from Petrolona (Greece), Arago (France), and Bodo. In the nomenclature widely adopted in the 1970s, the Group C fossils are referred to by many anthropologists as "early archaic" *Homo sapiens*. Their successors (the Neanderthals and present-day humans) are called "late archaic" and "modern" *Homo sapiens*, respectively.

Some anthropologists, in comparing Bodo to the Broken Hill cranium, refer to both as the subspecies *H. sapiens rhodesiensis*. In recent years, the trend has been to call the Group C hominids, Bodo included, *H. heidelbergensis* and the Neanderthals (Group D) *H. neanderthalensis*, even though these names originate from incomplete fossils.[15] The differences in nomenclature employed to distinguish one of these species, or subspecies, from another reflect fundamental differences of opinion among anthropologists over how much variation between forms warrants a taxonomic distinction and how much variation exists between sexes of the same species.

In terms of evolutionary relationships—and accepting on faith the current species definitions—we can now roughly divide the genus *Homo* into just three groups: "early" *Homo* (*rudolfensis* and *habilis*), "intermediate" *Homo* (*ergaster, erectus, heidelbergensis,* and *neanderthalensis*), and "modern" *Homo* (*sapiens*). The two end members, early and modern *Homo,* are more gracile than the intermediate forms and have the smallest and largest relative brain sizes, respectively, of the genus. The more robust, intermediate forms (including our original Group C) are distinguished by their larger relative body mass compared to early and modern *Homo.* Using cranial morphology to predict body mass, anthropologist John Kappelman of the University of Texas, suggests that African *heidelbergensis,* from Broken Hill and Bodo, may have weighed 100 kilograms (220 pounds) or more, which could make them the largest fossil hominids on record, or at least among the largest hominids of the intermediate *Homo* kind.[16] The figure for Bodo accompanies a new estimate by Glenn Conroy for its absolute brain size, 1313 cubic centimeters, the largest for the known crania commonly assigned to *heidelbergensis.*[17] Overall, it is apparent that the robustness and massiveness of these hominids represent the extreme of a trend that began with *ergaster* in the late Pliocene (c. 1.8 million years ago) and ended with the Neanderthals in the late Pleistocene (c. 40,000 years ago). With the appearance of *Homo sapiens,* came the decrease of relative body mass and the dramatic increase of relative brain size that characterize our species.[16]

The transition in the later Pleistocene from robust to gracile forms, from a large body mass to a smaller body mass, and from a moderate brain size to the very large brains of modern humans may have been due to various factors. First, it is very hard not to associate large body size and increased bone density with "big-game feeders,"[18] such as the Bodo people, who were able to transport core boulders and to wield slabs of cut stone to butcher animals much larger than themselves. In Africa at least, it is apparent that big-game consumption lasted for the duration of the Acheulean period, which coincided with the emergence of *erectus* (or *ergaster*) in the late Pliocene and peaked with *heidelbergensis* in the middle Pleistocene (c. 500,000 to 600,000 years ago). By the late Pleistocene (c. 100,000 to 125,000 years ago), perhaps with the appearance of the Andaleeans and their equivalents, major climatic shifts and faunal turnovers forced humans to use a more diversified toolkit to pursue a more diversified diet. Extreme robustness and body mass as a specialized asset for survival thus became secondary to increased relative brain size, a more gracile body, advanced tool use, and resourcefulness. Kappelman argues that it is likely that sophisticated group foraging, perhaps kin-based, and rapid language development also coincided with advances in technology and increased intelligence.[19]

Thus, what appears to be unique about "Bodo Man" is that because of his physique, strength, and skills, he represented the apogee of the killing and dismembering abilities of the Acheulean hunter and meateater. If Bodo was in fact cannibalized, as White suggests, then he was not only the hunter but also the prey. Or, as White proposes, "it is possible that the original Bodo individual was a victim of interpersonal conflict whose cranium was prepared as a meal, a meal and a trophy, or a trophy only. Alternatively, the specimen could represent the remains of a venerated ancestor who was ritually cannibalized or curated, or both."[20] White cautions that, excepting "verified human gnaw-marks on human bone," confirmation of cannibalism must rely on indirect evidence such as cut marks, breakage, or burned bone. D. Carl has also weighed in on this subject, suggesting that the defleshing and butchery of Bodo may have proceeded "before his death,"[21] adding sadism and mutilation to the potential list of Acheulean meat-carving activities. Surely this could be worthy of another headline:

FAMOUS ANTHROPOLOGIST PROCLAIMS EARLY MAN EATEN ALIVE

In view of the relationships that White and Johanson had with their colleagues, as these relationships would come to be known over the years, we can conclude that both were well versed on the subjects of "interpersonal conflicts." Added to this list would be fratricide, or "cannibalism": The two would eventually expend considerable energy trying to consume one another. But by mid-1978—following their encounter in Nice two years earlier after White's blowup with Richard Leakey— the bonding that developed between Johanson and White was a match made in heaven, both for their careers and, many would agree, for science. This was made possible by Johanson's role as caretaker of the Hadar hominids and by White's study of the hominids from Laetoli, Mary Leakey's prized site in Tanzania. Despite the distance between the two sites, White was the first to recognize that the Hadar and Laetoli hominids were remarkably similar.[22]

Following the burst of hominid discoveries by the IARE in late 1974, Johanson identified three hominid species at Hadar. In the March 1976 issue of *Nature,* he compared the jaws found by Alemayehu to *Homo,* Lucy to the gracile *Australopithecus africanus,* and a partial femur to *Australopithecus robustus* (see Appendix I).[23] Then, in the May 1977 issue of *Nature,* he also assigned the "First Family" to *Homo.*[24] In both papers he and his colleagues estimated the age of "*Homo*" from Hadar to be close to 3 million years. As we have seen, the identifications of early *Homo* fit well with the ideas of Richard and Mary Leakey, faithful to Louis's cherished belief in the great antiquity of humans. Early *Homo* fossils from Hadar also bolstered Richard's (contested) age of 2.6 million years for the 1470 *Homo* skull from Turkana (now dated at 1.9 million years old) and White's published descriptions of the Laetoli hominids, also assigned to *Homo.*[25] Because the Laetoli fossils were dated at about 3.7 million years, a 3-million-year-old "*Homo*" from Hadar argued for a 2.6- to 3.7-million-year time frame for the earliest humans.

There was more.

In early 1978 Mary announced the discovery of what proved to be the capstone of her career, a remarkable trail of human footprints found at Laetoli on the surface of a volcanic tuff.[26] The impressions came from the same beds as the hominid fossils and, following more excavations, suggested that the footprints were made by an adult male and female and a child.[27] One report even suggested the female may have been pregnant.[28] Thus, by the spring of 1978, the post-Louis years for the Leakeys, although not without their problems, were constructing a grand edifice of human evolution that would have warmed the old man's heart—except that Johanson and White soon provided convincing evidence that it was all a house of cards.

Following months of collaborative study on the Hadar and Laetoli hominids, the two first concluded, on the basis of the similarity and primitive nature of the fossils, that they all belonged to the same genus, *Australopithecus*. Furthermore, even though they were found 1700 kilometers apart, with an age difference of more than half a million years, the hominids were nearly identical and therefore had to belong to a single species. Previous identifications by Johanson of multiple species at Hadar were now attributed to age and sex differences between individuals. The diverse collection of males and females and infants and adults recovered from L 333 (the "First Family") provided ample data on size, sex, and age variation within a single population. In many ways the Hadar/Laetoli hominids resembled *Australopithecus africanus* from southern Africa, but because the former exhibited even more primitive and ape-like features than *africanus*, the Hadar/Laetoili hominids were placed in a new species, which Johanson and White named *Australopithecus afarensis* after the Afar Depression. In other features the fossils presaged *Homo* and suggested a close ancestral relationship to *Homo habilis* and to modern humans; therefore, Johanson and White argued, *afarensis* was *the* common ancestor of all other australopithecines and humankind.[29]

According to their provocative new model of human evolution, *Homo* diverged from this common ancestor about 2.6 million years ago (Appendix I); thus it was impossible for humans to have existed at Laetoli a million years earlier, as Mary Leakey still believed. Johanson and White also accepted an earlier date (1.8 million years ago) for the 1470 skull, thus implying that fossils of earliest *Homo* were yet to be found.

In 1978 Johanson and White, with Coppens as a third author, published a description of their new species, *H. afarensis,* in a bulletin of the Cleveland Museum of Natural History.[30] In early 1979 the two alone then published a follow-up article in *Science* on the broader implications of the new species to human evolution.[29] Their conclusions, although not without critics, were masterful and consistent with state-of-the-art knowledge of early hominids, from the cave sites of South Africa to the Afar lowlands. Among their detractors were the Leakeys. What drew the lasting enmity of Richard and Mary was not the conclusions reached by Johanson and White, but the manner in which they were presented. Although the name selected for the new species was *afarensis*, at Johanson's urging, the fossil chosen as the holotype—the reference specimen used to define the new species—was from Laetoli.[31] According to taxonomic convention, the holotype and its name, if that name is chosen for its geographical location, should have a common origin. Thus, if the Laetoli fossil, a partial mandible, was to be the holotype, the new species name should reflect where it was found, as would, for instance, "*laetoliensis*." Alternatively, the holotype could have been chosen from the much more complete collection of "*afarensis*" fossils from Hadar. Matched with comparable specimens from Hadar, the Laetoli mandible was a clearly inferior fossil, broken on both sides and missing anterior dentition.[31] In *Lucy* Johanson justified selecting the holotype from Laetoli, and the species name from the Afar, by saying this was the best way to tie fossils from the two sites together. Furthermore, "if future field work produced a whole avalanche of material from the Afar, people might lose sight of the fact that there was this other population [of hominids] in another part of Africa. . . . It might be overlooked."[32] I doubt that Mary Leakey viewed Johanson as the champion of the hominid underdog or was convinced that the Laetoli collection would be lost to obscurity without his heroic efforts. Rather, it is apparent that Mary interpreted Johanson's motives as an *in your face* taxonomic hijacking of one of her proudest discoveries.[33]

Not content to press his scientific differences with the Leakeys in scholarly journals, over the next few years Johanson repeatedly took his *afarensis* conclusions to the media, particularly to rub Richard's and Mary's noses in an upgraded model of human evolution that differed from their own.[34] Johanson's highly visible broadsides were tailor-made to cause contention, which no doubt enhanced his fundraising efforts to establish (in 1981) his own institute in Berkeley, the Institute of Human Origins.[35] In the course of his self-promotion, the Leakeys endured what they felt were calculated insults and damaging "rumors" aimed at undermining their accomplishments, professionalism, and good name. This is not to say that the Leakeys were shrinking violets, but, as Richard observed, paleoanthropology had stooped to a new low.[36] The breach that followed between the Leakeys and Johanson was irreparable. The break with White, who owed his career largely to the Leakeys, followed.

CHAPTER 25
National Security

At the end of March 1978, Cramer returned to New York, to the relief of the U.S. embassy in Addis Ababa. He had taken to drinking with the U.S. Marine guards during their off hours in the late afternoon; the parties convened at the Cottage, a bar and restaurant not far from our office. On several occasions Doug was the only one left standing, and the Marine commandant I spoke with early one morning insisted he was a security threat. I suggested that the real person to blame for such off-campus activities was Jimmy Carter, whose abstinence from spirits had been adopted by our embassy. I remember a reception given at the embassy shortly after President Carter took office. Only tea, soda pop, and coffee were offered, and the outraged diplomatic corps fled even before the chocolate cake was served.

By early 1978, Judy and I had been in Ethiopia for seven years and were considered something of a novelty. Not many Americans had remained in the country that long, and most of those who had were missionaries and teachers. There was an African-American, however, who according to embassy legend, showed up one day seeking help on the renewal of his visa. No one there had ever heard of him, but he gained instant fame when his passport revealed that he had been in Ethiopia for 12 years on 3-month tourist visa. He lived in the western part of the country, apparently a model member of the community, running a prosperous farm that supported his wife and many children.

Many foreigners had left the country by choice, had been expelled, or had had their businesses nationalized. Others—especially those serving with embassies, international organizations or companies—simply departed at the end of their tours leaving their positions unfilled. Many of the people who came for two-, three-, or four-year tours became our friends: We raised our children together, went to Entoto on picnics, organized campouts, or held volleyball games. In 1978, the fourth year of the revolution, one favorite meeting spot was the geothermally heated pool of the Hilton Hotel. With tourism at a lowpoint, the hotel offered pool memberships at cut-rate prices. I recall East Germans—said to be security advisors to the Derg—huddled in conversation in chest-deep water, no doubt with firsthand knowledge that the pool was one of the few meeting places in Addis Ababa that could not be bugged.

As the revolution progressed, friendships with Ethiopians became fewer, or at least less open. Many feared being accused of consorting with "undesirable foreign elements," particularly following the campaign of terror against dissidents that began in late 1977 and continued into 1978. The government-imposed segregation that resulted brought the diminished foreign community even closer together. Wives called one another to see where eggs or sugar was available in the city as the stocks of stores declined, or to find out which medical doctors were still in the country and had not been impressed into the military. When gunfire erupted at night, we would call each other to see what parts of the city were off limits or simply to check on one another's well-being. It would seem that such violence would force everyone to stay home at night, but the need to reduce tensions in the "stricken capital," as one journalist put it, was greater, or so we rationalized. Parties seemed to spring up at the worst times. A lot of drinking went on and probably more cross-marital relationships than usual, which were always a source of well-deserved gossip.

Then there were the unique moments brought on by the particular circumstances of urban warfare intersecting with the humdrum of everyday life. One night after a period of relative calm, Judy and I left the children with our housekeeper and went to a movie at the National Cinema in the center of the city. The film was *Waterloo*, with Rod Steiger playing a brooding Napoleon. It was a long movie, sparsely attended. About two-thirds of the way through, there was a loud explosion outside, and the film went dead, followed by absolute darkness.

Probably at one of the nearby ministries, I told Judy. Perhaps Defense.

We sat there.

Machine gun fire erupted. Not Defense, I thought, much too close.

After a few minutes, the film sputtered back to life. It was a battle scene with eighteenth-century cannons roaring away, interspersed with musket fire . . . and machine gun fire. We had no idea who was winning what battle, all the more so because in the confusion the projectionist mixed up the reels, and suddenly people killed earlier in the movie reappeared alive and well. To this day I could not tell you whether the British or the French were victorious at Waterloo.

On another occasion the municipal power plant was bombed and much of the city, including our house, lost its electricity. I was scheduled to go into the field the next morning. I had my entire crew sleep at the office overnight so that we could leave immediately after the curfew was lifted, at 5:00 A.M. I had a problem, however: My wristwatch was broken, and the only clock in the house was electric. All night long I kept waking up and looking out the window for the first light. Finally, I decided it must be about 4:30 A.M., so I called one of the large hotels to confirm the time. An obviously groggy desk clerk answered the phone.

"Could you please tell me what time it is?" I asked.

"Pardon. Please tell you *what* . . . ?"

"My clock is not working. Could you tell me what *time* it is?"

"Wait."

After a series of clicks the phone started ringing elsewhere in the hotel. After ten or so rings a husky voice finally answered, "Hallo."

"Yes, can you please tell me what time it is?"

"What tha . . . ?" a half-awake man answered.

"Yes, can you please give me the correct time?"

There was a long pause, followed by banging and crashing sounds. Finally, the voice came back on and said—in an emphatically British accent—"It's three-*bloody* thirty in the morning!"

"Thank you. Thank you *very* much."

The desk clerk had apparently connected me with one of the hotel guests.

From April into July, much of my time was spent trying to secure more scholarships. Because Tsrha's enrollment at New York University demonstrated that the Fulbright program could still function in the country despite the closing of the U.S. cultural affairs office, we were offered two more scholarships. One was for Dejene to pursue at the University of Pittsburgh a specially tailored graduate program in science administration in developing countries, and the other was for Sleshi to study vertebrate palaeontology at the University of Texas at Austin. After approval of these scholarships by the Ethiopian Commission for Higher Education, we secured two more: one for Assefa to study systematics, ecology, and vertebrate paleontology at the University of Kansas, and the other for Mesfin Asnake to study vertebrate paleontology, specializing in microvertebrates, at the University of Washington. Each of these scholarships took months of paperwork, meetings, and presentations with government offices. At one point I drew up a flowchart that outlined 33 steps required for the approval of just one scholarship, showing the trail of documents from office to office. Along the way files were lost, minutes of meetings remained untyped, officials were replaced or arrested, a signature was lacking, or offices were closed for demonstrations, political orientation sessions, or funerals. In truth, most of these delays were understandable and logical, given the events taking place in the country, and those concerned invariably went out of their way to be helpful, even at some risk.

In view of the political problems I had experienced with NSF, there was an added urgency to these efforts. It was apparent that the RVRME could not compete with the vested-interest network that I was certain permeated the foundation, so we had to seek other means to carry out our research. Just as Tsrha would study the Bodo skull for her Ph.D. dissertation, Sleshi would study the Middle Awash suids for his, Assefa the proboscideans, Mesfin the rodents, and Dejene the interaction (and funding) of scientific research and development in the Third World. In order to fill other roles, I employed three more assistants who worked alternately in the Paleobiology Research Laboratory, our offices, and the museum. I also convinced the Fulbright program to put up the funds for four new teaching positions at Addis Ababa University in the fields of paleontology, paleoanthropology, human anatomy, and geology to help the science faculty plan its first-ever master's degree program.

All of this was essential, but we still lacked money for our current operation, and I was going deeper and deeper into debt. I took delivery of the Land Rover acquired with my Wenner-Gren grant, and more small grants were in the works, but major funding was months off, at the least. I was also aware that the problems generated by the NSF spyfest the previous year were still very real. The latest rumor—passed on to me gleefully by an American busybody working as an architectural historian at the Ministry of Culture—was that the Institute of International Education of New York, which administers the Fulbright program, "had CIA connections." I also knew that Desmond Clark still had designs on the Middle Awash, and I assumed that this continued to involve Berhanu Abebe and NSF's John Yellen.

As more and more of each day came to be occupied with administrative matters, and as pressure mounted to find money for research and field work, I spent less and less time with my family and became insensitive to the problems of those around me. At times I felt I was being swallowed up by Africa. I remember going home one evening and finding the house and verandah awash in Ethiopians. There were our two housekeepers and four of their children; Kelati's widow and child, whom I was then supporting; the soldier living in our basement, who was somebody's nephew; our zebanya and his wife and child; our 80-year-old part-time gardener; and our neighbor's children. As I walked in the front door, I yelled to Judy, "Get all these goddamn Ethiopians out of here!" And there was the time that I stormed into the office one morning and complained to Liz that I was "sick of having that beggar woman and her child sit by our gate *every* morning, and I am not going to give her another goddamn cent!" She always had a plastic bag wrapped around her head for a hat and had her hand held out. "It's irritating. Surely she can find some kind of a job!"

Liz quietly pointed out that the woman had leprosy.

One afternoon Menda, still our maintenance man at the office as well as our field cook, came to me to say that his three-year-old son was ill. Menda and his wife and child lived at the office in an outbuilding (along with his wife's nephew and an orphan adopted by a New Zealand geologist who had then left the country). I told one of my staff to look into Menda's son's problem and thought nothing more of it. Such matters occurred almost daily. Then early the next morning, some time not long after midnight, there was a loud knock at the front door of my home. Now what? I jumped out of bed.

Middle-of-the-night emergencies were not uncommon. One night our zebanya was drunk and threatened to kill his wife with a butcher knife. Another night a different zebanya's wife was having a breach birth and I rushed her to the hospital, curfew or no curfew. Then there were the ubiquitous thieves and shootings. But this particular night it was Menda. He stood on our front steps in the moonlight with his head hanging and his arms wrapped around himself to tell me that his son was *yemote*—dead. "*Yemote?*" I exclaimed. I threw on some clothes, and together we ran to the office. His son, a fine, handsome boy with thick black, curly hair, was laid out on a cot illuminated by a lantern. His mother sat beside him moaning, rocking back and forth. The boy had no pulse, and his forehead was cold as the early morning air. Menda told me he had died several hours earlier. There was nothing to be done. I have no idea what he died of, unless it was asthma. The funeral, held at the office with much wailing, singing, and dancing—the custom of the Oromo—lasted three days.

With the Somali hold on the Ogaden broken, in the spring of 1978 the Soviet-backed Ethiopian army and air force launched a major counteroffensive against the Eritreans, who then controlled much of the north. Ferocious fighting continued into the summer, with heavy military and civilian casualties in and around Asmara. Although the Ethiopians regained much lost ground, the Eritreans built heavy fortifications at Nafta, in the mountains north of Asmara, and survived the onslaught.[1] Despite the large number of Ethiopian casualties suffered during this brutal and protracted war, one never, or rarely, saw a wounded soldier on the streets of Addis Ababa. Rumor was that amputees and the severely scarred were kept in special houses in the capital, or in camps, so that the civilian population would not be demoralized. It was also reported that thousands of political malcontents were now being shipped off to "indoctrination camps" in the countryside.[2] Despite these cosmetic attempts to alter the image of the military government, in late June Addis Ababa radio reported, oddly, that in the last ten months there had been nine attempts to assassinate Colonel Mengistu.[3] Perhaps he was being portrayed as invincible.

In mid-July word reached me that Berhanu was calling the U.S. embassy, asking it to pressure the Ministry of Foreign Affairs to expedite an entry visa for Clark. Also, Berhanu was pestering me with an uncharacteristic interest in a catalog of the Bodo skull fragments collected during our most recent field trip. In a meeting with the minister of culture over other matters, I expressed my concern that Berhanu might again attempt to interfere with the RVRME permit. The minister assured me that it was safe but advised me that "rumors" were once again floating around the ministry. He advised that I keep a "low profile."

At the time I was keeping anything but a low profile, because I was making frequent trips to the U.S. embassy on Fulbright matters and my days were filled with appointments with Ethiopian officials. Then one Sunday, at the request of the chargé of the embassy, a friend, I arranged a lunch at an Ethiopian restaurant for the newly appointed U.S. ambassador and the Canadian ambassador, whose tour was ending. The restaurant I chose was outdoors, a scenic spot on the top of a hill at the edge of town. We arrived conspicuously in two embassy vehicles, one a black limousine flying the American flag. I recall the new ambassador, Frederick Chapin, a career diplomat, as a modest man. Judy asked him if his appointment signified new policies toward Ethiopia. He responded, "I *am* the new policy"—meaning that the embassy had functioned for four years without an ambassador, and his appointment was intended to open new dialog between Washington and Addis Ababa.[4]

On July 19, Judy left for six weeks in England to attend courses in country dancing while collecting materials for teaching dance to children. She had raised the money for the trip from the dance lessons she gave in our home, and from an advance on her pending salary from the International School (formerly the American School), where she would teach on her return. Judy was concerned that something might happen in her absence. I assured her that "everything would be okay" and that I would pay special attention to the girls.

In early August 1978 I was served notice by the Department of Public Security that I was being expelled from the country. I was given six days to settle my affairs. The lieutenant colonel who gave me the order offered no explanation, but he did instruct me to return to the National Museum all the "antiquities"that I had taken to the university for study.

"Antiquities?" I asked.

"Yes, the ancient relics from our historic past."

I thought about asking him if he had heard of Claude Lévi-Strauss, but I knew the answer.

Until the collections were returned a few days later, the Paleobiology Research Laboratory was locked and sealed with an official notice pasted across the door frame. When Assefa and I later stripped the fossils from the tables and specimen drawers, it took us only a few hours to undo what it had taken us months to assemble. Most of the larger specimens were suid and elephant molars, but there were also hundreds of rodent teeth and jaws painstakingly glued to pinheads and arranged on foam rubber pads to facilitate study under a microscope. Following delivery of the collections to the front door of the museum, I was given a letter to present to Public Security confirming the deposition of the collections.[5] The fossils were then taken to the same tin shed behind the museum that contained the rest of our collections and the discarded plaster and bronze busts of Haile Sellassie. I wondered whether the closing of the laboratory was the first time in history that the study of evolutionary biology had been treated as a threat to national security.

On my last day in Ethiopia I gave generous severance pay to all my employees, using the proceeds from the quick sale of the two Toyota pickup trucks, and then spent the rest of the day and evening with the girls, saying good-bye to friends who came by our house. The next morning we boarded the plane for England, where we would join Judy. The plane lifted off, banked over the capital, and flew northeast toward San'a, Yemen, to pick up more passengers. Our route took us straight over the Awash Valley, a flight pattern I knew well. With Justine and Spring settled in their seats reading magazines, I looked out the window. I could see it all. It was roughly in the reverse order of that which I had first seen on the ground, like a movie reel played backwards: Ananu, Hatowie, Sagantoli, Aramis, Wee-ee, Belo-dolie, Maka, Matabaietu, Wilti Dora, Bodo, Hargufia, Meadura, Andalee, Talalak, Meshellu, Hadar, Geraru, Ledi, Kariyu, Tendaho, Lake Abhé . . .

Soon we passed over the Red Sea.

CHAPTER 26

African Cake

K ing Leopold II of Belgium, born in 1835, inherited his colonial ambitions
from his father, Leopold I.[1] Like those of his father, Leopold II's colonial
efforts were repeatedly frustrated. While other European powers, the British especially, expanded their empires throughout the world, Leopold II unsuccessfully
sought to establish colonies as far afield as the Caribbean, Borneo, and the Fiji
Islands. By the mid-1870s, however, his attention was focused on Central Africa,
stimulated by reports of the explorations of Livingstone, Stanley, and Verney Cameron. The king was particularly interested in the reports of Cameron, an Englishman,
who in 1875 was the first European to cross the continent from east to west. He
described "unspeakable richness" in the Congo basin: gold, ivory, coal, palm oil,
and rubber trees waiting to be tapped.[2] Leopold saw his future.

To pursue his African dream, the king organized the Geographical Conference
of Brussels in 1876, inviting experts to discuss the means to "open up to civilization the only part of the globe which it has not yet penetrated."[3] Leopold spoke
eloquently of explorers and scientists moving across the African countryside unfettered by nationalistic boundaries."[4] Privately he wanted to be sure Belgium got its
"share in this magnificent African cake."[5] To implement his scheme, the king urged
the conference delegates to establish an advisory body to oversee the exploration
enterprise.

Two weeks after I was expelled, on August 21, 1978, Desmond Clark arrived in Ethiopia and soon obtained Berhanu's blessing to work in the Middle Awash, an area that
promised unspeakable prehistoric richness. Clark wrote Yellen to say that he hoped
to seize the Bodo sites "before some Eastern bloc scientists take them over."[6] On his
return to the United States, Clark again wrote Yellen, this time urgently requesting
funds to begin field work in the Middle Awash as soon as possible: ". . . it is very desir-

able that American interests in paleoanthropological research there should not be jeopardized."[7] Yellen urged his superiors at NSF to approve Clark's request for financial support, because it was in "the interests of U.S. science that American researchers" maintain access to the "extremely important" sites in the Middle Awash.[8] Berhanu, however, was abruptly removed from his position as head of antiquities and assigned to a less influential position in the Ministry of Culture, and the new head of antiquities canceled Clark's plans to work in the Middle Awash.[9]

Clark wrote Yellen suggesting that the reason for this was that Maurice Taieb was trying to regain his original concession to the area.[10] In a memo to the deputy director of NSF, Yellen concluded that "the question of who would have the rights in the Middle Awash had not been settled."[11]

In May 1979 Clark laid the groundwork for a French–American scientific conference to be held in Addis Ababa at the end of the year. The goal would be to plan for the "coordination and integration of policy for future research projects in the Afar."[12] He wrote Taieb that "in the interests of the international and interdisciplinary groups" they should share the Middle Awash unfettered by "the kind of restrictions that rigid geographical study areas impose."[13] He proposed that an "advisory committee," composed of themselves, be formed to help the Ministry of Culture plan future field work in the Afar. Yellen justified using NSF expense money for the American representatives to the conference because they would be "acting for all American paleoanthropologists and will not [themselves] have exclusive U.S. rights" to the sites.[14]

Leopold's obsession with the Congo basin intensified after Stanley completed his second expedition into Central Africa in September 1877. After finding Livingstone in 1872 and then learning of the great explorer's death the following year, Stanley was determined to complete his mentor's life's work by tracing the source of the Congo. The feat, which Stanley and his Zanzibari porters and riflemen accomplished in 999 days, revealed the vast extent of the enormous waterway in the heart of the continent, opening up unlimited opportunities for exploitation in Leopold's visionary mind. The king soon replaced his "civilizing" objectives with more direct commercial goals and secretly began raising funds from investors, which led to the formation of the International Association of the Congo. The IAC was in fact neither international nor an association: It was a front to promote Leopold's monopolistic ambitions, while supposedly serving as a trading company that promised to open the Congo up to all countries. To help legitimize his plans for "free trade" in Central Africa, Leopold relied on an admirer and confidant, Henry Sanford, who was well connected in Washington and managed to persuade the U.S. Senate to recognize the "philanthropic" work of the IAC. The king considered such formal recognition essential to thwart the intentions of the French, who had territorial designs of their own on the lower Congo. The French interest was an outgrowth of the prodding of Pierre de Brazza, an Italian-born Frenchman who in 1880 explored the basin nearest the Atlantic and sought claims on the upper Congo on behalf of France. There he clashed directly with Stanley, who was then working for Leopold to establish trading stations along the route of his previous explorations.

In September 1979, Clark submitted a grant application to NSF entitled "Proposal for a Joint Seminar on the Organization of Research into Human Origins in East Africa with Special Reference to the Afar Sector of the Ethiopian Rift Valley."[15] In no sense did Clark plan on a *seminar* in Addis Ababa at the end of the year, nor was the proposal subjected to the scrutiny of NSF's peer-review system. Rather, the Addis Ababa meeting was organized to allow the Americans access to the Middle Awash sites discovered by the RVRME; the competitive review of Clark's grant application was waived because, as Yellen stressed to his superiors, the riches of the Middle Awash were "up for grabs."[16] All together, Yellen would approve four grant requests for Clark leading up to the Addis Ababa meeting, all of which were "secret" because none was subjected to peer review.[17] In late November 1979, Clark wrote Yellen to add urgency to his request for the meeting, passing on rumors that there were "some Japanese people now in Ethiopia picking up fossils(!)"[18]

By late 1884 the race between Leopold and the French for "territorial rights" to the Congo basin was in high gear, as was the scramble by other European powers for more pieces of the continent. The Germans coveted parts of western Africa, the British sought the lower Niger and Egypt, and the Portuguese also laid claim to parts of the Congo. To resolve these issues, on November 15, 1884, the Berlin West Africa Conference was convened at the behest of the German chancellor, Otto von Bismarck, who would later remark, "No one, not even the most malevolent democrat, has any idea how much nullity and charlatanism there is in this diplomacy."[19]

Although Leopold's International Association of the Congo was not officially represented at the conference because it was not a state, and the king himself was not present, his surrogate, Henry Sanford, and others attended on his behalf. By continuing to claim that the IAC would guarantee free trade in the Congo, and by masterfully playing off the interests of the European powers against one another, Leopold won international recognition for the association and its territory, while France, Germany, Portugal, and Britain gained lesser, though still important, concessions of their own.

The Berlin Act, signed by the conference delegates on February 26, 1885, combined the commercial interests of the Great Powers with their "civilizing" concerns. The central provisions of the *entente cordiale* were as follows:

1. An international commission should be established to oversee navigation of the Congo waterways.
2. The Congo and all its outlets should be a free-trade zone.
3. All powers should improve native conditions, suppress the slave trade, protect all scientific institutions organized to achieve these ends, and instruct the natives in the "the blessings of civilization."
4. The IAC and the territories claimed by it should be recognized as a sovereign state.[20]

The "Joint Seminar" organized to determine "territorial rights" in the Middle Awash Valley was convened at the Ministry of Culture on January 5, 1980. The meetings were nominally chaired by Berhanu Abebe, even though he was no longer head of antiquities. Clark, Johanson, and Yellen represented the American interests, Taieb and Raymonde Bonnefille the French. Taieb still hoped to see the Middle Awash reinstated to his original concession; Clark wanted his own concession on the Hararghe plateau extended to include the Middle Awash; and we can assume that Johanson wanted access to the Middle Awash any way he could get it.[21]

On January 10, after six days of maneuvering and haggling at the Ministry and in hotel rooms, a jointly written document was completed. Entitled "Recommendations to the Ministry of Culture and Sports on the Organization of Research in the Ethiopian Rift and Adjacent Areas,"[22] it was signed by all except Clark, who by then had left the country, Yellen signed as his surrogate. The "Recommendations," later described by Clark as an *"entente cordiale,"* carefully portioned out the respective interests of the competing parties.[23] Among the central provisions of the document were the following:

1. A "Working Group" should be established to help the Ministry of Culture develop a "coherent research program." The group would consist of five Americans and five Frenchmen.
2. Research "should not be constrained by definite geographic areas or concessions"; rather, scientists should be allowed to "move freely over the landscape."
3. Ethiopians should be trained and incorporated into the research. The Working Group would provide "some direct support to accomplish these goals."
4. The most immediate research priority of the Working Group should be "the area between Hadar and Gewani . . . bounded by the Awash-Assab road and the Western Ethiopian Escarpment"—the Middle Awash.[24]

Although Leopold's IAC gained the international recognition that it needed to be a sovereign state, Britain refused to accept the actual boundaries of the territory staked out by the association. When the details of Leopold's claims were delivered to the

British Foreign Office, however, a clerk mistakenly acknowledged the document by return mail, a "stupid blunder" that in diplomatic terms amounted to an irreversible acceptance of the king's frontiers.[25] As Leopold's biographer Barbara Emerson has noted, the king thus became the "absolute ruler of a country of over 1,000,000 square miles."[26] The name that Leopold gave his newly won kingdom, the "Congo Free State," was one of the great euphemisms of European colonization, as time would tell. On a grander scale, Leopold's extraordinary achievement was a victory over French claims to the Congo and greatly accelerated the partition of Africa.

Although Clark was to be a full beneficiary of the Addis Ababa accord, which would give him unfettered access to the Middle Awash, by the terms of the *entente cordiale* he agreed to share the prize with Johanson and the French. As fate would have it, however, a "stupid blunder" would give Clark control of the entire area. On visiting the National Museum at the end of the meetings, Johanson was accused of attempting to steal a hominid fossil from the collections and was ordered to leave the country.[27] Whatever the truth of the matter, he lost his chance to work in the Middle Awash, as did Taieb, because he too was somehow involved in the altercation that followed. Their permit to Hadar was also revoked.[28]

Clark was now heir apparent to the Middle Awash, an area he described as "probably the richest anywhere in the world for discovering the crucial evidence for the emergence of man."[29] On his return to Washington from the "seminar," Yellen described the Middle Awash to his superiors as having "enormous paleoanthropological potential."[30] The efforts of the two would have been the envy of King Leopold. Once again the French were out of the picture, and the only thing left to do was to devour the magnificent cake.

From London the girls and I traveled by train to the University of Durham to meet Judy, where she was taking her courses. Justine and Spring, now eight and four, had their first university experience: They slept in a dormitory and received ample attention from the "other" students. Justine took recorder lessons and on talent night played "Go Tell Aunt Rhody."

Judy was shaken by the turn of events; Ethiopia had been our home for more than seven years. As it happened, we lost most of the belongings we had accumulated. The local militia confiscated all the RVRME office and field equipment, much of the latter belonging to Clifford Jolly. With the help of the U.S. embassy and the U.S. Geological Survey, I did get most of my data out of the country, but not of course the fossil collections, hundreds of aerial photographs, and a number of blown-up site photos, all heavily marked with critical data.

From Durham we went to Bath, west of London, where the Romans had built a city of pleasure around the only hot springs in England. With its ancient ruins and stately homes, statues, and exquisite gardens, Bath must have been a curious place for Haile Sellassie and his family during the four years they lived there in exile. However, after fleeing Ethiopia before the advancing Italian armies in May 1936, the Emperor felt isolated and miserable in England; he hated the gloomy winters, made barely tolerable by the curative waters.[31] During the second winter, he was injured in an automobile accident, became seriously ill, and fell increasingly short of money. The worst indignity, however, was watching the British government give in to pressures to recognize Italy's annexation of Ethiopia.[32] Nevertheless, he remained in constant communication with partisan exile groups in the Sudan, and in June 1940 his fortunes turned when Italy declared war on the Allies. Suddenly, the British saw the liberation of Ethiopia as essential to the security of the Suez Canal.[33] A year later the Emperor was back in Addis Ababa, a hero.

Bath was also the stage of one of the great tragedies, and lingering mysteries, surrounding African exploration: the shooting death of John Hanning Speke, a British army officer turned explorer. The story began in 1857–1858 during Richard Burton and Speke's second expedition together in Africa.[34] The two had barely escaped with their lives in their first expedition to Somaliland in 1855, when they were savagely attacked by Issa. (Burton was speared through the face, and Speke was severely beaten and stabbed.) The second expedition took them across eastern Africa in search of the source of the White Nile, one of the great remaining scientific plums in world geography. Burton was particularly intrigued by the work of two German missionaries and respected explorers, Johann Rebmann and Johann Krapf, who reported stories from African and Arab sources about the existence of large "inland seas," lakes, south of the equator.[35]

From Zanzibar, Burton and Speke spent eight grueling months reaching the shores of the one of these "seas," Lake Tanganyika. The lake was itself a great discovery, but they found no evidence that it drained northward into the Nile. On their march back to the coast, Burton, seriously ill, called a halt to gain strength for the final push to the Indian Ocean. Speke took the opportunity, with Burton's mixed blessing, to travel north to investigate accounts of another large lake nearby. After a 23-day trek, Speke discovered "a vast expanse of . . . pale-blue waters," which he immediately assumed was the source of the White Nile. He named the lake in honor of Queen Victoria.[36]

On their return to England in March 1859, Burton stopped over in Aden to convalesce, while Speke continued on to London. Both agreed to wait for Burton's return to England before reporting the results of their explorations to the Royal Geographical Society, which had awarded Burton a grant for the expedition. Instead of waiting for Burton's return, however, and in collusion with the ambitious president of the RGS, Speke addressed a packed audience of the society, announcing that he—and he alone—had discovered the source of the White Nile. When Burton learned of Speke's perfidy, he challenged his subaltern's character and the nature of his claim, but the RGS ignored Burton and awarded Speke a grant to return to Africa the following year to confirm the discovery.

In 1860 and 1861, accompanied by James Grant, Speke mapped more of the lake, but his work was hasty and left unanswered questions about the river's source. Once again Burton challenged his findings. By then Speke had proved to be not just outspoken but also intolerant of criticism, and he had many detractors in British scientific circles. Both his report to the RGS and his best-selling popular accounts of his explorations were heavily criticized for lack of scholarship. Speke was deeply wounded by the treatment he received.[37] To help settle the issue of the Nile's source, the RGS—now in collusion with Burton—suggested that he and the beleaguered Speke debate their differences in a public forum, which arrangement Speke was pressured to accept as a matter of honor. The affair was to be held at Bath in the meeting hall of the Mineral Water Hospital on September 15, 1864. Many predicted that the brilliant Burton would humiliate and destroy Speke, who was by nature withdrawn and had a reputation for making reckless statements.[38]

The day before the scheduled "Nile Duel," Speke, an expert hunter, shot and killed himself while hunting partridges on his uncle's estate. An inquest concluded that it was an accident, although Burton and many subsequent commentators suspected suicide. It is also possible that Speke simply attempted to wound himself— the gunshot pierced his left side—to avert the public confrontation with Burton and gain sympathy. The tragedy is still debated.[39]

In 1874–1875 Stanley circumnavigated Lake Victoria and carefully mapped its northern outlet. He confirmed what Speke had said all along: The lake was indeed the source of the White Nile.

CHAPTER 27

The CIA Lives

O n returning to the United States from Addis Ababa in January 1980, Johanson and Clark applied to the NSF anthropology program for $570,000 to conduct a three-year research program in the Middle Awash and Hadar areas.[1] Johanson applied for an additional $58,000 to "stimulate scientific cooperation" between the French and Americans in the Awash Valley.[2] As he described in the proposal:

> In the history of the search for human origins much of the early research has been typified by an intensely individualistic and/or nationalistic flavor. Such a short-sighted and narrow approach has not always been conducive for the best scientific results.

> Nationalistic attitudes about the proprietorship of fossil-bearing deposits has sometimes permitted only scientists of a particular country to actually participate in fieldwork. The shortcomings of such an approach are obvious.[3]

In mid-1980, NSF awarded Johanson and Clark $259,000 with field work scheduled to begin in the fall.[4] Because Johanson and Taieb's field permit had been revoked six months earlier, however, apparently as a result of the "stupid blunder" at the National Museum, no field season took place.[5] Instead, much of Johanson's time was spent basking in the publicity generated by the publication of *Lucy*. In reviews of the best-seller, it was generally stated that field work in the Afar had been halted because of the "political situation" in Ethiopia, but which political situation was never made clear.[6]

In the last chapter of *Lucy,* Johanson described a three-day visit he and others had made to the Middle Awash after the Addis Ababa summit. He referred to his plans, in collaboration with Tim White, then at the University of California at Berkeley, to explore the older deposits in the region, saying, "What we find in them could well blow the roof off of everything."[7]

However, in 1981 the Ministry of Culture once again denied Johanson and Taieb permission to work in Ethiopia.[8] This time Clark applied for the permit on his own and got it.[9] Helping matters was another $70,000 awarded to Clark from Yellen's program, this amount for the construction of a laboratory/collection-storage facility at the National Museum.[10] Not surprisingly, the collaboration between the Americans and the French, Coppens included, never materialized, nor did the Wendorf family ever see the sun rise at Bodo. Instead, White became co-director of the Berkeley team that spent a full field season in the Middle Awash in the fall of 1981.[11] On his return from the field, Clark wrote Yellen, "It was by far the most successful field season in all my forty-odd years of work in Africa."[12]

Among their discoveries was a proximal hominid femur from Maka dated at 3.4 million years, a frontal cranium fragment from Belodolie dated at 3.9 million years, and a parietal cranium fragment of a second hominid found at Bodo.[11] The two older fossils were identified as *afarensis*. The leg bone provided additional evidence that *afarensis* walked upright, and the 3.9-million-year-old frontal extended the time range of the species by another 150,000 years (beyond its lower age of 3.75 million years at Laetoli, Tanzania) (see Appendix I on page 319). *Afarensis* was now known in the fossil record for a period of at least 700,000 years.

With the chancellor of the University of California beaming at their side, Clark and White announced their finds at a packed press conference in Berkeley on June 10, 1982. Their press release described the Middle Awash area as "the most important yet discovered for human origins studies" and their 1981 field season as "the first extensive survey of the [area]. In only two months the scientists had hit upon rich finds of prehistoric tool use extending back as far as 1.5 million years." Also described was "an almost continuous fossil record of many animal species also spanning the past six million years."[12]* Sitting on the table in front of Clark and White were the hominid fragments from Maka and Belodolie, and the Bodo skull, which had been taken to Berkeley six months earlier. White used the occasion to announce the presence of cut marks on the skull that were indicative of human defleshing.

The discoveries were widely reported in the press but were perhaps best summarized in *California Monthly,* a publication of the University of California. It noted that Clark "has searched the length of Africa for 42 years, studying man's evolution; but never before had he captured such a major prize."[15]

---

*Prior to their 1981 field work, Clark and White had available to them a considerable body of RVRME materials, including published reports and maps,[13] data contained in at least our declined NSF proposal, and apparently even the numerous unpublished reports that the RVRME had submitted to the Ministry of Culture.[14]

From England my family and I eventually made our way to Austin, Texas, where in the fall of 1978, I obtained a nonsalaried position as a museum research associate in the Vertebrate Paleontology Laboratory of the University of Texas. Over the next few years I co-authored ten papers with my former RVRME colleagues on the geology, paleontology, and archeology of the Middle Awash. Two of the papers appeared back-to-back in *Nature* in 1982, with a picture of the Bodo skull on the cover.[16] We described four new formations in the region, two below the Hadar Formation, late Miocene to early Pliocene in age, and two above the Hadar Formation, late Pliocene and Pleistocene in age (Appendix III). Accompanying a geological map of the area, we provided a detailed account of the fossils and artifacts in the different units, and their ages. I gave the London editor the final changes to the text over the telephone at three in the morning, Austin time. I had the galley proofs spread across the dining room table in the quiet of my home while hearing the distinct sounds of London's morning traffic in the background. I made some mistakes in my haste to get the papers completed, but they were published in July— just in time to reach Ethiopia before Clark and White's scheduled second field season in the Middle Awash.

In mid-June 1982 Sleshi Tebedge, then at the University of Texas on a Fulbright scholarship, returned to Ethiopia, using modest funds he had raised from several foundations. After going to the United States in mid-1978 (shortly before my expulsion), he had completed his master's thesis on the Middle Awash suids, using data and photographs he had taken with him.[17] He hoped to conduct his Ph.D. dissertation research on Bodo fossils. His office was next to mine in the Vertebrate Paleontology Laboratory, and we talked daily about his work and future plans. At noon we also ran the π-route (3.14 miles) through the woods next to the lab. Except for the deer and the occasional rattlesnake, the setting was not very different from the thickets next to the Awash, and the Texas summer was ample reminder of the Afar heat.

In Ethiopia Sleshi was joined by his graduate supervisor, paleontologist Ernest Lundelius Jr., a crusty native Texan specializing in Pleistocene faunas, who would accompany him to Bodo to see the area firsthand. Also joining them would be Tsrha Adefris and Clifford Jolly, who was still her graduate supervisor at New York University. Tsrha had completed her master's degree and returned to Ethiopia, where she hoped to begin her Ph.D. dissertation research on the Bodo skull. When she arrived in Addis Ababa, however, she learned that White and Clark had taken the skull to Berkeley, ostensibly so that an Ethiopian then studying with them could use it as a "training fossil."[18]

After a brief visit to the Middle Awash in mid-July, Lundelius and Jolly returned to the United States, while Sleshi and Tsrha remained in Ethiopia. At that time the two renewed a petition they had submitted to the Ethiopian government the previous year. It asked that the Middle Awash be set aside as a scientific reserve for Ethiopian scholars and students. The salient passage read

We plead with the Ethiopian Government to consider our request and its implications on Ethiopian development and pride. Set aside something for Ethiopians and Ethiopians alone. It is our country and we deserve this right. Give us and other Ethiopians the opportunity to complete our training and to return to Ethiopia to train others and to develop our institutions. At that time we can work with the Ministry of Culture, the National Museum, and Addis Ababa University in developing programs that are fully responsive to Ethiopia's development and prestige.[19]

Sleshi and Tsrha were joined in the petition by two other former RVRME associates—Assefa Mebrate, then at the University of Kansas, who was writing his master's thesis on Middle Awash elephants, and Mesfin Asnake, at the University of Washington, who was studying fossil rodents.

Clark and White and their team began assembling in Addis Ababa in late August 1982 for the fall field season, having received an additional $470,500 from NSF.[20] Johanson's permit was finally renewed by the Ministry of Culture, for Hadar only, so he too would be returning to the field on NSF money. This time, however, the Hadar work was to be conducted under the sponsorship of his newly established Institute of Human Origins in Berkeley. The IARE was long since history; Taieb was then working in Tanzania.

Just before the two Berkeley teams left for the field in mid-September, however, the Ministry of Culture informed Clark, White, and Johanson that their field permit was suspended. The reason given was Ethiopia's need to re-examine its policies regulating prehistory research.[21]

Over the next few weeks all hell and damnation broke loose in Addis Ababa. The three anthropologists used every means at their disposal to pressure and intimidate Ethiopian officials into reversing the decision. They singled out the commissioner for science and technology, a former supporter of the RVRME, as most responsible for the suspension. Clark and White's Ethiopian student circulated a letter to government officials laced with innuendo about the commissioner's support of "Kalb's CIA activities."[22] All together, the Berkeley group solicited the aid of six embassies, NSF, and the chancellor of the University of California, and they sent letters describing their plight to Colonel Mengistu, the Council of Ministers, the Ministry of State and Public Security, the Ministry of Foreign Affairs, the Ministry of Mines, the commissioner of higher education, the commissioner of tourism, and the president of Addis Ababa University.[23] Jolly later remarked, "What, they did not write the president of India?"

The three anthropologists overplayed their hand. In early October the central government, the Derg, issued a nationwide proclamation ordering the suspension of *all* prehistory research in the country.[24]

On their return to the United States, Clark and White wrote a lengthy postmortem report for NSF, giving the reasons for the suspension of prehistory research.[25]

1. There is recognition that students and staff of the Addis Ababa University need to be better integrated into palaeoanthropological and archaeological research. . . . The Ministry is and has been for nearly two years without a Minister and is, therefore, in a weak position and certain interests have seized this opportunity to try to obtain complete revision of procedures relating to antiquities. This power struggle could result in the University's taking over responsibility for research of this kind. . . .

2. Foreign expeditions have often been perceived as "ripping off" the Ethiopians, removing fossils and contributing little or nothing to local institutions or scholars. These accusations were most often leveled against the French. The Middle Awash team has taken major steps to correct this problem by building a laboratory with NSF funding. . . .

3. The action . . . in suspending research seems to have been precipitated by the actions of a group now based in the U.S. Jon Kalb . . . was expelled by Ethiopian Government security in 1978. The suspicions aroused by his activities continue to play a very negative role in relations with the Ethiopians: paleoanthropologists are mistrusted by officials. After his expulsion, Kalb arranged for several . . . undergraduates to come to the United States to train. . . . Two of these students [Sleshi and Tsrha] . . . returned to Ethiopia in the summer of 1982 and engaged in a campaign of propaganda against the Johanson and Clark/White expeditions. They contacted governmental and university officials and poisoned the atmosphere with a welter of false and ridiculous accusations. . . . Most regrettably it appears as if this campaign was carefully orchestrated by American "researchers" [Lundelius and Jolly] posing as advisors to these students. The students were sent by their advisors to perform "research."

At the end of October, the *Economist* reported on the moratorium, quoting from unidentified sources, "By common consent it was deemed to be all the fault of the French, who had put the government's back up by their secrecy in the field and their habit of publishing their findings, sometimes years late, in obscure journals rarely found in Addis Ababa."[26] Clark and White suggested to journalists that I had been expelled because I had violated Ethiopian security and that the Ethiopians must have known what they were doing.[27] Johanson was more explicit, claiming that my alleged CIA connection was raised with him again and again by Ethiopian officials, even by the Swedish ambassador.[28] In the *New Scientist,* Johanson allowed that "Kalb was obviously a CIA agent."[29] Appearing on National Public Radio with White, he elaborated: "Until this last visit, this last year 1982, that a number of officials said that he was expelled with 48 hours notice because he was a security risk to the country. And some people went so far as to say that it was almost certain that he was a CIA . . ., or that they thought he was a CIA, agent. And that is the first time I heard those allegations."[30]

It is hard to say which offense was the most serious in the eyes and ears of the Unidentified Sources and Johanson: the suspicions aroused by my "activities" or the French habit of publishing articles in French journals. But Johanson found more evil afoot before he left Addis Ababa in the fall of 1982. He discovered that his Land Rover, purchased with NSF money, which had been parked in a storage shed at the French embassy since 1980, had been stripped and looted. As he described the matter in an outraged letter to Taieb and several French officials, missing from the vehicle were the differential, U-bolts, two rear axles, drive shaft, distributor, battery, one rear mainspring, one shock absorber, three tire rims, oil cap, mirrors, horn, brake lights, light switch, and the dashboard. Johanson offered no explanation why they had overlooked the fuel pump.[31]

CHAPTER 28

Prehistoric Feast

In June 1982 I wrote to Vice President George Bush, asking for his assistance with NSF.[1] Because he was a former director of the CIA, I hoped my request would spark a special interest. Specifically, I wanted the foundation to publicly acknowledge the role that the CIA rumors had played in the evaluation of the three RVRME-related proposals in 1976–1977. Also, I hoped to see revealed the circumstances in which the rumors had arisen, which I thought would finally lay the matter to rest. On a broader plane, I hoped an inquiry might lead to a re-evaluation of NSF's peer-review system, which I felt was overly secretive and inherently corrupting.

In response to a request by Bush's office, NSF reinvestigated the affair but said there was "no evidence" that the CIA story had had any significant influence on the rejection of the proposals. The foundation also claimed that the files of the SMU proposal had been "routinely destroyed"—but not those of the NYU and Harvard proposals.[2] Later NSF would claim that the files of the Harvard proposal had also mysteriously disappeared.

Between late 1982 and early 1985, I submitted 115 requests to NSF for records under the Freedom of Information Act. In early 1986 I summarized my findings in two reports, which I submitted to the U.S. Congress's House Committee on Science and Technology,[3] the committee responsible for investigating how scientific research is selected for funding by the federal government. I described what I felt was a fundamental lack of due process in NSF's peer-review system and proposed a step-by-step examination of the legality of its procedures.

At the end of 1986 I then filed a lawsuit against NSF, claiming that the agency violated the Privacy Act and other federal laws by denying me the opportunity to challenge the veracity of derogatory information used in its decision making.[4] I was represented by Public Citizen of Washington, D.C., an organization that specializes in open-government issues. A year later I won a court-stipulated settlement in which NSF finally admitted that the CIA charges had indeed played a major role in the evaluation of the RVRME proposals. NSF also publicly apologized for its own part in spreading the charges and paid my legal fees and costs. Most important, NSF established a precedent by allowing me the right, prior to any decision on a proposal, to rebut damaging allegations that might arise during the review of any future applications I submit to NSF.[5]

Seven months later, in mid-1988, NSF implicitly agreed to give *all* grant applicants such rights by acknowledging for the first time that its peer-review records were subject to the Privacy Act, which it previously had denied. In doing so, NSF also implicitly acknowledged that since passage of the act in 1974, the foundation had unlawfully used exemptions to the Freedom of Information Act to selectively withhold supplementary peer-review records from applicants and, in the process, had maintained a secret filing system on tens of thousands of American scientists. These conclusions were supported by NSF's own inspector general.[6]

In 1989 the Ethiopian government finally lifted the moratorium on prehistory research put into effect in late 1982.[7] Concurrently, it announced a new proclamation approving guidelines to regulate the study and protection of antiquities.[8] The preamble of the proclamation read, in part,

WHEREAS antiquities constitute the imprints of a people's age-old way of life, labour and creativity . . .

WHEREAS antiquities make a major and a universal contribution to the development of science . . .

WHEREAS antiquities play a major role in imbuing the working people with a spirit of national pride and love for the Motherland commensurate with the span of their history and the profundity of their culture . . .

The document then formalized procedures regulating the use of antiquities in promoting science and education; the removal of antiquities from sites; the export of fossils, artifacts, or other antiquities; the purchase or sale of antiquities; the repatriation, protection, ownership, and registration of antiquities; and the all-important granting of research permits. Regarding the duties of permit holders, one provision required "the participation and training of Ethiopians in the exploration and research of antiquities."[8]

Although significant numbers of Ethiopians would eventually be trained by groups working in the country, the new regulations did little else to promote Ethiopia-based prehistory research, other than reaffirming that all matters antiquarian fell under the authority of the Ministry of Culture. Foreign-based expeditions would still be running the show, and Addis Ababa University would play virtually no operational role. In this respect Clark and White were correct in their report to NSF in late 1982: There had been an effort to involve the university directly in the sponsorship of prehistory research, and part of the reason for the moratorium was to thrash out this issue.[9] But the Ministry of Culture, with considerable support from the Berkeley team, won that power struggle. As a result, the ministry reissued the field permit to Clark and White to work in the Middle Awash in 1990, once again giving them exclusive control of nearly 4000 square kilometers of some of the most important and richest prehistoric sites in the world. (The area extends from Gewani north to the Talalak River/Wallo border on both sides of the Awash; see Map VII on page 140).[10] The difference in the significance of the permit when it was given to them and when it was issued to me 15 years earlier was that in 1975, no one knew what was *in* the Middle Awash. In 1990, Clark and White certainly understood there were enough sites in the region to keep 20 teams busy for 200 years.

Johanson once again lost out on the prehistoric feast to be had in the Middle Awash, and this set the stage for a behind-the-scenes battle with Tim White and others at the University of California at Berkeley that would severely alter D. Carl's career. Nonetheless, because the French had completely disappeared from the radar screen, Johanson and his Institute of Human Origins (IHO) were now the sole proprietors of the still fabulously rich Hadar area. To supplement the spectacular fossils recovered from the site from 1973 to 1976, a new crop of hominids would have eroded to the surface just waiting to be reaped.

In late 1990 the IHO team discovered at Hadar 18 more hominid bones, teeth, and jaws, which represented 15 individuals and expanded the sample of fossils assigned to *afarensis*.[11] By the end of the 1993 field season, 35 more specimens had been added to the *afarensis* inventory. The most dramatic of the finds by the IHO team, now led by anthropologist Bill Kimbel, was a partial skull of an adult male from the upper levels of Hadar. Its great size—it was reported to be the largest *Australopithecus* yet found—contrasted with smaller, fragmentary skull remains belonging to the female of the species, like Lucy.[12]

Another great find, in November 1994, came from the uppermost levels of Hadar in beds unconformably overlying the main Pliocene sequence. There the IHO team, again led by Kimbel, found an upper jaw of *Homo* that was dated at 2.3 million years (Appendix I).[13] Among the very oldest human remains known, the maxilla is close in age to a cranium fragment of *Homo* found in Kenya and to a mandible from Malawi, both dated at about 2.4 million years.[14] All three fossils are believed to fall near the split of the *Homo* lineage from *Australopithecus*, but more complete skull material is required for identification to species level (as *H. habilis*, *H. rudolfensis*, or possibly a new species).

What gives the Hadar *Homo* particularly significance is that the jaw was found with manufactured stone flakes and cores—the oldest association of human remains and artifacts known, and one that supports long-held views of a causal link between the appearance of our genus and stone tools. Fossils of other animals from the *Homo* locality indicate that humans living in the central Afar in the later Pliocene had adapted to a drier, more open habitat that contrasted with the wetter, more wooded environments of the underlying *afarensis*-bearing deposits. This shift in environment supports the view that human origins are also linked to the onset of global cooling in the later Pliocene (currently believed to be about 2.8 million years ago), which forced adaptations to drier, more demanding conditions.[15] The exact timing of this climatic shift and its relationship to the first appearance of humans and lithic technology are questions to be resolved.

Although the new discoveries at Hadar spelled a triumph for Johanson and his institute, much of 1994 was filled with controversy that pitted the University of California at Berkeley (UCB) and IHO-based teams in Ethiopia in a fight to the death. The origins of the conflict are apparently buried in the murky territorial intrigues among Johanson, White, and Clark over the "rights" to the Middle Awash carried over from the early 1980s. After the UCB reclaimed the entire permit in 1990, intense infighting developed between Johanson and White that involved Ethiopian officials and students (none formerly with the RVRME).[16]

Giving White a decided edge in this tragic bloodletting was Ann Getty, San Francisco-based arts patron and wife of billionaire Gordon Getty, heir to the John Paul Getty oil fortune.[17] Mrs. Getty was then taking courses from White at UCB and soon joined the Middle Awash team. She became an outspoken supporter of White and, during trips to Ethiopia (in her private Boeing 727), was in a position to make large donations to the National Museum, winning influence in the Ethiopian government. Worse for Johanson, Mr. Getty was the IHO's largest financial donor and a member of its board. In April 1994, he abruptly withdrew his support from the institute, charging Johanson with mismanagement of funds and personnel.[18] Because Getty was responsible for half of the IHO's nearly $2 million annual budget, his defection was widely reported in the press, from the *Wall Street Journal* to the *London Times*.[19] Adding substance to Getty's claims was a long-brewing schism in the 15-member IHO itself, including complaints that Johanson was more interested in popularizing science while promoting himself than he was in research. Dissatisfaction was especially strong among 10 members of the radiometric dating group of the IHO, who left the organization *en masse* following Getty's withdrawal of support. With a million-dollar start-up grant from Getty, the group immediately formed their own institute, the Berkeley Geochronology Center. In July, with an ever-widening circle of enemies surrounding Johanson, the IHO board voted to move the institute to a new location.[20] It is now located in Tempe, Arizona.

More controversy followed in 1995.[21] The affair surfaced at the March meeting of the American Association of Physical Anthropology in Oakland, California. An Ethiopian graduate student at Rutgers University, Sileshi Semaw, openly accused Johanson's team of stealing hominid teeth from an area he was excavating for his doctorate, an area he claimed was producing the oldest artifacts in the world.* The larger site, Gona, was the same one to which Johanson had guided

---

*Apparently the teeth came from the higher levels of the site and are believed to be in the time range of *H. erectus*. Evidently the fossils remain unstudied in the National Museum.[22]

Jack Harris in 1977, over the protests of a French archeologist working in the area. Harris had moved on to Rutgers from the UCB after receiving his Ph.D. and had become Semaw's graduate supervisor. At the Oakland meeting, Semaw was joined in his complaints by Berhane Asfaw—a member of Clark and White's team since 1982—who warned that Ethiopians would not tolerate "neocolonialist paleoanthropologists."[21]

Semaw's concerns about Gona were understandable, because the site had in fact produced the oldest stone tools known, dated securely at between 2.5 and 2.6 million years at the prospering Berkeley Geochronology Center (Appendix I).[23] The presence of artifacts at Gona, and the IHO discovery of artifacts and the *Homo* jaw dated at 2.3 million years at upper Hadar, suggest that the first appearance of humans and stone tool manufacturing is still at least several hundred thousand years earlier than present evidence allows. Semaw and his colleagues cite the sophisticated understanding of rock breakage that characterizes the Gona technology, implying that the hominids living there 2.5 to 2.6 million years ago "were not novices to lithic technology," and thus "even older artifacts will be found."[24]

While the IHO crowd was alternating between the higher reaches of Hadar and Gona and bad press, the Clark and White team was once again harvesting the Middle Awash sites. In 1990 they retrieved a partial hominid arm bone from the Oldowan-bearing Matabaietu beds and then found skull fragments in equivalent beds at Gemeda.[25] The fossils were of indeterminate identification but were dated at about 2.5 million years. In 1996–1997, the hominid hunting was even more productive: The team discovered a partial skull of a gracile australopithecine, cut bone, and Oldowan tools in comparable deposits in the western Awash. The find was made at Buri from deposits beneath the Acheulean-bearing Dakanihyalo beds.* The skull is distinguished from *A. africanus* and early *Homo* by having a small cranium and a projecting ape-like face. It is similar to *afarensis* but differs by having enlarged molars and flattened, human-like premolars. For these and other reasons, the UCB team named the fossil as a new species, *A. garhi* (which means "surprise" in Afar). In the same deposits were found long, human-like leg bones and long, ape-like fore-

---

*In 1982 I assigned the Matabaietu and Gemeda beds to the Matabaietu Formation (and dated it, using fauna, at 2.0 to 2.4 million years) (Appendix III); in 1999 the UCB team assigned equivalent beds at Buri to the newly designated Bouri Formation.[26] In 1993 I estimated the age of the Dakanihyalo beds to be 1 million years, as the UCB did in 1999; however, I placed these beds in the Acheulean-bearing lower Wehaietu Formation (*Wehaietu* is the Afar name for the Awash River), whereas the UCB included the Dakanihyalo beds and the Oldowan-bearing and Middle Stone Age–bearing units all within the Bouri Formation.[27] UCB spellings of place names also sometimes differ from those of the RVRME. Thus Buri, Hatowie, Gemeda, Belodolie, and Dakanihyalo in RVRME publications later become Bouri, Hatayae, Gemedah, Belohdelie, and Dakanihylo in UCB publications. Unfortunately, neither set of names comply with acceptable phonemic spellings of Afar names adopted by linguists.

arms that may belong to the same taxon. All together the evidence, including its 2.5-million-year age, suggests that the big-toothed hominid is a descendant of *afarensis* and possibly a direct ancestor of early *Homo* (Appendix I). Cut marks, chop marks, and hammerstone scars on associated animal bones, along with the presence of nearby Oldowan tools, support a close relationship of the Buri hominids to early humans. The scored bones, reportedly the oldest remains known of hominid butchery, offer compelling evidence that stone tool manufacturing may have originated not with *Homo* but with an immediate australopithecine ancestor.[25]

The UCB team had also renewed their hominid hunting at Bodo, finding a partial upper arm in 1990. In 1994 they dated a tuff layer at the site and concluded that the hominids and the associated Acheulean industry were about 600,000 years old. They reported Oldowan and Developed Oldowan tools from eastern Bodo, where Acheulean-bearing sediments are in fault-produced contact with Matabaietu-equivalent beds. The team then claimed that the Oldowan, Developed Oldowan, and Acheulean artifacts hominids were *all* about 600,000 years old and that the transition from the Oldowan to the Acheulean stone tool cultures in the Middle Awash took place nearly a million years later than the shift of these traditions elsewhere in Africa (Appendix I).[28]

At the time I left Ethiopia in mid-1978, Mengistu was at the peak of his power. The Somalis had been routed, the Ethiopians had regained lost ground in Eritrea, thousands of dissidents had been eliminated, and one opponent after another within the Derg had been killed. Of course, the military successes owed much to the USSR, but beginning in late 1979 the Soviets had their own Eritrea to deal with, in the mountains of Afghanistan, and before long their military aid to Ethiopia began drying up. When another devastating famine struck in 1984, it was predominantly the West that again came to the rescue.[29] Because the Mengistu government was preoccupied with celebrating the tenth anniversary of the revolution, the call for help was delayed, and tens of thousands of people perished needlessly. When the government tried to relieve the pressures on the limited crop production in the north by forcibly resettling 600,000 people in the south, thousands more died. The Afar area was again hit by the famine. At a lecture Johanson gave in early 1985 at Creighton University in Omaha, Nebraska—at the time, the moratorium was still enforced by the Ministry of Culture—he attributed the ban on field work at Hadar to the drought: "It is virtually impossible to work in that area. People are starving to death."[30]

By mid-1988 the Ethiopian army had suffered catastrophic defeats at the hands of the secessionist Eritreans and their Tigray neighbors, who sought the overthrow of Mengistu and power sharing in Addis Ababa.[31] In July of that year Mengistu went to Moscow, desperately seeking more military aid, just as Haile Sellassie had gone to Washington 15 years earlier, and with the same results.[32] Mengistu's request was denied in spite of his government's official conversion to communism in 1984 and its adoption of a Soviet-style constitution in 1987. What the Ethiopian leader could not have predicted was that the Soviet system of government was crumbling, despite the belated reform policies of perestroika and glasnost. By 1989 Soviet control of Eastern Europe was finished, and in 1990 the Berlin Wall was finally demolished. During this same period, the Ethiopian forces suffered crushing defeats from the then-united Tigray and Eritrean forces. By early 1989, Eritrea was essentially controlled by the rebels, and by September, Tigray Province was as well. To the south, Wallo and Gondar were overrun in 1990, Gojam and much of Shoa in the spring of 1991. By May Addis Ababa was completely surrounded by rebel armies, and on May 21 Mengistu fled to Zimbabwe. The conquering forces marched triumphantly into the capital on May 28, 1991.[33]

A year later I received an unsolicited letter from a newly appointed director of antiquities:

We all know that you started out as a pioneer of geological and paleoanthropological research in Ethiopia. You are now a distinguished scientist with significant contributions to the knowledge of mankind, particularly to our knowledge of Ethiopia. I am sure you will continue your illustrious career both abroad and in Ethiopia.[34]

The letter ended with an invitation to return to Ethiopia: "Please rest assured that you are always welcomed."

In follow-up telephone conversations the director gave me permission, subject to the necessary formalities, to work in the western Middle Awash. Since I had left Ethiopia 14 years earlier, no one had visited the area. The director assured me that the UCB team had permission to work in the eastern Middle Awash only.

Over the next six months I organized a large team of scientists that included six Ethiopians with advanced degrees. If it was expeditions the Ministry of Culture still wanted, I would give them a world-class expedition. In early December 1992, I sent the ministry an 18-page permit application.

A week later I learned that the UCB had just been granted permission to work in the western Awash. By long distance, the director of antiquities told me that he had been put under extreme pressure in ways that he "could not describe over the telephone. . . ." Tortured? Forced to watch Johanson's documentary, "How I Found Lucy"?

Of course I had been naive. There was no way I could compete with a team backed by a billionaire. Also, John Yellen, still program director of anthropology at NSF, had finally come out openly as a member of the UCB team.*

*Yellen was a member of the first team to survey the Middle Awash following the work of the RVRME, an opportunity arranged by Clark after the "seminar" held in Addis Ababa in early January 1979.[35]

To my surprise, six months later, in May 1993, I received a letter from the vice-minister of culture. "We all know that you started out as a pioneer of geological and paleontological research in Ethiopia. You are now a distinguished scientist. . . ."[36]

Strangely familiar words.

Once again the letter ended with an invitation to resume work in the country. By then I had learned that the UCB team had concentrated their 1992 field work in the lower Pliocene beds of the Central Awash Complex, apparently finding "sensational" hominids. This meant that the upper Miocene beds in the western foothills at Ananu and elsewhere still had not been mined. There were far fewer surface fossils there than in the central Awash, but I knew of localities to the west that would produce rich faunas upon excavation, and there were more areas to be surveyed.

In August 1993 I received another letter from the vice-minister, this time saying that my research application had been approved.[37] But no, in November I received a letter from the minister of culture, saying that Clark and White had been given the permit for the entire Middle Awash in 1990 and that their research was "still active."[38]

Obviously I am a slow learner, but this time I understood.

White's team did indeed make sensational hominid finds in the Central Awash Complex in late 1992, at Aramis (see Map VII on page 140). And a year later they recovered more hominids from the same locality. All together 17 specimens were found, mostly isolated teeth, including most of those of a single individual, but also jaw and cranial fragments and partial arm bones.[39] The fossils were dated at 4.4 million years: 500,000 years earlier than the oldest *afarensis* fossils found in the eastern Middle Awash.[40] The date also fell within the time period (4 to 6 million years ago) when australopithecines had long been predicted by molecular biologists to have diverged from the African apes. In White's announcement of the discovery, carried on the front page of the *New York Times* in September 1994, he and colleagues described the fossils as a new species of *Australopithecus,* which they named *ramidus* (from *ramid,* which means "root" in Afar).[41] An editorial in *Nature* described *ramidus* as *the* Missing Link that fills the morphological gap between the apes and all other known hominids.[42] Clark, though he was not directly part of the discovery, called *ramidus* the "find of the century."[43]

The hominids were recovered along the catchment of the Aramis stream. They came from the same fossiliferous sediments that the RVRME included (in 1982) in the Aramis Member of the Sagantole Formation (the latter named for the stream just south of Aramis) (see Appendix III on page 321).[30]* Faunal data and radio-

---

*A "member" is a stratigraphic subdivision of a formation.

metric dates show that the formation is pre-Hadar in age and spans a time period of more than a million years.

The first major hominid find at Aramis, discovered by the incomparable Alemayehu Asfaw, was a lower jaw fragment of a child possessing a single deciduous molar. White described the morphology of the tooth as more like that of a chimpanzee than that of a hominid, and by itself enough to signal a new species.[44] The thinness of the enamel of this and other *ramidus* teeth was also chimp-like and indicated that the early hominid lived on a soft diet, such as fruits and leaves. Other characters linked *ramidus* to later hominids, particularly *afarensis:* the relatively reduced canines, and a fragment of the basal cranium showing the downward position of the foramen magnum (the passageway for the spinal cord at the base of the skull), an indicator of some form of bipedalism. A partial forearm, reinforced with a bony ridge, bespoke a chimp-like dependence on arms when walking.

Overall, the features of *ramidus* immediately led experts to recognize a different and more primitive human ancestor. Its uniqueness was so apparent that in May 1995, White and colleagues published a 225-word "correction" to their original *Nature* description of the new hominid by placing it in a new genus, *Ardipithecus* (*ardi* means "earth" in Afar) (Appendix I).[45] They also reported that startling discoveries made at Aramis in late 1994, including a jaw and a skeleton, would offer major new insights about the new genus. As of this writing (April, 2000), descriptions of the new fossils still have not been published, apparently because of the effort devoted to excavating and describing the skeleton. However, reports have described the skeleton as 70 percent complete (30 percent more so than Lucy), with long, curved, ape-like finger bones.[46]

Hundreds of other vertebrate fossils were identified at Aramis, including mole rat, bush rat, squirrel, bat, mongoose, otter, bear, monkey, baboon, kudu, bovini, giraffe, horse, rhino, proboscideans, hippo, turtle, crocodile, and fish.[43] Also identified were thousands of *Canthium* seeds, a common plant in African woodlands and forests. Combined with colobine monkeys (30 percent of the total fauna), this evidence indicates that *ramidus* lived in a forested setting, which led White and colleagues to suggest that ideas linking human origins and bipedality with adaptations to open, savanna habitats require re-evaluation. The recovery of more hominids from other parts of the Sagantole section and the underlying Adu-Asa Formation (named by the RVRME for sediments in the Hatowie graben, the western foothills, and elsewhere) (Appendix III), all exhibiting a range of paleohabitats, will certainly elucidate such ideas further. According to a report from Addis Ababa in early 1999, such discoveries may have already been made by the UCB team.

Today no one disputes that the Awash Valley in the Afar contains one of the most complete records of human habitation in the world. The only question is whether it contains *the* most complete record. I think the answer will prove to be yes. Over the past few decades, discoveries in the region have produced.

- The putative ancestor of all hominids, *Ardipithecus ramidus,* 4.4 million years old, from Aramis.
- The putative ancestor of all later hominids, *Australopithecus afarensis,* ranging from 2.9 to 3.9 million years old, from Hadar, Maka, and Belodolie.
- The earliest known stone tools, 2.6 million years old, from Gona.
- A possible direct ancestor of *Homo, Australopithecus garhi,* 2.5 million years old, from Buri.
- The oldest known evidence of hominid butchery, 2.5 million years old, from Buri.
- The earliest known *Homo* directly associated with stone tools, 2.3 million years old, from upper Hadar.
- An early Acheulean industry, approximately 1.0 million years old, from Dakanihyalo.
- The largest single occurrence of later Acheulean industries, from Bodo to Alkanasa.
- A *Homo* skull intermediate between *erectus* and later humans, 600,000 years old, from Bodo.
- The first skull of *Homo* in Africa found in direct association with an in situ Acheulean industry, from Bodo.
- A "terminal" Acheulean industry (Sangoan), of later Pleistocene age, from Andalee.
- Extensive but as yet largely undescribed Middle Stone Age industries, overlying the middle Pleistocene gravels from Buri to the Talalak River to Meshellu.
- Late Stone Age industries surrounding Lake Yardi, at Nemay Koma, and throughout the Kariyu and Tendaho grabens.
- Holocene to recent ceramics at Buyelle and south of Lake Yardi.
- Historic stonework and tools throughout the region.

We know that if all the known sedimentary strata in the Awash Valley, just in the area from Gewani to Hadar, could be piled on top of one another, they would reach at least 1 kilometer in total thickness.[47] On the basis of strata still to be explored and mapped in the region, I predict that in the decades to come, new pieces of the regional section will be found in this area to nearly double this thickness. In the process, these and currently described strata will yield scores of "sensational" additions to the hominid fossil and archeological records. Older and younger remains of *ramidus* and *afarensis* will be discovered, as may pre-*ramidus* hominids, or hominoids, that will be yet more chimp-like, perhaps even indistinguishable from *Pan,* the living chimpanzees. Other species of australopithecines that are contemporaneous with both *ramidus* and *afarensis* will also be described. Older stone tools will be found, including *manuports,* unmodified rocks used as crushing instruments, that will probably antedate stone tool manufacturing by several million years. Older and more complete fossils of early *Homo* will be recovered. Both *Homo* and *Australopithecus* will be identified in the Matabaietu Formation in the eastern Awash, along with more Oldowan tools in stratigraphical context. Similar finds will be made at Geraru or elsewhere north of Hadar. More post-Matabaietu and pre-Bodo lithic industries will be discovered, as will their toolmakers. More fragments of the Bodo skull will be recovered, and when these and the many pieces that still remain unattached to the cranium are reconstructed to the skull, it is likely to be the most complete early hominid skull yet recovered from Ethiopia.

Bodo and the rest of the Acheulean sites will literally take at least 100 years to excavate to the satisfaction of archeologists, and when this is done, entire settlements—"cities"—of Acheulean peoples will be revealed in the form of seasonal base camps and communities of base camps. Acheulean tools will be documented in several dozen levels throughout a time range of hundreds of thousands of years, and they will prove to be the handiwork of several species of toolmakers.

The effect on human populations of the devastating climatic event that produced the middle Pleistocene plateau gravels will become better understood, as will the timing of the rehabitation of the Afar by the intrepid Andaleeans and others. A panoply of later Pleistocene humans will be discovered, as will multiple levels of Middle Stone Age industries—particularly in the area from Issie to the Talalak River to the upper Meshellu basin—and they may prove to be among the most extensive stone tool occurrences of this age in Africa.

Studies of the wanderings of the Afar nomads—as determined by their abundant stonework and their (rapidly disappearing) present-day life patterns—may serve as an analog to studies of the distribution of the abundant Late Stone Age sites that surely exist throughout much of the Afar Depression. The "enigma" of the origin of the Afar people will also become understood through a combination of oral histories and linguistic, genetic, and archeological studies. Finally, the scramble for concessions in the Awash Valley by scientific teams in the latter part of the twentieth century, and into the new millennium will offer historians rich and provocative insights to the colonization of Africa at the end of the nineteenth century.

Once the lengthy stratigraphical and fossil record of the Awash is fully pieced together, it will prove to be one of the most valuable paleoclimatic records in Africa—one that will further elucidate the relationships between global climatic events and faunal and floral turnovers over the last 6 or more million years. This will be true particularly for the periods that coincide with the emergence of the Hominidae in the early Pliocene and late Miocene and with the appearance of *Homo* in the later Pliocene. The migrations of hominid populations in the Afar since the late Miocene will also become better understood in the context of lake basins and river systems migrating as a consequence of shifting plates, the unraveling triple junction, sea-floor spreading, and oceanization. Specific spreading events in the Red Sea, the Gulf of Aden, and the Afar itself will be found imprinted in sedimentary successions at places like Ananu and Hadar, as revealed by discordant and distorted bedding, faults, and volcanic events.

By 1995 enough was known about the geology of the Awash Valley for me finally to publish, in a Dutch journal, a working model of the movements of lakes and animals in the sinking and expanding grabens of the western and central Afar (see Appendix II on page 320). To illustrate the dispersal of animals throughout the basins within these grabens, I used a monograph that Assefa Mebrate and I had prepared describing the Middle Awash proboscideans and that was published in 1993 by the American Philosophical Society.[48] The model depicts how more and more evolved species of *Anancus* and elephantids, comprising 17 identifiable taxa, followed the lakes over millions of years. In essence, one herd followed the other deeper and deeper into the subsiding lowlands of the triple junction as the surrounding plates inched and jerked farther and farther apart. Certainly, these migratory movements of animals are applicable to those of our ancestors. They must also be interconnected with broader dispersal events of mammals brought on by climate swings and the presence or absence of landbridges in the areas of the Sinai Peninsula and the Bab el Mandab Straits at the southern end of the Red Sea. I never did find the "Fossil Man Ripped Apart by Sea-floor Spreading" that I had expected to find, but I know it is only a matter of time before scholars realize that the birth of an ocean in the Afar coincided with the birth of humankind.

This is what I now think about the most. Not of discoveries made, but of those still to be made, and especially those not made. I think about those I missed that late afternoon in early 1972 at Meshellu, when I turned back to Camp 270 instead of going farther west; about those I failed to find east of the Magenta Gap in early 1976, when the military officer at Dubti advised us to stay out of the area; and those I missed along the Borkana River that same year, when I heeded the warnings about rebels along the western escarpment. I think about the pottery we found at Buyelle in 1975 with the handaxe lying in it and wonder who those late-comers were that chose to make ceramics on top of an Acheulean site. I particularly think about the time when I pushed on to the Busidima River at night instead of camping at the Talalak River and surveying the area as I should have. There is a lesson here: Discoveries are not made by the weak, timid, and feeble-minded. This is what I think about—the hill not climbed, the stream not crossed.

The more pregnant question is who will carry on with the work? Anyone who has stood on the north rim of the Hadar basin or the south rim of the Meshellu basin, or anyone who has flown the length of the Middle Awash Valley, knows that this unique region is an invaluable scientific resource that demands long-term planning, a world-class research facility, and a means of sustaining an ongoing, bona fide *Ethiopia-based* research program. Although increasing numbers of highly qualified Ethiopians who have received doctorate degrees abroad could contribute to this effort, many do not remain in Ethiopia because of limited salaries and positions, or they become hangers-on at the National Museum. One way to secure this talent, and to build a sound research apparatus in the country for pre-history studies, is for the Ethiopian government to take the initiative in raising research dollars. As Johanson has demonstrated, Ethiopian hominid discoveries can attract millions of dollars for institution building. And as Clark and White have demonstrated, these fossils can also attract millions for sustaining a long-term research program for one's own institution. Despite many lost opportunities, there is no reason why Ethiopian scholars and institutions cannot independently seek these same opportunities.

I propose that the Ethiopian government organize a touring exhibit of its fossil hominids for fund-raising purposes, in a manner not unlike that of the Tutank-hamen and Rameses II exhibits that have toured the United States and major Euro-pean cities in years past, or the Human Ancestors exhibit that was held at the Ameri-can National History Museum in New York in 1984. There are qualified Ethiopians to help prepare and organize such an exhibit—to be displayed in cities such as Paris, London, New York, and Tokyo—and I have no doubt that museums, organizations like UNESCO, corporate and private sponsors, and scientists the world over would contribute their time and expertise to making the undertaking a success. The effort would allow people of other nations to see *their* ancestors firsthand and to contrib-ute individually to an extraordinary chapter of scientific discovery.

Epilog

January 25, 1994. We took the early morning Ethiopian Airlines flight from Djibouti to Addis Ababa. We flew a somewhat northerly course, staying well clear of Somali airspace, over the stretched-out and ripped-up terrain in the southern Afar that connects the westward extension of the Gulf of Aden rift with the Red Sea rift. This is the rugged landscape that Werner Munzinger crossed during his disastrous invasion of Ethiopia in 1875 and that Wilfred Thesiger crossed in 1934 after exploring Lake Abhé. As we flew over the Djibouti border into Ethiopia, I got a full view of the lake itself, which I had first seen with Taieb during that crazy field trip in late 1971.

As our flight continued southwest toward Dire Dawa, I saw deeply eroded sediments in the lowlands north of the Hararghe escarpment, none of which have been explored for fossils so far as I know. The area has always been described as contested by Issa and Afar and frequented by smugglers—and therefore too dangerous to work in. But you never know. I stuck my camera to the window and snapped some pictures.

From Dire Dawa we flew west toward Addis Ababa. We passed just north of Awash Station and south of the former Trapp Company base camp at Arba, where I had spent pleasant evenings talking with the German road engineers and drinking beer. Beyond, I could see that the plantations along the Awash had been greatly expanded. The irrigated fields spreading over the once wildlife-rich Alledege plains were testimony to the five-volume master plan to develop the Awash Valley drawn up 30 years earlier by the UN FAO in endless charts, maps, and tables.[1] I guessed that plantations farther north had been similarly expanded, absorbing thousands of hectares and drawing more and more water from the Awash. *That* was why Lake Abhé had looked so diminished from the airplane. I wondered what had become of the scheme to build a dam at Tendaho and flood the Kariyu plain, depriving the Afar of yet more pasturage. And what about the magnificent wildlife that once had filled that landscape?

Even though I knew how desperately agricultural development was needed in Ethiopia—there was another famine in 1992–1993—I felt sick. One report described the nomads as increasingly "marginalized" and "peripheralized" with the loss of more and more grazing lands.[2] Its author, an Ethiopian and obviously courageous, recommended that the land taken away from the Afar be given back. Also, I heard that the sultan Ali Mirah had returned from exile and was once again calling for self-determination of his people and the return of traditional lands. Fat chance. However, along with other factions of Afar, his former Afar Liberation Front was now a political party in the restructuring allowed by the new government. Nominally, at least, the Afar people had a voice in the country's affairs.

Soon our plane reached the familiar eucalyptus-covered green hills and lofty volcanoes surrounding Addis Ababa. Once again I could see the thriving city sprawled across the upper Awash basin.

We were met by U.S. embassy personnel and quickly shuffled through customs to a waiting van that took us into the city. This time I was part of a three-man technical team sponsored by the U.S. Geological Survey. My two colleagues were Fred Simon, chief of the African branch of the USGS Office of International Geology, and Mike Foose, an economic metals specialist. Our purpose was to sell the Ethiopian government on a multinational gold exploration program in the Afar to be carried out in cooperation with Djibouti as well as Eritrea, which had become independent the previous year. The idea had been spawned several years earlier when Djibouti geologists asked the USGS to analyze a collection of volcanic rock samples and scales from hydrothermal wells. Some of the samples showed anomalous concentrations of gold—a special type called "epithermal" gold, which is formed near the surface by hot solutions rising from magma sources. The gold had not been reported previously because it is generally found only in microscopic concentrations, but the ore becomes commercial when mined in bulk. As Fred and Mike explained to the Djibouti geologists during the week we were there, such gold has been recognized only quite recently and is commonly found along plate boundaries. Djibouti, as we well know, is literally cooking on top of a spreading zone between plates and is porous with geothermal centers.

The best-known epithermal gold deposit is the site of a multibillion-dollar operation in the United States in Nevada, where the ore is mined from the surface and the gold is leached out of the rock by solution methods. The extraction process is relatively easy but can be environmentally messy unless carefully carried out. Because most of Djibouti lies within the Afar Depression, near the triple junction, it made sense that similar levels of gold might be found in the rifts across Djibouti's western border in Ethiopia and its northern border in Eritrea. The USGS was interested in the project because of what it might reveal about the genesis of this type of geological occurrence and for its development potential. The USGS was not promising that the Djibouti gold anomalies would prove economical, only that they were worth investigating. I was recruited because of my knowledge of Afar geology and a previous interest in the metalliferous brines in the Red Sea. (I had always believed there was gold in the Afar besides fossil hominids.)

I felt disoriented my first few days in Addis Ababa. I had known the city so well that I thought it would be like 1978 all over again. But 16 years had passed. What threw me, besides new buildings, was the growth of the trees lining some of the main streets—and certainly the incredible traffic and mobs of people. When I went to Ethiopia in 1971, Addis Ababa had a population of about 1.5 million. Its population had now tripled and included thousands of homeless people. Beggars were everywhere. My colleagues and I stayed at the Hilton, where tourists were ambushed by street people as soon as they stepped off the hotel grounds, away from the protection of the cane-wielding hotel guards.

One late afternoon I went to meet a friend at the Ethiopia Hotel and found the lobby and bar packed with people just off work. Someone had laid a young boy next to the entrance on a large piece of cardboard with a small bowl beside him for donations. To me the boy looked like he was dying. His jaundiced yellow skin was covered with perspiration, and he was breathing deeply and unevenly. As he exhaled he would yell out in a croaking voice, yet people bustled around him, greeting friends and hurrying to make engagements; occasionally someone dropped change in the box. Callous, you say? Had I remained in the country those last 16 years, I would hardly have noticed him. One becomes inured to suffering when there is so much of it.

Twenty years of revolution, war, and civil strife had left its mark. Poverty was everywhere, and there were few in the country who had not experienced some tragedy or another. It seemed that most of the people walking the streets were wearing the same clothes they had been wearing before the revolution. On the other hand, beginning in the late afternoons, the bars and restaurants of the Hilton quickly filled with Ethiopians, something I had never seen in the 1970s under Mengistu, or under Haile Sellassie. Smartly dressed, animated young couples enjoyed the simple pleasures of one another's company in the new openness the country was experiencing. In the evenings, the banquet rooms of the hotel were filled by wedding parties and other revelers, as music, rhythmic clapping, ululating, and singing echoed down the hallways. Once before dawn, I went to the pool for a solitary warm-water swim. I was sure the pool would be empty, but it was filled with Ethiopians swimming laps.

The buildings of the Ministry of Mines and Geological Survey looked hardly different to me from the day I first walked into the compound in early 1971. Of course, most of the faces had changed. My friend and former colleague Mesfin Asnake, who had worked with the RVRME, was there, as were some others I had known, including the librarian, Metasabia. She reminded me that her daughter had taken dance lessons from Judy.

The Ethiopian geologists whom my USGS colleagues and I met during our stay were extremely cordial and receptive to the prospect of exploring for a new resource. However, they were uncertain how a joint project with Djibouti and Eritrea would work. This surely would have to be cleared at the highest levels of the government. They were also hesitant to use their limited resources to pursue a type of gold that could not be seen with the naked eye, when there were known visible deposits in the Precambrian rocks of western Ethiopia.

At the end of January we went to Asmara to meet geologists with the hastily formed Eritrean Department of Mines. En route we flew over the Blue Nile basin, one of the most breathtaking sites in the world, with its chasms, winding canyons, soaring peaks, and isolated mesas. I have since learned more about the mysterious earthwork in the basin mapped by an American survey team, which I had read about years before in Alan Moorehead's book *The Blue Nile*. I recall that the author described the feature as "an immense ditch too wide for a horse to jump across [that] . . . winds its way for hundreds of miles." He had suggested that it was the ancient boundary between two tribes.[3] He was right, except that the ditch is only 50 miles long. It lies just west of the Didessa River at the southern margin of the Blue Nile basin, approximately 250 kilometers southwest of Addis Ababa (see Map II). It was mapped in 1960 by Darwin Jepsen and Murray Athearn, geologists with the U.S. Bureau of Reclamation, and was labeled "Ancient Tribal Boundary Trench." Their field work was part of a five-year pioneering study of the land and water resources of the Blue Nile that had been conducted in cooperation with the Ethiopian Ministry of Public Works. The effort resulted in a 4053-page report that was made public in the United States only in 1987.[4] By comparing Jepsen and Athearn's geology map showing the location of the trench with topographic and historical sources, I found that its position falls almost exactly along the boundary separating the former Oromo kingdoms of Gomma and Limmu.[5] Both were established in the early nineteenth century and were later conquered by Menelik. The kingdoms were also located due west of the former Oromo kingdom of Janjero, which lay in the catchment of the Omo River. It was their descendants who had served my family and friends that memorable feast after our outing into the gorge in 1974.

———

When we landed at the Asmara Airport in Eritrea, the first things I noticed were military tanks positioned around the airstrip, some covered with camouflage netting. I was told that they had been captured from the Ethiopians but were of Soviet origin, like most of the newer equipment used by the Eritrean army. In downtown Asmara, however, direct evidence of the 30-year war for independence—the longest waged in modern African history—was little in view, as was the case in Addis Ababa. It was as though each side had honored an unwritten agreement not to harm the other's capital. Fortunate, because Asmara is one of the showpieces of the continent, although much of what is attractive about the city is hardly African. Rather, you would think you were in a town in the Italian alps, except that the main streets are lined with palm trees. The colonial-period architecture is a combination of the art deco fashionable in the 1930s and classical Italian, which includes several ornately built, stone Catholic cathedrals.

Asmara lies 700 kilometers due north of Addis Ababa at a similar elevation, about 2500 meters. It is connected to the port of Massawa by a precipitous road and a narrow-gauge railroad that covers most of this elevation over a horizontal distance of a mere 45 kilometers. During the week that Fred, Mike, and I were in Asmara, banks of clouds rolled over the city in the late afternoon and covered the slopes of the surrounding mountains, making them look like islands in a field of snow. By midmorning the next day the clouds had burned off, revealing a brilliant blue sky.

I saw very little of the poverty that was so evident in Addis Ababa, even though the per capita income of Eritrea, like that of Ethiopia, was less than $300 per year. Only once was I approached by a beggar, a boy eight or nine years old, who was immediately grabbed by a passer-by who said to the boy, "Who do you think you are, an *Ethiopian?*" Such pride was noticeable everywhere, with patriotic posters, slogans, and flags adorning buildings. A war museum located in a large warehouse in the central city was devoted to the numerous victories the Eritreans had won over the successive might of the Lion of Judah and Colonel Mengistu. Stores had their doors open, people walked arm in arm down the sidewalks, and in a city of only 400,000 people there was none of the congestion that clogged Addis Ababa. Still, Eritrea's road to survival was in many ways just beginning and depended on its ability to do what neither the Italians nor the Ethiopians had been able to do before them: sustain a secure country, build a secure economy, and restrain themselves from overreaching. At the same time, Ethiopia had to recognize that the only way to win back a meaningful relationship with Eritrea was through trade and commerce, not war. Fighting would only consume the resources of both countries, and destroy countless lives.

Once again our days were filled with meetings. As it turned out, I knew the two senior geologists with the Department of Mines, Alem Kibreab and Tesfaye Michael Kelets. Both had worked for the Ethiopian Geological Survey when I was there but had fled the country in 1977. They had spent the next 17 years as freedom fighters, but Tesfaye did show me a wrinkled geological map that he had made during the war. Both geologists were interested in our ideas about epithermal gold in the part of Eritrea that stretched into the Afar, but like their neighbors, they were uncertain how a multinational exploration project would work. Also like the Ethiopians, the Eritreans were more interested in developing known prospects. We visited one such site south of Asmara, an important copper deposit that had been partially developed in the early 1970s by a Japanese–Ethiopian mining effort. As we saw, however, the entire operation had been blown up by the rebels, lest the profits benefit the Ethiopians.

As we passed through the countryside and small towns during that field trip, we saw evidence of fighting almost everywhere we went: blown-up tanks and armored personnel carriers; buildings shot up, burned out, and bombed out. But it was nothing like what we saw in the port city of Massawa the next day, a Sunday. Evidence of what was to come was first visible during our plummeting drive down the escarpment. At one point I was perplexed by white blotches along a hillside. They looked like mining test pits, but I learned they were bomb craters left by the Ethiopian air force when it had tried to break through the guerrilla forces on the road to the coast. Then we saw the Eritreans' trenches that had been pounded during bombing runs. When we reached the outskirts of Massawa, the destruction increased. As a result of ten months of continuous bombing at the end of the war, which reportedly included cluster bombs and napalm, whole stretches of buildings were destroyed.

At one point we stopped at the edge of the city so that our driver could fix something on our Land Rover. I got out and walked across the road to take a picture of a partially destroyed armored personnel carrier. As Fred was yelling at me to be careful of possible land mines, I noticed a few army helmets in the field next to the road. As I walked around the carrier and looked again, I realized that I was looking at a former battlefield. It was smothered with the debris of fighting and carnage. Bones of shattered bodies lay everywhere amid helmets, cartridge cases, boots with leg bones protruding from them, smashed skulls, canteens, spent shell casings, broken rifles, and pieces of uniforms. I took more pictures and had started back to the Land Rover when an Eritrean came running up and said he wanted to show us something. Our driver and Fred agreed, so we drove up the road a short distance, the Eritrean running alongside us, until we came to a small compound enclosed by a fence made of barbed wire and sheet metal. He took us inside through a makeshift gate. The compound was stacked with wooden ammunition boxes filled with skeletons of soldiers. Lying on top of some of the boxes were rows of skulls. All had bullet holes in the head. Our guide told us the remains were those of Ethiopians who had been executed by their own officers because they refused to continue fighting. Their remains had been gathered by the Eritreans and placed in the compound as a memorial.

During the late 1970s we had all heard stories of such atrocities. One was that Mengistu ordered the execution of a number of generals for questioning the continuation of the war, or for losing the war. Stories varied, and it was hard to say what was true, but one account in the New York Times states that on November 18 and 19, 1977, truckloads of Ethiopian officers and soldiers were executed on the outskirts of Asmara for failing to break through the Eritrean forces on the road to Massawa.[6] Reportedly the men were all handcuffed, forced to sit beside mass graves, and shot.

In the center of Massawa—a city consisting in part of sixteenth-century buildings divided by narrow lanes and arched passageways—was more devastation. Yet amid all the destruction, there was an automobile rally that day. Souped-up Fiats and Peugeots raced around the port city, blaring their horns and sporting Eritrean flags. Because it was a Sunday, we went to the beach for a swim in the tepid waters of the Red Sea, amid picnicking families and young people. Many of the men and some of the women were scarred with battle wounds, some terribly, but a holiday atmosphere prevailed, and the beach and waters were filled with people.

The evening before we left Asmara, the U.S. ambassador, Robert Hodeck, a hefty and bearded career diplomat, invited us to a reception at his home overlooking the city. The event was attended by relief workers, U.S. AID and Peace Corps representatives, embassy personnel, and Eritrean officials wearing short-sleeved shirts and sandals. At one point the ambassador and I were standing on the front porch talking, when we both looked up and saw an extraordinary sight: a meteorite shower moving ever so slowly across the horizon. It looked as if it were not more than 30 or 40 kilometers away, although it was surely much farther distant. Soon the drivers of the official vehicles waiting nearby were pointing to the sky and talking and yelling excitedly. The shower must have consisted of thousands of fragments burning up in the atmosphere, each glowing and leaving an incendiary trail. It was moving to the southeast—toward the Afar Depression; after several minutes it disappeared below the horizon. Some day a geologist, or more likely a goatherd, will find some of the scattered remains and regard them as just another mystery of the region.

The next morning, accompanied by the two Eritrean geologists, whose travel expenses we had paid, we flew back to Addis Ababa to meet again with the Ethiopians. The Eritreans were nervous but eager to see their former Ethiopian friends with the Geological Survey after so many years. Also flying to Addis Ababa that morning were the two geologists from Djibouti, whose costs our project also covered. Over the rest of the day and the next, we all sat around a conference table at the Geological Survey and thrashed out a joint exploration program for the area in the Afar that might contain epithermal gold where the borders of the three countries come together, at a triple junction. Although the project still has not materialized as a joint undertaking, and may not come to fruition for many years, if ever, at least we all made the effort.

Fred and Mike returned to the United States the next day, and our Eritrean and Djiboutian associates returned to their countries. I remained in Addis Ababa a while longer. I visited some old friends and missed seeing others. Dejene Aneme was now vice-minister of labor, and Assefa Mebrate was with the Commission for Science and Technology. Dejene had never made it to the United States for graduate studies, but Assefa had attended the University of Kansas, obtaining two master's degrees and a Ph.D. He wrote his dissertation on North American fossil proboscideans.[7] Tsrha was then in the United States, having only recently completed her Ph.D. in anthropology at New York University. Her dissertation was on the Bodo skull.[8] Mesfin had obtained a master's degree at the University of Washington, specializing in microvertebrate paleontology, before returning to Ethiopia and joining the Geological Survey. Sleshi Tebedge had completed his Ph.D. at the

University of Texas, writing his dissertation on a Pleistocene cave fauna from New Mexico. He never returned to Ethiopia again after his visit there in 1982 and his heroic standoff with the Berkeley teams, but he always hoped to. Instead he taught biology and geology at two small colleges in Austin. He died of complications from diabetes in 1995, leaving behind a widow and two young sons.

I looked up an Italian friend of a friend, Carlos, who offered to drive me up to the Entoto Hills overlooking the capital. We were accompanied by an Australian geologist I had met and by an Ethiopian journalist, Tefera Ghedamu. The last time I had been to Entoto, on a Sunday outing in 1978, we had been forced to leave by soldiers for some vague security reason. This time we came across groups of joggers in colorful outfits, their voices echoing up and down the mountain trails. Tefera had looked me up concerning several articles he had written for an Amharic magazine about fossil-hunter intrigues at the National Museum.[9] One article said favorable things about the RVRME but was critical of Tim White and Ann Getty. It had created a local stir and had also elicited a five-page diatribe to the editor from White, saying the article was "full of lies and willful distortions."[10] He noted, in a fashion I recognized, that copies of his letter had been sent to the prime minister, the minister of foreign affairs, the minister of security and internal affairs, the Ethiopian ambassador to the United States, the U.S. ambassador to Ethiopia, and so on.

I also took the time to peek over the gate of our former home near Teodros Circle, which was now occupied by a government official. The house looked as though it had not been painted or cared for since we left the country. The once beautiful gardens were largely destroyed and overgrown with weeds. I walked down the lane and crossed the street to what had been the RVRME office and found it to be in similar condition, except that the big old house had been divided up into tiny apartments. I guessed that 30 or more Ethiopians lived there. When I walked into the compound, accompanied by Mesfin, a dozen or so children, from tots to teenagers, came running up. I told them the house was once my office, and they told me that they all lived there. They seemed pleased by the association, and so was I.

The next day I took the early flight to Khartoum. I had always wanted to visit Sudan, the largest country in Africa, one-third the size of the contiguous United States. The land where the Blue Nile and White Nile converge at Omdurman. The Mahdist revolt that defeated the British. The mystical Sufi sect of the whirling dervishes. And I had some ideas about where fossils might be found, but I dare not make this known, lest the area soon be crawling with Berkeley types. They have an international reputation for being followers, you know.

Just about everybody I talked to warned me about visiting Sudan. High on the U.S. State Department's list of terrorist nations. Hamas, the Hizbollah, the PLO. Also the dreadful civil war and famines in the south. But I had letters of introduction to the Sudan Geological Research Authority, and I planned to spend a month exploring the country. Geologists have a common bond with the earth, and I was not worried about terrorists.

Metric Equivalents
Glossary of Terms and
Abbreviations

millimeter: 25 mm = 1 inch
centimeter: 1 cm = 0.39 inch; 2.5 cm = 1 inch
kilometer: 1 km = 1000 meters = 0.62 mile; 1.6 km = 1 mile
square kilometer: 1 km^2 = 0.39 mi^2; 2.6 km^2 = 1 mi^2
hectare: 1 ha = 2.47 acres
kilogram: 1 kg = 2.2 pounds

Acheulean. An Early Stone Age culture, lasting from about 1.7 million years ago to about 150,000 years ago, characterized by large, teardrop-shaped handaxes apparently first manufactured by *Homo erectus*.

AGIP. Azienda Generale Italiana Petroli, the national petroleum company of Italy.

Badlands. A deeply eroded and dissected area, commonly where there is little vegetation and rainfall is infrequent but intense.

Basalt. A dense, usually dark, fine-grained volcanic rock that commonly erupts along oceanic and continental rifts and on adjacent plateaus.

Biface. A rock core that is completely or nearly flaked on both sides forming a sharp edge around its periphery, such as a handaxe or a cleaver.

Birr. Ethiopian currency equivalent in name to the dollar.

Cleaver. Large, bifacial stone tool with a cutting edge transverse to the long axis, similar to an adze.

Derg. Amharic for "committee" or "a group of equals," a term applied to the central governing body of Ethiopia from 1974 to 1991, which was composed predominantly of military officers.

ESA. Early Stone Age. In Africa this period lasted from the earliest recognized manufactured stone tools, of Oldowan type, through the Acheulean culture. In Europe the ESA is referred to as the Early Palaeolithic.

Exposures. Sedimentary strata exposed to the surface by erosion or faulting.

FORGE. Fund for Overseas Grants and Education, founded by Alfred Kelleher in Stamford, Connecticut.

Firenji. Amharic for foreigner.

Graben. An elongated trough, formed by extension or "stretching," bounded on either side by faults and raised margins. Grabens are also known as rifts, and those of great length as rift valleys.

Handaxe. Pointed, teardrop-shaped, bifacial stone tool most commonly used for butchery purposes.

IHO. Institute of Human Origins. Founded by Don Johanson in Berkeley, California. Currently based in Tempe, Arizona.

LSA. Late Stone Age. Characterized by microlithic tools made of obsidian or quartzitic stone, notably hafted into wood or bone handles, by *Homo sapiens*. Believed to have lasted from about 40,000 to the near present.

MSA. Middle Stone Age. Characterized by light-duty flaked tools manufactured by early *Homo sapiens* between about 185,000 and 35,000 years ago.

Oldowan. An Early Stone Age industry, first described from Olduvai Gorge, lasting from about 2.6 to 1.0 million years ago and characterized by scrapers made from flakes, by hammerstones, and by crude choppers made from cobbles—hence the name "pebble tools." These tools are believed to have been manufactured by *Homo habilis* and possibly other early hominids.

NSF. National Science Foundation. A nominally regulated agency of the U.S. government that dispenses grants for scientific research via a secret decision-making process centered on anonymous peer reviews.

Radiometric dating. Clock-like dating methods based on the measurement of the constant rate of decay of naturally occurring radioactive materials.

RVRME. Rift Valley Research Mission in Ethiopia.

Sangoan. A "terminal" Acheulean or post-Acheulean stone tool tradition characterized by picks, small scrapers, core axes, and some handaxes.

Section. As used here, a vertical exposure of strata or a drawing or diagram of such an exposure.

Tuff. Consolidated volcanic ash and related materials.

UCB. University of California at Berkeley.

Zabanya. Amharic for "guard." Here refers to guards of private homes.

Zemacha. Amharic for "campaign." An abbreviated reference to the Development Through Cooperation Campaign, a national service program established under the Mengistu government in Ethiopia.

Appendices

Appendix I

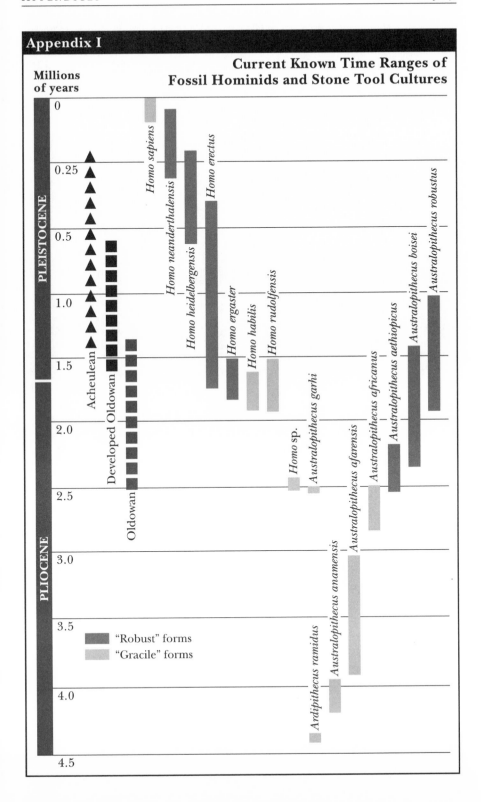

Current Known Time Ranges of Fossil Hominids and Stone Tool Cultures

Millions of years

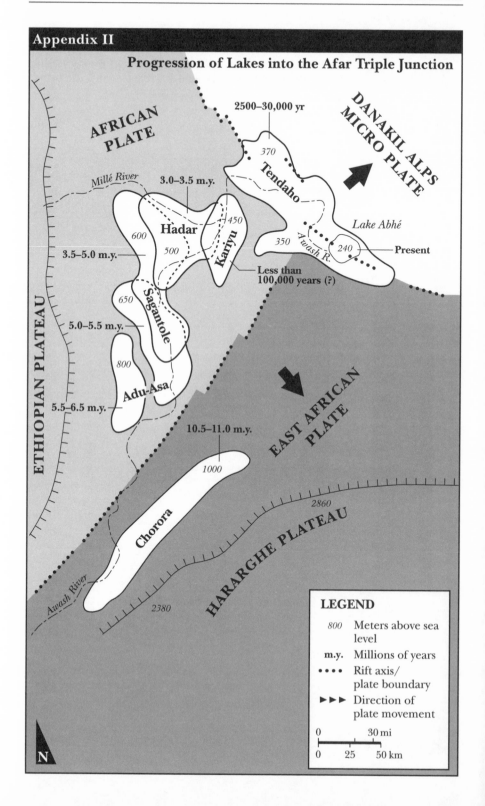

Appendix II

Progression of Lakes into the Afar Triple Junction

AFRICAN PLATE

DANAKIL ALPS MICRO PLATE

2500–30,000 yr

370

Tendaho

Millé River

3.0–3.5 m.y.

Hadar

450

Kariyu

Lake Abhé

600

500

350

240 — Present

Awash R.

3.5–5.0 m.y.

Less than 100,000 years (?)

650

Sagantole

5.0–5.5 m.y.

800

Adu-Asa

5.5–6.5 m.y.

ETHIOPIAN PLATEAU

10.5–11.0 m.y.

1000

EAST AFRICAN PLATE

Chorora

2860

HARARGHE PLATEAU

Awash River

2380

LEGEND

800 Meters above sea level

m.y. Millions of years

•••• Rift axis/ plate boundary

▶▶▶ Direction of plate movement

0   30 mi

0   25   50 km

N

Appendix III

Stratigraphy, ages, and highlights of Middle Awash/Hadar areas

Millions of years

Millions of years	Highlights	Formation
0	Later stone tool industries unconformably overlying plateau gravels.	**UPPER WEHAIETU FORMATION**
0.25		
0.5 — H	Middle Acheulean in eastern Middle Awash. *Homo* at Bodo.	**LOWER WEHAIETU FORMATION**
0.75		
1.0 — H	Early Acheulean at Dakanihyalo.	**DAKANIHYALO BEDS**
2.0 — H	Oldowan at Matabaietu, Gona, Buri; *Homo* and tools at Hadar; *Australopithecus garhi* at Buri.	**MATABAIETU FORMATION**
3.0 — H H	*A. afarensis* at Hadar/Meshellu.	**HADAR FORMATION**
4.0 — H	*A. afarensis*, late Anancus and early elephants in eastern Middle Awash.	
4.5 — H	*Ardipithecus ramidus* at Aramis in Central Awash Complex.	**SAGANTOLE FORMATION**
5.0	Early *Anancus* and elephants in Hatowie graben.	
5.5		
6.0	Early *Anancus* and elephants at Ananu and elsewhere in western foothills.	**ADU-ASA FORMATION**
10.0	*Hipparion primigenium* at Chorora on Hararghe escarpment.	**CHORORA FORMATION**
11.0		**H** = Hominids

PLEISTOCENE · PLIOCENE · MIOCENE

Notes

For a comprehensive overview of the geology, prehistory, geography, and exploration of the Afar Depression and surrounding areas, from classical times to the present, see *Bibliography of the Earth Sciences for the Horn of Africa: Ethiopia, Eritrea, Djibouti, and Somalia—1620–1993* by Jon Kalb, Dejene Aneme, Ketema Kedane, Mohamed Santur, and Afifa Kechrid. This work, with over 10,000 citations, has been "in progress" since 1971 and will be published in 2000 by the American Geological Institute (Alexandria, Virginia). Accompanying the paper copy of the bibliography is a CD-ROM with a component that enables historians, historians of science, and other researchers to search through the data base chronologically as well as by author, subject, and geography.

Much of the documentation used in this book, particularly in Chapters 11, 23, and 25 through 28, in the form of correspondence, grant proposal records, and memoranda, has been made available to me by the U.S. National Science Foundation through the Freedom of Information Act. Many of these same documents were also made available to author Robert Bell and are cited in his book *Impure Science: Fraud, Compromise and Political Influence in Scientific Research*. A number of related documents are cited in court records stemming from my lawsuit with NSF and are available from the U.S. District Court for the District of Columbia (Civ. No. 86-3557 and 86-3557SSH, 1986 and 1987, respectively).

Finally, much of the source material used to describe the geology and prehistory of the western and central Afar comes from my unpublished field notes, maps, and other data. The corpus of this material, particularly the discoveries by the RVRME, was first submitted to the Ministry of Culture from 1975 to 1978 in 22 formal reports, which were made available to researchers soon after I left Ethiopia. My former colleagues and I published a number of other reports in international scientific journals from 1978 to 1982, prior to the first reports by others working in the Middle Awash. From 1982 to 1996 we published yet more works that describe specific topics of RVRME research in more detail. Most of this material is cited in the Notes and the Bibliography that follow.

Chapter 1: An Oceanic Desert

1. Moorehead, *The Blue Nile,* 287.
2. Tazieff, "The Afar Triangle," 32–40.
3. Wegener, *The origin of continents and oceans.*
4. Wegener, *Die Entstehung.*
5. K. Wegener, "Alfred Wegener," iii–v; Georgi, "Memories of Alfred Wegener."
6. Wegener, *The origin of continents and oceans,* 190.
7. Balk, "Memorial to Hans Cloos (1886–1951)."
8. Cloos, "Hebung-Spaltung-Vulkanismus."
9. Cloos, *Conversation with the earth,* 395.
10. Ibid., 396–402.
11. Cloos, "Tektonische Bernerkungen."
12. Pettijohn, "Memorial to Ernst Cloos."
13. King, "Obituary notice"; S. I. T., "Obituary notice of Arthur Holmes."
14. Duff (1993), *Holmes' principles of physical geology,* iv.
15. Holmes, *The age of the earth.*
16. Holmes, "Radioactivity and earth movements."
17. Holmes (1944), *Principles of physical geology,* 506, fig. 262.
18. Note 16, 595–598; note 17, 438–441.
19. Holmes (1965), *Principles of geology,* 1080–1081.
20. James, "Harry Hammond Hess," 109–125. Much of the chronology and discussion that follows on the discoveries that highlight the "plate tectonic revolution" is based on Menard (1986) and Glen (1982).
21. Hess, "History of ocean basins"; Menard, *The ocean of truth,* 152–153.
22. Hess, "History of ocean basins," 618.
23. Dietz, "Continent and ocean basin evolution."
24. Dietz, "The spreading ocean floor," 34–35.
25. Menard, *The ocean of truth,* 216–221; Vine and Matthews, "Magnetic anomalies."
26. Glen, *The road to Jaramillo,* chap. 7; M. N. Hill, "A discussion concerning the floor"; Falcon et al., "A discussion on the structure"; Menard, *The ocean of truth,* chap. 17; Laughton, Whitmarsh, and Jones, "The evolution of the Gulf of Aden"; Allan, "Magnetic and gravity fields"; Drake and Girdler, "A geophysical study."
27. Matthews, "A major fault scarp"; Matthews, Vine, and Cann, "Geology of an area"; Matthews, "The Owen Fracture Zone"; Cann and Vine, "An area on the crest."
28. Wilson, "A new class of faults;" Glen, *The road to Jaramillo,* 305.
29. Menard, *The ocean of truth,* 247.
30. Wiseman and Sewell, "The floor of the Arabian Sea." Notes 28–33, 43–46.
31. Vine, "Spreading of the ocean floor"; Girdler, "The relationship of the Red Sea"; Girdler, "Initiation of continental drift"; Drake and Girdler, "A geophysical study."
32. Rothé, "La zone seismique."
33. Isacks, Oliver, and Sykes, "Seismology."
34. Irvine, *The world rift system;* Pilger and Rösler, *Afar depression of Ethiopia;* and note 29.
35. Menard, *The ocean of truth,* 284–286.
36. Morgan, "Rises, trenches, great faults and crustal blocks."
37. McKenzie and Parker, "The North Pacific."

Chapter 2: Afar Gold

1. McKenzie, Davies, and Molnar, "Plate tectonics."
2. Mohr, *The geology of Ethiopia.*
3. Mohr, "Major volcano-tectonic lineament"; Mohr, "Plate tectonics."
4. Mohr, "Notes on the Afar."
5. Bonatti, "Submarine volcanoes"; also see Bonatti and Tazieff, "Exposed guyot."
6. Howell, "Omo Research Expedition"; Howell, "Remains of Hominidae."
7. R. Leakey, "In search of man's past."
8. Mohr, *Geological map of horn of Africa.*
9. Harland et al., *A geologic time scale.*
10. Mohr, *The geology of Ethiopia,* 200–202.
11. Taieb, "Differents aspects du quaternaire."
12. D. Dow, letter to author, December 17, 1970, with attached proposal, "Research project south-eastern Afar," dated December 10, 1970.
13. Degens and Ross, *Hot brines.*
14. Falcon et. al., "A discussion on the structure and evolution of the Red Sea."

Chapter 3: The Awash Valley

1. United Nations, *Geology, geochemistry and hydrology.*
2. Taieb, "Évolution quaternaire du bassin de l'Aouache."
3. J. Chavaillon, "Les habitats acheuléen"; N. Chavaillon, "Les habitats oldowayens."
4. Chavaillon, Brahimi, and Coppens, "Premiere decouverte"; Howell, "Omo Research Expedition."
5. J. Chavaillon, "Prehistorical living-floors."
6. Thesiger, "The Awash River."
7. Taieb, "Évolution quaternaire du bassin de l'Aouache," fig. 87.
8. Levine, *Greater Ethiopia,* 48–51.
9. Buxton, *Travels in Ethiopia,* 154.
10. Schönfeld later revised the age of the fossils to 9 to 10.5 million years on the basis of radiometric dates. See Sickenberg and Schönfeld, "The Chorora Formation."
11. Tiercelin, Michaux, and Bandet, "Le Miocène superieur."
12. Christiansen, Schaefer, and Schönfeld, "Southern Afar and adjacent areas."
13. Nesbitt, *Hell-hole of creation,* 249.
14. Winid, "Comments."
15. Karl Butzer, personal communication, November 16, 1998.
16. Rubenson, *The survival of Ethiopian independence.*
17. Marcus, *The life and times of Menelik III,* 37–40.
18. Wylde, *Modern Abyssinia.*
19. Rubenson, *The survival of Ethiopian independence,* 142, 301.
20. Munzinger, "Narrative of a journey."
21. Johnson, *Travels in southern Abyssinia.* The following maps, based largely on second-hand information, also show the terminal flow of the Awash River into lakes, but with less accuracy than Johnson's: Rochet-D'Hericourt, *Voyage sur la cote orientale;* Cecchi, *Da Zeila alle Frontiere del Caffa;* Wylde, *Modern Abyssinia;* United Kingdom, War Office, *Abyssinia, map no. 2319.*
22. Johnson, *Travels in southern Abyssinia,* 150, 195.

23. Thesiger, "The Awash River," 16; Nesbitt, *Hell-hole of creation*, 218, 247.
24. Salt, *A voyage to Abyssinia*.
25. Ibid., 137, map on facing page.
26. Nesbitt, *Hell-hole of creation*, 203, 238.
27. Thesiger, "The Awash River," 13; also see Barberi et al., "Geology of the northern Afar," 448.
28. Thesiger, "The Awash River," 13.
29. Thesiger, *The life of my choice*, 151.
30. Marcus, *The life and times of Menelik II*, 135–139.
31. Background information on Taieb is from *Who's who*; Nelson, *Tunisia*; Perkins, *Tunisia*; and Jean-Pierre Slakman, personal communication, New York, April 1997.
32. Taieb, *Sur la terre*.
33. Scientific American, *Continents adrift*.
34. Gasse, "Les diatomées Holocenes."
35. Notes 36–38.
36. Rognon and Gasse, "Depots lacustres Quaternaires."
37. Varet, comp., *Geological map*.
38. Gasse, "Sedimentary formations."
39. Faure et al., "Les formations lacustres holocenes."
40. Butzer et al., "Radiocarbon dating"; Gasse and Street, "Late Quaternary lake-level fluctuations."
41. Johnson, *Travels in southern Abyssinia*, 208.
42. Holmes, *Principles of physical geology*, 1081.
43. Taieb, "Evolution quaternaire du bassin de l'Aouache," chap. 2.
44. Ferrara and Santacroce, "Dama' Ale."
45. Gallagher, "A preliminary report"; Gallagher, "Ethno-archaeology"; Gallagher, "Contemporary stone tools."
46. Bonatti, "Submarine volcanoes"; Bonatti and Tazieff, "Exposed guyot."
47. References 36 and 48.
48. Harrison, Bonatti, and Stieltjes, "Tectonism of axial valleys."
49. Taieb, "Évolution quaternaire du bassin de l'Aouache," 303.

Chapter 4: Louis and Mary Leakey

1. McCann, *People of the plow*, 23.
2. Mohr and Zanettin, "The Ethiopian flood basalt province"; Zanettin, "Evolution of the Ethiopian volcanic province."
3. Hofmann et al., "Timing of the Ethiopian flood"; Kreuzer et al., "Age of basalt."
4. Merla et al., *A geological map of Ethiopia*.
5. Mortan, Geological map of the Addis Ababa area.
6. Abebe, 1976, *Proceedings of the VIIth (1971) Pan-African Congress*.
7. "Scientist says man originated in Ethiopia."
8. Dietz and Holden, "Reconstruction of Pangaea."
9. Much of the general discussion in this chapter on primate evolution is based on relevant chapters in Fleagle, *Primate adaptation and evolution;* Jones, Martin, and Pilbeam, *The Cambridge encyclopedia of human evolution;* Day, *Guide to fossil man;* Martin, *Primate origins and evolution;* and Stanley, *Earth and life through time.*

10. Wallace, *The geographical distribution of animals,* 288–290.
11. Heybroek, "The Red Sea Miocene."
12. Patterson, et al., "Geology and fauna"; A. Hill, "Late Miocene," 131; Leakey et al., "Lothagam"; McDougall and Feibel, "Numerical age control."
13. Girdler and Styles, "Two stage Red Sea floor spreading"; Girdler, "The evolution of the Gulf"; Girdler, "The Afro-Arabian rift system"; Whybrow, "Geological and faunal evidence from Arabia"; Whybrow, "Land movements."
14. Dart, *Adventures with the missing link.*
15. Kazmin, *Explanation;* Merla et al., *A geological map.*
16. Kreutzer et al., "The age of the Red Beds."

Chapter 5: Looking for Leadu Man

1. L. Leakey, endorsement letter for J. Kalb and M. Taieb, January 27, 1972.
2. U.S. Army, *Joint operations graphic,* NC 37-4, ND 37-1C.
3. J. Chavaillon, "Prehistorical living-floors."
4. Deino and Potts, "Single-crystal"; Clark et al., "African *Homo erectus.*"
5. Clark, "The Middle Stone Age."
6. Fisher and Schmincke, *Pyroclastic rocks,* 264.
7. Klingel, "The Somali wild ass," 110; Klingel, "Social behavior of African Equidae"; also see Bauer et al., "The historic ranges of three equid species."
8. Daniel Barker, personal communication, October 1997.
9. Hans Klingel, University of Braunschweig, Germany, personal communication July 3, 1996; also see reference 7.
10. Thesiger, "The Awash River," 13; Thesiger, *The life of my choice,* 141.
11. Taieb et al., "Depots sedimentaires et faunes"; Varet, *Geology of central and southern Afar.*
12. Churcher and Richardson, "Equidae"; Ray Bernor, Howard University, personal communication, December 4, 1995.
13. Gasse and Street, "Late Quaternary lake-level fluctuations."
14. Roberts, "A discussion."
15. Tazieff et al., "Tectonic significance."
16. United Nations, *Geology, geochemistry, and hydrology.*

Chapter 6: Hadar

1. Fernández-Armesto, *The Times atlas,* 139, 188.
2. Rice, *Captain Sir Richard Francis Burton,* chap. 21.
3. McLynn, *The making of an explorer,* chapters 15 and 16.
4. McLynn, *Hearts of darkness,* 356; Fernández-Armesto, *The Times atlas,* 138, 194, 197; National Portrait Gallery, *David Livingstone,* 46, 49.
5. Suess, *The face of the earth;* Suess, "Di Bruche des ostlichen Africa,"
6. Gregory, *The Great Rift Valley,* 9, 220, 231; Gregory, *The rift valleys,* 18; Rotberg, *Africa and its explorers;* Fernández-Armesto, *The Times atlas,* chap. 39.
7. Kent, "Historical background," 1–4.
8. Also see Brown, *Where giants trod.*
9. Gregory, *The rift valleys,* 332–333.

10. Nesbitt, *Hell-hole of Creation*, 48, 290, 295–296, 301–303; Marcus, *The modern history of Ethiopia*, nos. 827, 829, 834; Marcus, *The life and times of Menelik II*, 62.

11. Jones and Monroe, *A history of Ethiopia*, 137–138; R. L. Hess, *Ethiopia*, 54–55; Marcus, *The modern history of Ethiopia*, no. 710.

12. Nesbitt, *Hell-hole of creation*, 301–302, map insert; U.S. Army, *Joint operations graphic*, NC 37-4, ND 37 1C.

13. Franchetti, *Nella dancalia Etiopica*.

14. Varet, comp., "Geological map of the central and southern Afar."

15. Franchetti, *Nella dancalia Etiopica*, 174.

16. Vinassa de Regny, *Dancalia*; U.S. Army, *Joint operations graphic*, ND 37 16; United Nations, *Geology, geochemistry and hydrology*.

17. Jones and Monroe, *A history of Ethiopia*, 187–192; Gigli, *Strade imperiali*.

18. Willis, *The hominid gang*, 45; Jennings, "The daybreak inquirey."

19. Maglio, "Early Elephantidae"; Maglio, "Vertebrate faunas"; Cooke and Maglio, "Plio-Pleistocene stratigraphy in East Africa."

20. Cooke and Maglio, "Plio-Pleistocene stratigraphy in East Africa"; Maglio, "Origin and evolution of the Elephantidae."

21. Coppens, "Decouverte d'un australopithecine"; Howell, "Hominidae," 199, table 10.6.

22. Morell, *Ancestral passions*, 398–400.

23. Brown, *Where giants trod*; Imperato, *Quest for the Jade Sea*.

24. Barkeley, *The road to Mayerling*, 2.

25. Marcus, *A history of Ethiopia*, 97–100.

26. The background and research in the Omo Valley can be found in Cole, *The prehistory of East Africa*, 271, 287–288; Coppens, "Earliest man and environments," v, xviii–xix; Johanson and Eddy, *Lucy*, chap. 5; Willis, *The hominid gang*, 38–39; Boaz, *Quarry*, 53–54; and Morell, *Ancestral passions*, 272 and chap. 20.

27. In the late 1960s archeological research in Ethiopia was administered through the "Institut Ethiopien d'Archéologie," as it was then known, in Addis Ababa under the Ethiopian Antiquities Administration. The "chef de la mission française" was Francis Anfray; representing the institute at the 1967 Pan-African Congress of Prehistory held in Dakar was Jean Chavaillon, who became field director of the French team in the Omo.[28]

28. Arambourg, Chavaillon, and Coppens, "Expedition internationale de recherches paleontologiques," 139.

29. Morell, *Ancestral passions*, 288.

30. Ibid., 385–388, 397.

31. Ibid., 396–397.

32. Ibid., 351–352.

33. Johanson and Eddy, *Lucy*, 127.

34. For descriptions of the area east of Magenta Gap, see Rognon and Gasse, "Depots lacustres Quaternaires," 310–313; Gasse and Rognon, "Le Quaternaire des bassins lacustres," 407–408; and Gasse, "Sedimentary formations," 1978, 46.

35. Parker and Hayward, *An Afar–English–French dictionary*, 36; R. Hayward, personal communication, April 12, 1999.

36. Taieb, "Aperçus sur les formations"; Taieb et al., "Depots sedimentaires et faunes."

37. Cecchi, *Da Zeila alle frontiere del Caffa*; Nesbitt, *Hell-hole of creation*; Badoglio, *The war in Abyssinia*; United Kingdom, War Office, *Abyssinia*, map no. 2319.

38. Nesbitt, *Hell-hole of creation*; Thesiger, "The Awash River."

39. U.S. Army, *Joint operations graphic*, NC 37-4; Kalb et al., "Stratigraphy of the Awash Group," fig. 3.

40. In his book *Sur la terre des primiers hommes* (published in 1985), Taieb says he discovered Hadar in 1968.[a] In several activity reports written in 1968 and 1969, however, he describes field work carried out during those years in the upper Awash and on the Hararghe plateau, with no mention of Hadar.[b] In his book he also refers to field work carried out in the central Afar in 1969 with the UN geothermal team,[c] but they did not begin their work in Ethiopia until mid-1970.[d] At a congress in 1969, Taieb gave papers describing field work carried out in 1968 along portions of the Ledi and Millé rivers, again with no mention of Hadar.[e] In a paper published in 1971, however, he mentions Hadar and its abundant fossils for the first time, which he says were collected in December 1970.[f] In an activity report published in 1971 on research during 1970–1971, he again mentions Hadar.[g] If Taieb did discover Hadar in late 1970 instead of 1968, why would he claim the earlier date? It may be that he feared I might ultimately claim discovery or co-discovery of the site, or of its counterpart, the Meshellu basin, and that he wanted to put distance between the alleged discovery date (1968) of Hadar, our joint field work in late 1971, and my independent explorations of the Afar beginning in early 1972.
   a. Taieb, 1985, 26, 284;
   b. Chavaillon and Taieb, "Stratigraphie du quaternaire"; Alimen et al., "Les etudes françaises"; Taieb, "Differents aspects"; Taieb, "Melka Kontouré"; Chavaillon, *Les civilisations Ethiopiennes*, 20–22, 47–49; Anfray, Bonnefille, and Chavaillon, *Documents pour servir à l'histoire*, 23, 90, 103;
   c. Taieb, *Sur la terre*, 33–35;
   d. United Nations, *Geology, geochemistry and hydrology*, 1;
   e. Taieb, "Les études Françaises"; Taieb, "Les dépôts quaternaires;"
   f. Taieb, "Aperçus sur les formations quaternaires," note 5; also see Taieb, "Recherches sur le quaternaire," 112;
   g. Taieb, "Compte-rendu des travaux."

41. "International geological and paleontological expedition to the South and Central Afar (Ethiopia)" [original in French] May 11, 1972. Signatories: Maurice Taieb, Head of Mission; Yves Coppens; Carl Johanson; Jon Kalb.

42. Greenfield, *Ethiopia*, 124–127.

43. Badoglio, *The war in Abyssinia*, chap. 12; Barker, *The civilizing mission*, 261; Marcus, *A history of Ethiopia*, 146.

44. Rice, *Captain Sir Richard Francis Burton*, 238, 250–252; Burton, *First footsteps*, 55.

45. Hill, "Cat (*Catha edulis* Forsk); Murphy, "The Qat culture," C20.

Chapter 7: Transition

1. "International Awash-Afar research," 1972. Permit map boundaries drawn on 1:2,000,000 *Addis Ababa* sheet 20, series 2201, edition 6-AMS, U.S. Army Map Service, Corps of Engineers, Washington, D.C.

2. Morell, *Ancestral passions*, 401–402.

3. Ibid., 242–243, 250.

4. Ibid., chap. 4.

5. Marcus, *A history of Ethiopia*, 142.

6. Marcus, *Haile Sellassie*, 1.

7. Marcus, *A history of Ethiopia,* chap. 11.
8. Ibid., 178
9. Taieb et al., "Dépôts sédimentaires."
10. Kalb, "Preliminary bibliography of the geology"; Kalb, "Recent geological research in Ethiopia"; Kalb, "Special report: recent geologic research."

Chapter 8: Famine and Fame

1. Shipman, "Implications of drought for fossil vertebrate assemblages."
2. Mariam, *Rural vulnerability to famine in Ethiopia 1958–1977;* Keller, *Revolutionary Ethiopia,* 166.
3. Varet, comp., *Geological map of central and southern Afar.*
4. Kalb, "Geological, paleontological, and anthropological investigations."
5. Taieb, "Evolution quaternaire du bassin de l'Aouache," 266; also see Taieb et al., "Découverte d'Hominidés."
6. Taieb et al., "Geological and palaeontological background."
7. Morell, *Ancestral passions,* 409.
8. Lewin, *Bones of contention,* chap. 9; Johanson and Edey, *Lucy,* 170–171, 176–178.
9. Johanson and Taieb, "Plio-Pleistocene hominid discoveries in Hadar."
10. Johanson and Edey, *Lucy,* 156.
11. Ibid., 156–159.
12. Corvinus, "Palaeolithic remains"; Corvinus, "Prehistoric exploration at Hadar, Ethiopia."
13. For a review of the work at Gona and its dating, see Corvinus and Roche, "Prehistoric exploration at Hadar in the Afar"; Harris, "Cultural beginnings: Plio-Pleistocene archaeological occurrences," 7–11; Kalb, "Refined stratigraphy of the hominid-bearing Awash Group," 39, 51, fig. 14; Semaw et al., 2.5-million-year-old stone tools from Gona, Ethiopia."
14. Johanson and Edey, *Lucy,* 371–372; reference 12.
15. Bonnefille, Vincens, and Buchet, "Palynology, stratigraphy and palaeoenvironment of a Pliocene hominid site."
16. "3-million-year-old human fossils found," *Ethiopian Herald.*
17. Johanson and Edey, *Lucy,* 156, 161–162.

Chapter 9: Storm Clouds

1. *The unknown famine: A report on famine in Ethiopia.*
2. Mariam, *Rural vulnerability to famine in Ethiopia, 1958–1977,* chap. 2.
3. Marcus, *A history of Ethiopia,* 180–182; Schwab, *Ethiopia: politics, economics and society,* 15; Nelson and Kaplan, *Ethiopia, a country study.*
4. Ajayi and Crowder, *Historical atlas of Africa,* maps 14, 18, 20, 27, 38, 48; Clay and Holcomb, 1986, *Politics and the Ethiopian famine 1984–1985,* maps 2, 4, and 5; Holcomb and Ibssa, *The invention of Ethiopia,* maps 3–17.
5. Marcus, *A history of Ethiopia,* chap. 5; Marcus, *The life and times of Menelik II.*
6. Keller, *Revolutionary Ethiopia,* 65.
7. Marcus, *A history of Ethiopia,* chap. 11; Mockler, *Haile Sellassie's war;* Thesiger, *The life of my choice,* part v.
8. Tiruneh, *The Ethiopian Revolution, 1974–1987,* 29.

9. Marcus, *A history of Ethiopia*, chap. 11.
10. Ibid., chap. 12; Sellassie, *Conflict and intervention in the horn of Africa*, chap. 5; Holcomb and Ibssa, *The invention of Ethiopia*, chap. 7.
11. "5 male and 2 female skyjackers slain in gun duel on jetliner."
12. Ottaway and Ottaway, *Ethiopia: empire in revolution*, 154; Cobb, "Eritrea wins the peace," 84–103.
13. Tiruneh, *The Ethiopian revolution, 1974–1987*, 26.
14. Marcus, *A history of Ethiopia*, 181.
15. "Ethiopia—man of the year."
16. Marcus, *A history of Ethiopia*, 156; Marcus, *The politics of empire*, 48–50.
17. Marcus, *A history of Ethiopia*, 161.
18. Sellassie, *Conflict and intervention in the Horn of Africa*, 158; Sorenson, *Imagining Ethiopia*, 114; Tiruneh, *The Ethiopian revolution, 1974–1987*, 32.
19. Sorenson, *Imagining Ethiopia*, 36.
20. Tiruneh, *The Ethiopian revolution, 1974–1987*, 32–33, 217; Nelson and Kaplan, *Ethiopia*, 221; Ottaway and Ottaway, *Ethiopia*, 166–167.
21. Sellassie, *Conflict and intervention in the Horn of Africa*, 23.
22. Young, "Meet an ancestor, age 3 million."
23. The precipitous events of the revolution are described in Ottaway and Ottaway, *Ethiopia*, chap. 1; Keller, *Revolutionary Ethiopia*, chap. 6; and Tiruneh, *The Ethiopian revolution, 1974–1987*, chap. 2.
24. Taieb, "Évolution quaternaire du bassin de l'Aouache," fig. 97; Taieb et al., "Descoverte d'hominides dans les series plio-pleistocene d'Hadar."
25. Nesbitt, *Hell-hole of creation*, 207, 300.
26. In the spelling and translation of Afar names, I have been generously assisted by Loren Bliese with the Lutheran Mission, Addis Ababa, personal communication, 1996–1999; Richard Hayward, School of Oriental and African Studies, University of London, personal communication, October 22, 1997; and Parker and Hayward, *An Afar–English–French dictionary*.

Chapter 10: Desert Origins

1. Bliese, "Afar"; Loren Bliese, personal communication, June 1999; Richard Hayward, personal communication, October 1997: Newman, *The peopling of Africa*, 51–54, 89–91, 203; Harlan, "Indigenous African Agriculture"; Trimingham, *Islam in Ethiopia*.
2. Anfray, "The civilization of Axum"; Glen Brown, U.S. Geological Survey (retired), personal communication, July 1996; de Contenson, "Pre-Axumite culture"; Doe, *Southern Arabia*; Pankhurst, *Ethiopia: a cultural history*, 106–107; Sellassie, *Ancient and medieval Ethiopian history*, 39–40; Verin, "Madagascar."
3. Kobishchanov, *Axum*, 134–135.
4. Richard Hayward, personal communication, October 22, 1997.
5. Doe, *Southern Arabia*, 17, 73.
6. Burton, *First footsteps*, 52; I. M. Lewis, *Peoples of the horn*, 155–156; Munzinger, "Narrative of a journey through the Afar country," 209–210; Thompson and Adloff, *Djibouti and the horn*, 4–5, 23; Klemp, *Africa*.
7. Richard Hayward, personal communication, July 26, 1999.
8. Kobishchanov, *Axum*, 38–39, 45, 82.
9. Michels, "Axumite archaeology," map 1.

10. Huntingford, *The historical geography of Ethiopia*, 54–55. Kobishchanov, *Axum*, 67, 82, 132, 230, 233; Trimingham, *Islam in Ethiopia*, 36.

11. Huntingford, *The historical geography of Ethiopia*, 47; Mekouria, "Christian Axum," 227.

12. Shehim, "The influence of Islam on the Afar," 54–58; Neuman, *The peopling of Africa*, 93–94; Butzer, *Archaeology as human* ecology, 93–97, 142–145, 156.

13. Trimingham, *Islam in Ethiopia*, 32–33.

14. Marcus, *A history of Ethiopia*, 17–19; Sellassie, *Ancient and medieval Ethiopian history*, 37–38; Kobishchanov, *Axum*, 198–199; Neuman, *The peopling of Africa*, 95; Tamrat, "The Horn of Africa," 438; Perham, *The government of Ethiopia.*, chap. 12; Shillington, *History of Africa*, 166; R. Pankhurst, *A social history of Ethiopia*, 49–55, 64–66.

15. Trimingham, *Islam in Ethiopia*, 66–72; Cerruli, "Ethiopia's relations," 280–281; Neuman, *The peopling of Africa*, 94, 97; Huntingford, *The historical geography of Ethiopia*, 88–89; Ajayi and Crowder, *Historical atlas of Africa*, map 28; Tamrat, "Ethiopia, the Red Sea and the Horn," 142–147.

16. Tamrat, "Ethiopia, the Red Sea and the Horn," 145–146; Shehim, "The influence of Islam on the Afar," 4; Neuman, *The peopling of Africa*, 95.

17. Cerulli, "Ethiopia's relations," 283–284; Shehim, "The influence of Islam on the Afar," 64, 69–71; Trimingham, *Islam in Ethiopia*, 85–86.

18. Tamrat, "The Horn of Africa," 172–173; Huntingford, *The historical geography of Ethiopia*, 120–121, 125–126.

19. Trimingham, *Islam in Ethiopia*, 87.

20. Doresse, *Ethiopia*, 128, 145, 147.

21. Trimingham, *Islam in Ethiopia*, 91–92; Doresse, *Ethiopia*, 147–150; Haberland, "The Horn of Africa," 716–721; Ajayi and Crowder, *Historical atlas*, map 39.

22. Trimingham, *Islam in Ethiopia*, 125–128; Abir, *Ethiopia and the Red Sea*, 140–141; Shehim, "The influence of Islam on the Afar," 94, 98; Marcus, *The life and times of Menelik II*, 66–69.

23. Ajayi and Crowder, *Historical atlas of Africa*, map 48.

24. Kobishchanov, *Axum*, 112; I. M. Lewis, *Peoples of the Horn of Africa*, 145, 155–156; Shehim, "The influence of Islam on the Afar," 19, 58, 145–148; Trimingham, *Islam in Ethiopia*, 45, 60–61.

25. Barberi et al., "Evolution of the Danakil depression," 445.

26. Goss, John, 1993. *The mapmakers art*, plate 5.16, 131.

27. Shehim, "The influence of Islam on the Afar," 93–98.

28. Trimingham, *Islam in Ethiopia*, 93–96.

29. Dilebo, Getahun, "Emperor Menelik's Ethiopia"; Thompson and Adloff, *Djibouti and the Horn of Africa*, 4; Abir, *Ethiopia and the Red Sea*, 76–77.

30. Tiercelin, "The Pliocene Hadar Formation," fig. 3.

31. Kalb, "Refined stratigraphy," 41, 52, 61, fig. 20.

32. *The forbidden desert of Danakil.*

33. Thesiger, "The Awash River"; Thesiger, *The life of my choice*, part 2.

34. Thesiger, "The Awash," 1; Thesiger and Maynell, "On the collection of birds."

35. Thesiger, "The Awash," 15.

36. Ibid., 16.

37. Shehim, "The influence of Islam on the Afar," 101–102.

38. Thompson and Adloff, *Djibouti and the Horn*, 5, 9; R. Pankhurst, *A social history of Ethiopia*, 101, 244–246; Doe, *Southern Arabia*, 13, 51; R. Pankhurst, *Economic history of Ethiopia*, 393–394; R. Pankhurst, *A social history of Ethiopia*,

chap. 7; Thompson and Adloff, *Djibouti and the Horn,* 5; Shehim, "The influence of Islam on the Afar," 15, 103.

39. Shehim, "The influence of Islam on the Afar," chaps. 4–5; Trimingham, *Islam in Ethiopia,* 171–173; Thesiger, "The Awash River," 5–6.

40. Pankhurst, *A social history of Ethiopia,* 65, 84, 157.

41. Marcus, *The life and times of Menelik II,* 62; Rubenson, *The survival of Ethiopian independence,* 152–153; Marcus, *The modern history of Ethiopia,* items 427–428; Perham, *The government of Ethiopia,* 341.

42. Marcus, *The life and times of Menelik II,* 36–40, 60–62; Marcus, *The modern history of Ethiopia,* items 183–185; Shehim, "The influence of Islam," 103–113.

43. Marcus, *The life and times of Menelik II,* chaps. 2–3; Holcomb and Ibssa, *The invention of Ethiopia,* 305; Maknun and Hayward, "Tolo Hanfade's song."

44. Marcus, *The life and times of Menelik II,* 62–63, 109; Perham, *The government of Ethiopia,* 341; Greenfield, *Ethiopia,* 120–121.

45. Marcus, *The life and times of Menelik II,* 260, chap. 9; Greenfield, *Ethiopia,* chap. 8; Marcus, *The life and times of Menelik II,* chap. 9; Perham, *The government of Ethiopia,* 341.

46. Greenfield, *Ethiopia,* 137–144; de Prorok, *Dead men do tell tales,* chaps. 17 and 20; Thesiger, *The life of my choice,* 48–49.

47. Del Boca, *The Ethiopian war,* 16; Baeu, *The coming of the Italian–Ethiopian war,* 249; Martelli, *Italy against the world,* 69, 190; De Bono, *Anno XIII,* 51–52; Tom Ofcansky, U.S. Department of State, personal communication, Feb 23, 2000.

48. Nesbitt, *La Dancalia,* X; Nesbitt, *Gold Fever,* VII–XIX; Nesbitt, "From south to north"; L. E. J. 1936, "Review of *Desolate marches, by L. M. Nesbitt*," 84.

49. Martelli, *Italy against the world,* 94; Barker, *The civilizing mission,* 114.

50. Greenfield, *Ethiopia,* 198, 218; Barker, *The civilizing mission,* 242; Badoglio, *The war in Abyssinia,* 113–132, map 8; Mockler, *Haile Selassie's war,* 111–112; Nesbitt, *Hell-hole of creation,* 237; Thesiger, "The Awash," 12; Thesiger, *The life of my choice,* 141; Shehim, "The influence of Islam on the Afar," 108, 207; Ibrahim, "Politics and nationalism," 598–599.

51. Thompson and Adloff, *Djibouti and the Horn,* 108; Barker, *The civilizing mission,* 169; Greenfield, *Ethiopia,* 286–287; Lewis, *Peoples of the Horn of Africa,* 157; Shehim, "The influence of Islam on the Afar," 109.

52. Burton, *First footsteps in East Africa,* 53.

53. Nesbitt, *Hell-hole of creation,* 56.

54. Waugh, *Waugh in Abyssinia,* 49.

55. de Prorok, *Dead men do tell tales,* 239.

56. Diel, *Behold our new empire,* 147.

57. Trimingham, *Islam in Ethiopia,* 172.

58. Loren Bliese, personal communication, July 18, 1999.

59. Keller, *Revolutionary Ethiopia,* 184.

Chapter 11: Dawn of Humanity

1. Request for Authorization of Archaeological Research, submitted to the Ethiopian Antiquities Administration, J. Kalb, August 3, 1974.

2. B. Negussie to M. Taieb and D. Johanson, August 9, 1974.

3. D. Johanson to B. Negussie, August 16, 1974.

4. M. Taieb to B. Negussie, August 17, 1974.

5. M. Taieb to Tekle Tsadik Mekuria, August 23, 1994.
6. M. D. Leakey to J. Kalb, September 17, 1974.
7. B. Negussie to M. Taieb, August 28, 1974.
8. M. Mercouroff to B. Negussie, August 28, 1974.
9. M. Taieb to B. Negussie, September 2, 1974.
10. B. Negussie to M. Taieb (telegram), September 10, 1974; B. Negussie to the directors of CNRS, NSF, and FORGE, September 10, 1974. In response to Negussie's request, the director of NSF's Anthropology Program sent him a blanket endorsement of Johanson's research (I. Ishino to B. Negussie, September 13, 1974).
11. M. Taieb, D. Johanson, and other IARE members to B. Negussie, September 10, 1974.
12. Marcus, *A history of Ethiopia*, 187–188.
13. Keller, *Revolutionary Ethiopia*, 186–187.
14. Ottaway and Ottaway, *Ethiopia*, 58–59.
15. Bernstein and Woodward, *All the president's men*, 461.
16. Ottaway and Ottaway, *Ethiopia*, 54–58.
17. J. Kalb to B. Negussie, September 22, 1974.
18. Bell, *Impure science*, 29. FORGE was based in Stamford, Connecticut; its president was Bentley Glass, of the State University of New York at Stony Brook, also a former president of the American Association for the Advancement of Science. FORGE dissolved with the death of its founder and director, Alfred Kelleher, in July 1976 (Jeffery Kelleher, Cleveland, Ohio, personal communication, February, 1998).
19. Agreement for [the] International Afar Research Expedition 1974. Signatories: M. Taieb, D. Johanson, Y. Coppens, J. Kalb.
20. Johanson and Taieb (1976), "Plio-Pleistocene hominid discoveries."
21. International Afar Research Expedition, press release, October 25, 1974; see "Ethiopian fossil hominids."
22. Johanson and Edey, *Lucy*, chaps. 13–15; Lewin, *Bones of contention*, chaps. 11 and 12.
23. "A leading scientist hails fossil find"; Lewin, *Bones of contention*, 168–170, 174.
24. Ottaway and Ottaway, *Ethiopia*, 59; Tiruneh, *The Ethiopian Revolution*, 72–77.
25. Keller, *Revolutionary Ethiopia*, 185; Tiruneh, *The Ethiopian revolution*, 65, 78; Marcus, *A history of Ethiopia*, 189; Ottaway and Ottaway, *Ethiopia*, 59–60.
26. Marcus, *A history of Ethiopia*, 189.
27. "Ethiopian officials executed."
28. "60 former Ethiopian leaders executed"; "Ethiopia victims list."
29. Ottaway and Ottaway, *Ethiopia*, 61; Marcus, *A history of Ethiopia*, 189–190; Tiruneh, *The Ethiopian revolution*, 78; Thomson, *Ethiopia*, 119–124; Harbeson, *The Ethiopian transformation*, 124–125.
30. Johanson and Taieb (1976), "Plio-Pleistocene hominid discoveries," table 1.
31. "Skeleton may be 3 million years old"; "40 percent of Lucy."
32. Morell, *Ancestral passions*, 466.
33. Johanson and Taieb (1976), "Plio-Pleistocene hominid discoveries," fig. 2; Taieb et al., "Expedition internationale de l'Afar," figs. 1 and 2; Corvinus, "Palaeolithic remains," fig. 1; Kalb and Peak, "Documentation of fossil sites," Hadar locality map.
34. Johanson and Edey, *Lucy*, 196–198; Johanson and Taieb (1978), "Plio-Pleistocene hominid discoveries," fig. 2; Johanson et al., "Geological framework of the Pliocene Hadar Formation," fig. 34:3, profile 22.

35. Johanson, Progress report; also see Taieb et al., Découverte d'Hominides."
36. Taieb et al., "Geological and palaeontological background."
37. Rensenberger, "Skeleton fossils."
38. Renne et al., "New data"; Kalb, "Refined stratigraphy," 50–51.
39. Johanson and Taieb (1976), "Plio-Pleistocene hominid discoveries."

Chapter 12: A New Mission

1. Kalb, "The potential for prehistory research."
2. Kalb and Peak, Documentation of fossil sites; Oswald, *Geologic and paleontologic investigations.*
3. Ottaway and Ottaway, *Ethiopia,* 69–71.
4. Fred Wendorf, personal communication, October 1975 and September 3, 1982.
5. Johanson and Edey, *Lucy,* 168.
6. Jack Edwards, Tenneco (retired), personal communication, June 19, 1997; D.O. Nelson, Chevron (retired), September 19, 1997; "5 male and 2 female skyjackers slain."
7. Bill Cayce, Tenneco (retired), personal communication, June, 29, 1997.
8. Kalb et al., "Geology and stratigraphy of Neogene deposits," fig. 3.
9. Nesbitt, *Hell-hole of creation* (map insert); Thesiger, "The Awash River"; Thesiger, *The life of my choice.*
10. Dal Piaz et al., *Illustrazione geologica,* xi–xv, xxiv–xxviii, xxxv.
11. Gortani and Bianchi, "Itinerario Mieso-Gauani-Tiho"; Gortani and Bianchi, "Carta geologica della Dancalia meridionale." For an overview of the work of Gortani and Bianchi, see Kalb et al., *Bibliography for the earth sciences for the Horn.*
12. Mohr, *Geological map of the Horn of Africa.*
13. United Nations Food and Agricultural Organization, *Report on survey of the Awash River basin,* vol. 2, "Geomorphology map" (insert) and 4–5.
14. Taieb, "Évolution quaternaire du bassin de l'Aouache"—"Carte morphostructurale" (insert).
15. Taieb et al., "Dépots sédimentaires" (with map insert); United Nations, *Geology, geochemistry and hydrology*—"Photogeologic map of the southern and central Afar" (insert).

Chapter 13: An Acheulean "City"

1. Wendorf and Schild. *A Middle Stone Age sequence.*
2. Kalb et al., "Preliminary geology, paleontology and paleoecology."
3. Sampson, *The Stone Age of archaeology,* 222–229.
4. Schick and Toth, *Making silent stones speak,* 293–302.
5. For an overview of the archeological discoveries in the Middle Awash by the RVRME, see Kalb et al., "Early hominid habitation in Ethiopia."

Chapter 14: Rebellion

1. For a current overview of this work, see Zanettin, "Evolution of the Ethiopian volcanic province."
2. Fleagle, *Primate adaptation and evolution,* 387–390.

3. Schick and Toth, *Making silent stones speak,* 35–36.
4. For general discussion of the background to and events of the 1975 Afar rebellion, see Beshah and Harbeson, "Afar pastoralists in transition"; Bondestam, "People and capitalism"; Harbeson, "Territorial and development politics"; Harbeson, *The Ethiopian transformation,* 48–68; Ottaway and Ottaway, *Ethiopia,* 82–97; Shehim, "The influence of Islam on the Afar"; and Winid, "Comments on the development."
5. These are the author's own rough estimates based on Ottaway and Ottaway, *Ethiopia,* 202, footnote 18; Said, *Pastoralism and the state policies,* 1; and Ofcansky and Berry, 248.
6. Ottaway and Ottaway, *Ethiopia,* 17.
7. Winid, "Comments on the development," 154–156.
8. Beshah and Harbeson, "Afar pastoralists in transition," 256; Ottaway and Ottaway, *Ethiopia,* 84; Said, *Pastoralism,* 8–9.
9. Ali Said, *Pastoralism,* 12; Bondestam, "People and capitalism," 423.
10. It is worth noting that a five-volume report by the UN FAO, completed in 1965, hardly mentions the effects on the Afar people of agricultural development in the Awash Valley. See UN FAO "Report on survey."
11. Beshah and Harbeson, "Afar pastoralists in transition," 258; Shehim, "Ethiopia, revolution," 336–337.
12. Said, *Pastoralism and the state policies,* 10.
13. Vivó, *Ethiopia's revolution,* 120.
14. Ottaway and Ottaway, *Ethiopia,* 96.
15. Ibid., 96; Harbeson, "Territorial and development politics," 485.
16. Selassie, *Conflict,* 113; Shehim, "Ethiopia, revolution," 342–343.
17. Ottaway and Ottaway, *Ethiopia,* 96; Shehim, "The influence of Islam on the Afar," 43; "Fighting erupts in Ethiopia"; Richard Hayward, personal communication, October 22, 1997.

Chapter 15: The Oldowan

1. *Who's who in the world, 1976–1977.*
2. Clark, "Fred Wendorf"; Close, "Preface."
3. Wendorf and Schild, *A Middle Stone Age sequence.*
4. M. D. Leakey, *Excavations;* L. Leakey, Evernden, and Curtis, "Age of Bed I."
5. Walter et al., "Laser-fusion ^{40}Ar/^{39}Ar dating."
6. Clark, *The prehistory of Africa,* 68; Morell, *Ancestral passions,* 352, 398; Feibel, Brown, and McDougall, "Stratigraphic context."
7. Howell, Haesaerts, and de Heinzelin, "Depositional environments."
8. R. Leakey and Lewin, *Origins,* 84–85.
9. Ottaway and Ottaway, *Ethiopia,* 97–98, 156, 163–164.
10. Tiruneh, *The Ethiopian revolution,* 103.
11. Ottaway and Ottaway, *Ethiopia,* chap. 7.
12. Ibid., 108.
13. Hall et al., "Geochronology, stratigraphy."

Chapter 16: The "First Movie"

1. Ottaway and Ottaway, *Ethiopia,* 109–111, 203, note 22; Ofcansky and Berry, *Handbook of Ethiopia,* 56.

2. Kapuscinski, *The emperor*, 163.
3. Ciochon, "A remarkable lineage," 33.
4. D. Johanson to M. Green, November 11, 1975; NSF grants listing year 1975.
5. Abebbe [Abebe], *Evolution de la propriété*.
6. Greenfield, *Ethiopia*, 168–70; Marcus, *A History of Ethiopia*, 134; Clapham, *Haile Selassie's government*, 34–36; Nahum, *Constitution*.
7. Perham, *The government*, 70; Marcus, *A history of Ethiopia*, 17–18, 33, 38.
8. "Emperor presents prizes."
9. Tiruneh, *The Ethiopian revolution*, 69.
10. Johanson and Edgar, *From Lucy to language*, 23–24, 213–214.
11. Radosevich, Retallack, and Taieb, "Reassessment," 22–26.
12. Johanson, Taieb, and Coppens, "Pliocene hominids," 380.
13. Johanson and Edey, *Lucy*, 212–213.
14. For current dating of these localities, see Kimbel et al., "Late Pliocene *Homo*," fig. 1.
15. Johanson, "Ethiopia yields 'First Family.' "
16. Aronson and Taieb, "Geology and paleogeography," 187–191; Gray, "Environmental reconstruction," 211–216.
17. Stern and Sussman, "The locomotor anatomy."
18. Johanson and Eddy, 216. A brief examination of the L333 collection by taphonomist Pat Shipman revealed that nonhominid remains were recovered from the locality (P. Shipman, personal communication, June 16, 1997).

Chapter 17: The Miocene

1. Assefa Mebrate, "A primitive fossil elephant"; Assefa Mebrate, Fossil Probscidea from the Middle Awash."
2. Sleshi Tebedge, "Genus *Sus* (Suidae) from northwest Hararghe"; Sleshi Tebedge, "Fossil Suidae"; Kalb et al., "Catalog (revised) of fossil fauna."
3. Kalb, Oswald, and Mosca, "Spatial and temporal distribution, table 4."
4. White et al., "New discoveries," fig. 1; Clark et al., "Old radiometric ages," 1907.
5. A. Azzaroli to J. Kalb, September 16, 1975.
6. Berggren and Van Couvering, *The late Neogene*, 101–103, 172, fig. 11.
7. R. Bernor, personal communication, July 9, 1997; Woodburne, Bernor and Swisher, "An appraisal of."
8. J. Van Couvering, personal communication, July 28, 1997.
9. Berggren and Van Couvering, *The late Neogene*, 172.
10. Sickenberg and Schönfeld, "The Chorora Formation," 1975; Kunz, Kreuzer, and Mueller, "Potassium-argon age determinations."
11. Tiercelin, et al., "Le Miocène supérieur"; Dalrymple (1979), *Geology 7*: 558–560.
12. Petrocchi, "Il giacimento fossilifero di Sahabi"; Petrocchi, "Il giacimenta fossilifero di Sahabi. Parte 2."
13. Maglio, "Four new species"; Smart, "The Lothagam 1 fauna"; M. Leakey, et al., "Lothagam"; and McDougall and Fiebel, "Numerical age control."
14. Kalb and Mebrate, "Fossil elephantoids."
15. Maglio, *Origin and evolution*, 89–90.
16. Saunders, "North American Mammutidae"; Agenbroad and Mead, "Distribution and palaeoecology"; Dudley, "Mammoths, gomphotheres."
17. Shoshani and Tassy, "Summary, conclusions"; Webb, Hulbert, and Lambert, "Climatic implications."
18. Kalb and Mebrate, "Fossil elephantoids"; Kalb, Froehlich, and Bell, "Palaeobiogeography," 117.

19. Shoshani et al., "The earliest proboscideans," 70; Tassy, "Who is who."
20. Harris, J. M., "Deinotherioidea," 327.
21. Murray, *The love of elephants,* 76.
22. Shehim, "The influence of Islam on the Afar," 52–62.
23. McLynn, *Hearts of darkness,* chap. 8.
24. Kalb and Mebrate, "Fossil elephantoids," 86.
25. Powell-Cotton, *A sporting trip through Abyssinia,* 41–53, 483; Harrison, "A journey from Zeila," 263.
26. Thesiger, "The Awash River."

Chapter 18: Crossing the Talalak

1. Tiruneh, *The Ethiopian revolution,* 157–158.
2. Shipman, *Life history of a fossil,* 5–6.
3. Ibid., 3.
4. Folk and Hoops, "An early iron-age layer."
5. Andrews, *Owls, caves and fossils,* 6.
6. Kloos and Lemma, "Bilharziasis in the Awash Valley III."
7. Tazieff et al., "Tectonic significance."
8. Barberi et al., "Volcanism in the Afar Depression," 19–29; Barberi et al., "Evolution of the Danakil depression"; Black et al., "Sur la decouverte."
9. Varet, comp., *Geological map.*
10. Brinkman and Kürsten, *Geological sketchmap of the Danakil.*
11. Coppens, "Un nouveau proboscidean"; Kalb and Mebrate, "Fossil elephantoids."
12. Unfortunately, limited space in our Land Rover and lack of adequate preparation materials prevented us from collecting the foot. Plans to collect it later were cut short.
13. Wood, "A detailed report," 14.
14. Nesbitt, *Hell-hole of creation,* 178–179.

Chapter 19: The Central Awash

1. Kalb, "Annual report 1975–1976."
2. Hill, "Late Miocene and early Pliocene hominoids," 131–132; McDougall and Feibel, "Numerical age control."
3. Alan Patterson, personal communication, October 1997; Turnbull, "Bryan Patterson."
4. Kalb et al., "Stratigraphy of the Awash Group."
5. The *Anancus* "stages" went through several permutations: Kalb, "Mio-Pleistocene deposits," 92–93; Kalb et al., "Vertebrate faunas from the Awash"; Mebrate and Kalb, "Anancinae"; Kalb and Mebrate, "Fossil elephantoids."
6. Taieb et al., "Geological and palaeontological background."

Chapter 20: Baboons and Pigs

1. Ottaway and Ottaway, *Ethiopia,* 126, 135, 140; Tiruneh, *The Ethiopian revolution,* 182; "Ethiopia reports 2 fugitives slain," *New York Times,* August 13, 1976, A3.
2. Ottaway and Ottaway, *Ethiopia,* 132–133; Keller, *Revolutionary Ethiopia,* 197; Tiruneh, *The Ethiopian revolution,* 180–181.
3. Harbeson, *The Ethiopian transformation,* 155–156.

4. Jolly, "The classification and natural history of *Theropithecus.*"
5. Jolly and Ucko, "The riddle of the sphinx-monkey."
6. Jolly, "The seed eaters"; Jolly, "[Letter]."
7. Jolly and Plog, *Physical anthropology and archeology,* 86; Jablonski, "Evolution of the masticatory apparatus," 114, 142, 283–286.
8. Jablonski, "Evolution of the masticatory apparatus," 299, 300, 383, 418, 419.
9. Lewin, *Bones of contention,* 98–99.
10. C. Jolly, personal communication, August 1976.
11. Jolly and Plog, *Physical anthropology and archeology,* 68–69.
12. C. Jolly, personal communication, November 5, 1997.
13. Jolly and Plog, *Physical anthropology and archeology,* 69; Mariam, *Rural vulnerability to famine,* table 3.
14. Shipman, "Implications of drought."
15. Cooke, *Changing perspectives,* 1–3; Cross and Maglio, *A bibliography of the fossil mammals*; Simpson, *The Canadian who's who.*
16. Lewin, *Bones of contention,* 189–252, 273–277; Morell, *Ancestral passions,* 352–358, 388, 398, 399, 409, 420, 428–432, 478.
17. Cooke and Maglio, "Plio-Pleistocene stratigraphy"; Cooke, "Suidae from Pliocene–Pleistocene strata"; Cooke, "Faunal evidence."
18. Johanson and Edey, *Lucy,* 137–145, 170–177, 238–243, 353–358; Johanson and Shreeve, *Lucy's child,* 83, 91–101, 106
19. Fitch and Miller, "Radioisotopic age determinations."
20. Brown, "Development of Pliocene and Pleistocene chronology," 295–298.
21. Complete bibliography of Cooke's publications provided to the author; H. B. S. Cooke, personal communication, December 12, 1997.
22. Maglio, "Early Elephantidae"; Lewin, *Bones of contention,* 197–201.
23. Morell, *Ancestral passions,* 420.
24. Ibid., 430; D. Johanson, personal communication, September 1973. Although Cooke presented a paper at the 1975 meeting of the Geological Society of London, he chose not to offer it for publication because of the bitter feelings it raised; H. B. S. Cooke, personal communication, December 19, 1997.
25. Morell, *Ancestral passions,* 429.
26. Cooke, personal communication, December 19, 1997.
27. Tim White, personal communication, June 1976.
28. Harris and White, "Evolution."
29. Lewin, *Bones of contention,* 238.
30. Ibid., 238–239; Johanson and Shreeve, *Lucy's child,* 99–100; Morell, *Ancestral passions,* 478.
31. Morell, *Ancestral passions,* 477–478.
32. Lewin, *Bones of contention,* 239; Morell, *Ancestral passions,* 478.
33. Johanson and Shreeve, *Lucy's child,* 100.
34. Johanson and Edey, *Lucy,* 224; Lewin, *Bones of contention,* 280.
35. Morell, *Ancestral passions,* 478.
36. Lewin, *Bones of contention,* 240–241; White and Harris, "Suid evolution"; Harris and White, "Evolution."
37. Brown, "Development of Pliocene and Pleistocene chronology," 296; Lewin, *Bones of contention,* 245–252.
38. J. M. Harris to Soloman Tekalign, vice-minister of culture, June 9, 1975; Tekalign to J. M. Harris, September 3, 1975. Harris complained to the vice-minister accordingly but was assured that the RVRME was fully authorized.

39. Harris and White, "Evolution," 11, 37, 83, 88, 97.
40. See Coppens et al., *Earliest man.*
41. Cooke, "Pliocene–Pleistocene Suidae."

Chapter 21: The Artifact Trail

1. Alfred Kelleher began his career as an instrument maker at MIT and eventually became a grants officer for the Research Corporation of New York. After his retirement there he founded FORGE (J. Kelleher, Cleveland, Ohio, personal communication, February 2, 1998).
2. Discussion based on Singleton and Larson, "Catalog of artifacts"; Singleton, "Archeology report"; Larson, "Matabaietu"; Kalb et al., "Stratigraphy of the Awash Group"; and Kalb, "Refined Stratigraphy."
3. Kalb et al., "Preliminary geology."
4. Imbrie and Imbrie, *Ice ages,* chap. 15, fig. 40.
5. Corvinus, "Palaeolithic remains."
6. Ottaway and Ottaway, *Ethiopia,* 126–127.

Chapter 22: Human Origins

1. Whitehead, "Hominid discovery."
2. White, "Acheulian Man," 20; White, "Cut marks," 508–509.
3. Kalb et al., "Preliminary geology and palaeontology."
4. This age was our first impression, given in a late 1976 grant proposal submitted to NSF by Conroy and Jolly, although our later estimates would vary widely.
5. Clark et al., "Old radiometric ages."
6. Smart, "Report on the Bovidae"; Smart, "Report on the preliminary paleoecology."
7. Whitehead, "Summary of primate fossils." The remaining fossil identifications were included in Kalb et al., "Catalog (revised) of fossil fauna."
8. Singleton, "Archeology report for the Middle Awash"; Singleton, "Preliminary archeology report on the Hounda Bodo"; Larson, "Matabaietu."
9. Singleton and Larson, "Catalog of artifacts."
10. Kalb and Mabrete, "Fossil elephantoids."
11. Ottaway and Ottaway, 134; Lefort, *Ethiopia,* 287.

Chapter 23: Espionage Science

1. Marsh, "Curious dealer paid expedition."
2. Callison, "For 2 SMU students."
3. Alle Feleghe Selam to J. Kalb, February 1, 1977.
4. Morell, *Ancestral passions,* 347–349.
5. Lewin, *Bones of contention,* 197–198.
6. "Ethiopian students attack U.S. and British centers."
7. "Throngs in Addis Ababa condemn regime's foes."
8. "Ethiopian head and 6 in capital reported slain"; Ottaway and Ottaway, *Ethiopia,* 142–143.
9. Johanson and Edey, *Lucy,* 222.

10. N. Gonzalez to D. Johanson, November 11, 1976.
11. D. Johanson to N. Gonzalez, November 29, 1976.
12. Johanson and Edey, *Lucy,* 228–232; Lewin, "Ethiopian stone tools are world's oldest"; Corvinus and Roche, "Prehistoric exploration at Hadar"; Harris, "Cultural beginnings," fig. 4, 11–12.
13. Gibbons, "Claim jumping"; Johanson, Kimbel, and Walter, "Fossil collecting."
14. "Human skull, tools discovered in Awash."
15. Bell, *Impure science,* 12–13.
16. Ibid., 14.
17. Ibid., 16–18.
18. Ibid., 19–20.
19. N. Gonzalez, personal communication, September 8, 1978.
20. Bell, *Impure science,* 18.
21. F. Wendorf to Kalb, May 3, 1977 with NSF panel summary of Wendorf proposal 77-07964; F. Wendorf to N. Gonzalez, May 4, 1977.
22. Clark and Kurashina, "Hominid occupation."
23. J. D. Clark to N. Gonzalez, May 15, 1977.
24. Cooke, Harris, and Harris, "J. Desmond Clark."
25. J. D. Clark, "Glynn Isaac."
26. Clark, J. D., *Kalambo Falls,* vols. 1 and 2.
27. Clark, J. D., "Palaeoanthropological investigations in east central Ethiopia." Research proposal submitted to NSF, March 1973, with study area map enclosed.
28. Bekele Negussie to J. D. Clark, August 1, 1973; J. D. Clark to M. Taieb, August 14, 1973; M. Taieb to J. D. Clark, August 23, 1973.
29. NSF grants listing, 1974–1977; note 22.
30. "Panel summary," Fall 1978, NSF proposal application BNS 7826025, J. D. Clark, principal investigator; R.T. Louttit to J. D. Clark, January 16, 1979.
31. "Panel summary," May 5, 1977, NSF proposal application BNS 7708948, G. Conroy and C. Jolly, principal investigators.
32. Ottaway and Ottaway, *Ethiopia,* 168.
33. Marcus, *A history of Ethiopia,* 167–173; Ottaway and Ottaway, *Ethiopia,* 167; Kaufman, "Surrounded by enemies."
34. Korn, *Ethiopia, the United States and the Soviet Union,* 28.
35. Ibid., 26.
36. "Ethiopian leader visits USSR."
37. J. Kalb to N. Gonzalez, June 10, 1977, with statement (dated May 28, 1977).
38. N. Gonzalez to J. Kalb, July 8, 1977.
39. Bell, *Impure science,* 21.
40. J. D. Clark, proposal with budget to NSF, "Palaeo-anthropology and prehistoric environments in east-central Ethiopia," August 15, 1977; Bell, *Impure science,* 21–23.
41. J. D. Clark, permit application submitted to the Ministry of Culture, August 23, 1977.
42. Kalb, "Annual report 1976–1977."
43. Kaplan et al., *Area handbook for Ethiopia,* 194; Nelson and Kaplan, *Ethiopia,* 127.
44. Kaufman, "Ethiopian official is believed to have been executed."
45. J. Yellen to J. D. Clark, September 20, 1977.
46. "High Ethiopian casualties reported."
47. J. Yellen telephone memoranda, December 6 and 7, 1977.
48. Agreement of Affiliation of the RVRME and the Science Faculty of Addis Ababa, September 15, 1977.

49. Haile W. Mikael to Mamo Tesema, October 25, 1977.
50. J. Kalb to H. Guyford Stever, December 14, 1977.
51. J. Yellen memorandum to E. E. Clark, January 1978.
52. U.S. House Committee, "War in the Horn of Africa."
53. Ibid., 14.

Chapter 24: Big-Game Feeders

1. Tiruneh, 220–221; Ofcansky and Berry, 311–314.
2. Jaeger et al., "Evidence of itinerant breeding"; Mundy and Jarvis, *Africa's feath-ered locust.*
3. Isaac, *Olorgesailie,* 57–58, 91–92, 217.
4. Shipman, Bosler, and Davis, "Butchering of giant geladas," 260–263.
5. Potts, *Humanity's descent,* 215–216.
6. Jolly, "The classification and natural history."
7. White, *Prehistoric cannabalism.*
8. Claud Bramblett, personal communication, May 1, 1998.
9. Estes, *The safari companion,* 84.
10. Kingdon, *East African mammals,* 391.
11. Darnton, "Ethiopia drives Somalia's forces"; Hovey, "Brzezinski asserts"; Kaufman, "Somalis abandoning North Ogaden"; Nelson and Kaplan, *Ethiopia,* 270–271; Troy Dietz, personal communication, July 12, 1999.
12. Ofcansky and Berry, *Ethiopia,* 313–314.
13. Conroy et al., "Newly discovered fossil."
14. For general discussion, see Stringer, "Evolution of early humans"; Wolpoff, *Human evolution,* 478, 498–509, 524–525, 580–583; Rightmire, *The evolution of Homo erectus,* 204–218; Johanson and Edgar, *From Lucy to language;* Tattersall, *The fossil trail.*
15. Tattersall, "Species recognition"; Tattersall, "Species concepts."
16. Kappleman, "The evolution of body mass," table 4, fig. 3; John Kappleman, personal communication, August 25, 1999.
17. Glenn Conroy, personal communication, June 29, 1999.
18. Groves, "Comment on Shipman, Bosler, and Davis," 265.
19. Kappleman, "The evolution of body mass," 271–272.
20. White, *Acheulian man,* 20–21.
21. Johanson and Edgar, *From Lucy to language,* 93, 194.
22. Johanson and Edey, *Lucy,* 218; Morell, *Ancestral passions,* 475.
23. Johanson and Taieb, "Plio-Pleistocene hominid discoveries"; also see Johanson et al., "Geological framework."
24. Aronson et al., "New geochronologic and palaeomagnetic data," fig. 1.
25. Brown, "Development of Pliocene and Pleistocene chronology," 293–294; M. Leakey et al., "Fossil hominids from the Laetoli beds."
26. M. Leakey, "Pliocene footprints"; Morell, *Ancestral passions,* 475, 482, 485–487.
27. Morell, *Ancestral passions,* 501–502.
28. Johanson and Edey, *Lucy,* 247.
29. Ibid., chaps. 13 and 14; Johanson and White, "A systematic assessment."
30. Johanson, White, and Coppens, "A new species of the genus *Australopithecus.*" As a rule, in *Lucy* and his other writings Johanson gives Coppens minimal credit for his contributions to the Afar work. Lewin (*Bones of contention,* 286) states that Coppens was listed as a coauthor of the new species paper primarily because of the

1972 International Afar Research Expedition agreement "that gave [Coppens] the option of appearing as a coauthor on major publications." There was a clause in the agreement to the effect that credits of discoveries should be shared among members of the team, but there was no specific provision regarding shared authorships. In any case, by 1978 the agreement had long since lost any meaning.

31. Note 30, fig. 2; Johanson and White, "A systematic assessment," fig. 4.
32. Johanson and Edey, *Lucy*, 288–289; Morell, *Ancestral passions*, 487–488.
33. M. Leakey, *Disclosing the past*, 181.
34. Lewin, *Bones of contention*, 13–18, 288–298; Morell, *Ancestral passions*, 491–502, 520–524, 529, 537–538, 550.
35. Johanson and Shreeve, *Lucy's child*, 28–29, 32.
36. Morell, *Ancestral passions*, 550. For an in-depth view of the climate of paleoanthropology research in East Africa and Ethiopia in the 1970s and early 1980s, see Jennings, "The daybreak inquirey."

Chapter 25: National Security

1. Ofcansky and Berry, *Ethiopia*, 302–303.
2. Darnton, "Ethiopia uses terror to control capital."
3. "Ethiopia tells of 9 tries on Mengistu's life."
4. Korn, *Ethiopia, the United States and the Soviet Union*, 51.
5. Mamo Tesema, general manager National Museum, "To Whom It May Concern," August 4, 1978; letter copied to Zewde Girmu, Ministry of Culture permanent secretary.

Chapter 26: African Cake

1. For the discussion of King Leopold II and his ambitions, see Emerson, *Leopold II of the Belgians*, 73–121; Pakenham, *The scramble for Africa*, 239–255; Reader, *Africa*, 525–543; Hochschild, *King Leopold's ghost*, part I.
2. Reader, *Africa*, 529.
3. Ibid., 531.
4. Ibid., 532.
5. Emerson, *Leopold II of the Belgians*, 79.
6. J. D. Clark to J. Yellen, August 31, 1978; J. D. Clark to Berhanu Abebe, August 31, 1978.
7. J. D. Clark to J. Yellen, November 6, 1978; Bell, *Impure science*, 27.
8. Bell, *Impure science*, 28.
9. Tadessa Terfa to J. D. Clark, telegram, December 14, 1978.
10. J. D. Clark to J. Yellen, March 7, 1979.
11. J. Yellen to deputy director of NSF, November 20, 1979.
12. J. D. Clark to J. Yellen, May 7, 1979.
13. J. D. Clark to M. Taieb, May 7, 1979.
14. J. Yellen to Dick [Richard Louttit], July 26, 1979.
15. Bell, *Impure science*, 28.
16. J. Yellen to R. Louttit and E.E. Clark, memorandum, November 20, 1979.
17. Bell, *Impure science*, 27–28.
18. J. D. Clark to J. Yellen, November 24, 1979.

19. Reader, *Africa*, 551.
20. Chamberlain, *The scramble for Africa*, 124.
21. J. Yellen, Report on Paris and Addis Ababa conferences concerning future research in the Ethiopian Rift Valley, c. January 15, 1980 [undated]; D. Johanson and J. D. Clark, grant proposal (award no. BNS-17724), March 28, 1980, 42–52, Appendices A–C; Johanson and Eddy, *Lucy*, 367–376.
22. "Recommendations to the Ministry of Culture and Sports on the Organization of Research in the Ethiopian Rift and Adjacent Areas." D. Johanson, J. Yellen [for J. D. Clark], M. Taieb, and R. Bonnefille to Berhanu Abebe, January 10, 1980.
23. J. D. Clark to J. Yellen, January 25, 1979.
24. Note 22.
25. Emerson, *Leopold II of the Belgians*, 120; Reader, *Africa*, 542.
26. Emerson, *Leopold II of the Belgians*, 121.
27. Assefa Mebrate, personal communication, June, 1981; Sleshi Tebedge, personal communication, June and October, 1982; Tefera Ghedamu, Inspections officer, Ethiopian National Museum, personal communication, January 25, 1994; Ghedamu, "Pandemonium at the National Museum"; T. White, Letter to Editor, *EFOYTA Monthly*, February 4, 1993, 3 (English version of article that later appeared in the magazine in Amharic; see White, "For sure there is pandemonium.")
28. NSF Anthropology Program, file note, June 18, 1980.
29. J. D. Clark to H. Uznanski, January 25, 1980; J. D. Clark to NSF U.S.–France Program, report on January 1980 meeting in Addis Ababa, January 25, 1980.
30. Bell, *Impure science*, 29.
31. Marcus, *A history of Ethiopia*, 146, 151; Mockler, *Haile Selassie's war*, 150–152, 415–416; Selassie, *My life and Ethiopia's progress*, vol. 2, 36.
32. Mockler, *Haile Selassie's war*, 415.
33. Marcus, *A history of Ethiopia*, 151.
34. Rice, *Captain Sir Richard Francis Burton*, chap. 19–25; Imperato, *Quest for the Jade Sea*, 13–20.
35. Rice, *Captain Sir Richard Francis Burton*, 219; Imperato, *Quest for the Jade Sea*, 8–10; Pavitt, *Kenya*, 68–71.
36. Rice, *Captain Sir Richard Francis Burton*, 309.
37. Imperato, *Quest for the Jade Sea*, 18–20.
38. Severin, *The African adventure*, 229.
39. Rice, *Captain Sir Richard Francis Burton*, 384; Casada, 102–103.

Chapter 27: The CIA Lives

1. D. Johanson and J. D. Clark, grant proposal BNS 80-17724, March 28, 1980, abstract, 42–51, Appendices A–C.
2. D. Johanson grant proposal INT 80-24656, August 6, 1980, abstract.
3. Note 2, prologue.
4. J. L. Bostick to Director, Cleveland Museum of Natural History, August 4, 1980; J. L. Bostick to A. Bowker, August 4, 1980.
5. NSF Anthropology Program file note, c. mid-October 1980 [undated].
6. Tuttle, "Palaeoanthropology without inhibitions"; J. D. Clark to J. Yellen, September 9, 1981; D. Johanson to M. Green, January 14, 1982.
7. Johanson and Edey, *Lucy*, 375.

8. M G [Mary Green], file note, NSF Anthropology Program, August 3, 1981.

9. J. D. Clark to J. Yellen, September 9, 1981.

10. J. Yellen to Division of Grants and Contracts, memorandum re: grant award BNS 79-08342, University of California at Berkeley, September 29, 1981.

11. Clark et al., "Paleoanthropological discoveries"; Asfaw, "A new hominid parietal"; Asfaw, "The Belohdie frontal."

12. Office of Public Information, University of California at Berkeley, press release, File No. 8252, June 10, 1982.

13. Kalb, "Mio-Pleistocene deposits"; Conroy et al., "Newly discovered fossil hominid skull"; Kalb et al., "Preliminary geology and paleontology."

14. Bell, *Impure science,* 32.

15. "Awash in glory."

16. Kalb et al., "Geology and stratigraphy of Neogene deposits"; Kalb et al., "Fossil mammal and artefacts."

17. Tebedge, "Fossil Suidae from the Middle Awash."

18. Post, "The skeleton's in our closet"; J. D. Clark, D. C. Johanson, and T. White, report to Zewde Girma, acting permanent secretary, with attachments, Ministry of Culture, October 5, 1982.

19. Sleshi Tebedge, Tsrha Adefris, Mesfin Asnake, and Assefa Mebrate to Ministry of Culture, petition for an Ethiopian prehistory reserve, September 15, 1981.

20. J. Rom to N. Caputo, July 23, 1982.

21. J. D. Clark and T. White to S. Brush, with attachments, November 7, 1982; D. Korn to Secretary of State, U.S. State Department, cable, September 30, 1982.

22. B. Asfaw to Ethiopian Commissioner of Higher Education, October 11, 1982.

23. D. Korn to Secretary of State, U.S. State Department, cable, October 7, 1982; note 21; Clark, Johanson, and White under note 18; C. Redman to E. E. Clark, under note memorandum, item r, October 29, 1982.

24. To Whom It May Concern from Ministry of Culture, October 3, 1982.

25. J. D. Clark and T. White to S. Brush, November 7, 1982.

26. "Don't dig here."

27. Petit, "A fight for bones and money"; C. Petit, personal communication, February 1, 1983; Cherfas, "Grave accusations."

28. L. Garrett, "All Things Considered," National Public Radio, December 8, 1982. Interviews with D. Johanson, T. White, and J. Kalb; "Cold war, commerce, and the Lucy trail"; Redman to E. E. Clark, under note 23; Bell, *Impure science,* 29.

29. Cherfas, "Grave accusations."

30. Garrett under note 28.

31. D. Johanson to M. Taieb, H. Faure, J. Mirabel, and the French Embassy, October 11, 1982.

Chapter 28: Prehistoric Feast

1. J. Kalb to George Bush, June 23, 1982; Bell, *Impure science,* 19.

2. W. C. Atzert to director, NSF, September 10, 1982; J. Fregeau to "File" [Office of the Vice President], memorandum, November 19, 1982.

3. Kalb, "A case study of the peer review process"; Kalb, "A 'Covert' Process."

4. U.S. District Court for the District of Columbia (Dec. 31) 1986; U.S. District Court for the District of Columbia (Dec. 8) 1987.

5. "NSF admits spreading spy rumor"; "Suit on rumor of tie to C.I.A. brings apology to geologist"; Marshall, "Gossip and peer review at NSF"; Cordes, "NSF apolo-

gizes to geologist"; Agres, "NSF pushed to open up peer review"; "Suit on rumor of tie to C.I.A"; Bell, *Impure science*, 24.

6. Office of Inspector General, *Semiannual report to the Congress;* "NSF erred on Privacy Act." In mid-1989, Public Citizen and I submitted a lengthy rule-making petition to NSF, as allowed by the First Amendment.[a] Citing various federal statutes, and using *Kalb* v. *NSF* as a case study, we argued that NSF's peer-review system was fundamentally unfair by denying grant applicants basic Privacy Act rights of *timely* access to records that may be inaccurate, false, or even libelous, and the opportunity to amend such records *prior* to a decision on a proposal. We also argued that NSF's conflict-of-interest regulations were grossly inadequate, by allowing "archrivals" who serve as reviewers access to the research designs, data, and strategies of their competitors.

In early 1990, NSF agreed to further reforms of its peer-review system in virtually all areas we requested.[b] Also as a result of our petition, the chairman of the U.S. Senate Government Affairs Committee, Senator John Glenn, requested that the General Accounting Office (GAO) determine the degree to which open-government laws, such as the Privacy Act, are followed by federal agencies that use peer-review procedures. Second, he requested that the GAO investigate the degree to which abuses are perpetrated in NSF's peer-review system, as well as in those of other selected federal agencies.[c] Finally, the Administrative Conference of the United States, a government policy think-tank, commissioned a study to determine ways of improving the fairness of peer-review procedures used throughout the federal government, which at the time was using the procedure to award some $9 billion of annual research spending.[d] Although the reforms that were actually adopted by NSF and other federal agencies as a result of these studies did not go as far as we had hoped, for the first time the legal avenues that scientists can pursue to confront abuses of federal peer-review procedures were spelled out and publicized. Amazingly, despite the high cost and five years of effort that went into the GAO studies the Office failed to investigate the single greatest potential abuse of the peer-review system: direct and blatant conflicts of interests that anonymous reviewers may have with respect to a proposal or a grant applicant.[a]

    a. Public Citizen, Jon Ervin Kalb, petitioners. "Petition for rulemaking"; Agres, "NSF pushed to open up peer review"; Booth, "Change sought in rules"; Marshall, "NSF peer review under fire; Glitzenstein and Kalb, "NSF peer review"; Kalb and Glitzenstein, "Rejected applicant's petition says agency kept 'secret' filing system"; McCullough, "NSF official scoffs at allegations,"

    b. Mervis, "NSF makes it easier to appeal"; Raloff, "Revamping peer review."

    c. U.S. General Accounting Office, "Peer review: compliance"; U.S. General Accounting Office, "Peer review: reforms,"

    d. McGarity, "Bias in awarding"; McGarity, "Peer review in awarding federal grants"; Macilwain, "US agencies urged to tighten up peer review."

7. Gibbons, "Anthropology goes back to Ethiopia"; Gibbons, "First hominid finds."

8. "A proclamation to provide for the study."

9. J. D. Clark and T. White to S. Brush, November 7, 1982.

10. Tadessa Terfa to J. D. Clark and T. White, May 12, 1990; see permit map boundary in WoldeGabriel et al., "Ecological and temporal placement," Fig. 1 and in "Oldest human ancestors" *Ethiopian Herald,* Sept. 22, 1994, and compare with Map VII, page 140 of this book.

11. Gibbons, "First hominid finds"; Bower, "Fossil finds expand early hominid anatomy"; "Anthropologists take the measure of humanity."

12. Kimbel et al., "The first skull"; "Variable but singular"; " 'Lucy,' crucial early human ancestor"; Johanson and Edgar, *From Lucy to language,* 128–129.

13. Wilford, "2.3-million-year-old jaw extends human family"; Kimbel et al., "Late Pliocene *Homo* and Oldowan tools."

14. Kimbel, Johanson, and Rak, "Systematic assessment of a maxilla"; Hill et al., "Earliest *Homo*"; Feibel, "Earliest *Homo* debate"; Hill et al., "Reply to Feibel"; Schrenk et al., "Oldest *Homo* and Pliocene biogeography"; Wood, "Rift on the record"; Johanson and Edgar, *From Lucy to language,* 168–169.

15. deMenocal and Bloemendal, "Plio-Pleistocene climatic variability"; Vrba, "The fossil record of African antelopes."

16. Ghedamu, "Pandemonium at the National Museum"; Ghedamu, "Causes of the pandemonium"; White, "For sure there is pandemonium at the National Museum"; Jefferson, "This anthropologist has a style."

17. Fitzgerald, "Rift valley."

18. Gibbons, "Clash with billionaire."

19. Johanson, Kimbel, and Shea, "Institute of human origins"; Gibbons, "Response to Johanson, Kimbel and Shea"; Smith, "[Letter]"; Renne, "Institute of Human Origins breakup"; Apsell, "[Letter]"; Gibbons, "Lab custody fight"; Jefferson, "This anthropologist has a style"; "Paleo 'divorce' final"; note 21.

20. Gibbons, "Lab custody fight."

21. Gibbons, "Claim-jumping charges"; Johanson et al., "Fossil collecting."

22. Jack Harris, personal communication, September 28, 1998.

23. Wilford, "Human ancestors' tools are found in Africa"; Semaw et al., "2.5-million-year-old stone tools"; Wood, "The oldest whodunnit in the world."

24. Semaw et al., "2.5-million-year-old stone tools," 336.

25. Asfaw et al., *Australopithecus garhi;* de Heinzelin et al., "Environment and behavior"; Culotta, "A new human ancestor"; Bower, "African fossils."

26. Kalb et al., "Stratigraphy of the Awash Group"; de Heinzlein et. al., "Environment and Behavior."

27. Kalb, "Refined stratigraphy," figure 21.

28. Clark et al., "Palaeoanthropological discoveries"; Harris, "Cultural beginnings," 19; Asfaw et al., "The earliest Acheulean."

29. Korn, *Ethiopia,* chap. 7; Marcus, *A history of Ethiopia,* 205–209.

30. Day, "Drought halts anthropological dig."

31. Marcus, *A history of Ethiopia,* 211–212; Ofcansky and Berry, *Ethiopia,* 64–65.

32. Marcus, *A history of Ethiopia,* 207–212.

33. Ofcansky and Berry, *Ethiopia,* 64–66; Marcus, *A history of Ethiopia,* 211–217; Tiruneh, *The Ethiopian revolution,* 383.

34. Kassaye Begashaw to J. Kalb, May 8, 1992.

35. J. D. Clark to Tedessa Terfa, November 16, 1979; J. D. Clark, Request for Authorization for Archaeological Research Authorization, Ministry of Culture, November 14, 1979.

36. Tesfaye Fichala, vice-minister to J. Kalb, May 10, 1993.

37. Tesfaye Fichala, vice-minister to J. Kalb, August 13, 1993.

38. Leule Selassie, minister to J. Kalb, November 8, 1993.

39. White, Suwa and Asfaw, "*Australopithecus ramidus,*" Table 1.

40. WoldeGabriel et al., "Ecological and temporal placement; Kappelman and Fleagle Letter to Editor, [*Nature*], "Age of early hominids," August 17, 1995; in same issue, reply by WoldeGabriel et al.

41. Wilford, "New fossils take science close to dawn of humans."

42. Gee, "Uprooting the human family tree."
43. Clark, "Recent developments."
44. "Putting our oldest ancestors in their proper place."
45. White et al., "Corrigendum: *Australopithecus ramidus.*"
46. Johanson and Edgar, 116.
47. Notes 30 and 31.
48. Kalb and Mebrate, "Fossil elephantoids."

Epilog

1. United Nations Food and Agriculture Organization, "Report on the survey."
2. Gebre-Mariam, "The alienation of land rights among the Afar in Ethiopia."
3. Moorehead, *The Blue Nile,* 287.
4. U.S. Bureau of Reclamation, Department of Interior, *Land and water resources of the Blue Nile basin, Ethiopia.*
5. Ibid., D. H. Jepsen and M. J. Athearn, "General geology map of the Blue Nile," scale 1:500,000, plate I; United Kingdom, War Office, *Lechemt,* sheet NC37/4; United Kingdom, War Office, *Soddu,* sheet NB37/1; Trimingham, *Islam in Ethiopia,* maps 1 and 6; H. S. Lewis, *A Galla monarchy,* maps 3 and 4.
6. "Ethiopian deserters talk of massacres."
7. Assefa completed his master's thesis on the Middle Awash proboscideans. See Mebrate, "Late Miocene–middle Pleistocene proboscideans."
8. Adefris, "A description of the Bodo cranium."
9. Ghedamu, "Pandemonium at the National Museum"; Ghedamu, "Causes of the pandemonium."
10. White, "for sure there is pandemonium."

Appendices

I. Johanson and Edgar, *From Lucy to Language,* 38; Asfaw et al., "A new fossil hominid," Fig. 5; Feibel, "Debating the environmental factors," Fig. 3; Asfaw et al., "The earliest Acheulean"; Harris, John W. K., "Cultural beginnings," Fig. 11.
II. Adapted from Kalb, "Fossil elephantoids, Awash paleolake basins." Dates supplemented from Appendix III. Present-day elevations (in meters above sea level) reflect relative differences in elevation, not actual elevations at the time the lakes were formed.
III. The regional stratigraphy as it is known today and is presented here differs little from that depicted in Kalb et al., "Stratigraphy of the Awash Group," Figs. 2 and 6; additional information concerning the Dakanihyalo beds comes from Kalb, "Refined stratigraphy," 52 and Fig. 21. For refined dates, see Asfaw et al., "*Australopithecus garhi,*" 629–630; Clark et al., "African *Homo erectus*"; de Heinzelin et al., "Environment and behavior"; Kimbel et al., "Late Pliocene *Homo,*" Fig. 8; and Renne et al., "Chronostratigraphy," Fig. 8.

Bibliography

3-million-year-old human fossils found in Wollo governorate. 1973. *Ethiopian Herald* (Dec. 16), 1.

5 male and 2 female skyjackers slain in gun duel on jetliner. 1972. *International Herald Tribune* (June 25), sec. 9–10:1.

40 percent of Lucy after 3 million years. 1975. *Science News* 107:4.

60 former Ethiopian leaders executed. 1974. *Washington Post* (Nov. 25), A1, A24.

A leading scientist hails fossil find in Ethiopian valley. 1974. *New York Times* (Oct. 27), 9.

A proclamation to provide for the study and protection of antiquities. 1989. *Negarit Gazeta of the People's Democratic Republic of Ethiopia* (Oct. 7), no. 36:47–54.

Abbebe [Abebe], Berhanou. 1971. *Évolution de la propriété foncière au choa (Éthiopie): Du regne de Ménélik à la constitution de 1931.* Paris: Imprimerie Nationale.

Abebe, B., J. Chavaillon, and J. E. G. Sutton, eds. 1976. *Proceedings of the VIIth (1971) Pan-African Congress of Prehistory and Quaternary Studies.* Addis Ababa: Ethiopian Antiquities Administration.

Abir, Mordechai. 1980. *Ethiopia and the Red Sea.* London: Frank Cass.

Adefris, Tsrha. 1992. A description of the Bodo cranium: An Archaic *Homo sapiens* cranium from Ethiopia. Ph.D. diss., New York University.

Agenbroad, L. D., and J. I. Mead. 1996. Distribution and palaeoecology of central and western North American *Mammuthus.* Chap. 28 in *The Proboscidea: Evolution and palaeoecology of elephants and their relatives,* edited by J. Shoshani and P. Tassy. Oxford: Oxford Univ. Press.

Agres, Ted. 1988. NSF pushed to open up peer review. *The Scientist* (Jan. 11) 2(2):6.

Ajayi, J. F. Ade, and Michael Crowder, eds. 1985. *Historical atlas of Africa.* Cambridge: Cambridge Univ. Press.

Alimen, H., H. Faure, J. Chavaillon, M. Taieb, and R. Battistini. 1969. Les études françaises sur le quaternaire d'Afrique. In *Études françaises sur le quaternaire, Supplement au Bulletin de l'AFEQ,* n.s., pp. 201–214.

Allan, T. D. 1970. Magnetic and gravity fields over the Red Sea. *Philosophical Transactions of the Royal Society of London* 267:153–180.

Andrews, Peter. 1990. *Owls, caves and fossils: Predation, preservation, and accumulation of small mammal bones in caves, with an analysis of the Pleistocene cave faunás from Westbury-sub-Mendip, Somerset, U.K.* Chicago: Univ. of Chicago Press.

Anfray, F. 1990. The civilization of Axum from the first to the seventh century. Chap. 14 in *Ancient civilizations of Africa,* edited by G. Mokhtar. London: James Currey.

Anfray, F., R. Bonnefille, and J. Chavaillon, comps., 1970. Documents pour servir à l'histoire des civilisations Éthiopiennes. Massy, France: Centre National de la Recherche Scientifique.

Anthropologists take the measure of humanity. 1994. Science 264:350.

Apsell, P. S. 1994. "[Letter]." Institute of Human Origins breakup. Science 265:722.

Arambourg, C., J. Chavaillon, and Y. Coppens. 1972. Expédition internationale de recherches paléontologiques dans la vallée de l'Omo (Éthiopie) en 1967. In Actes du 6e congrès Panafricain de préhistoire et d'études du Quaternaire, Les Imprimeries Réunies de Chambery, pp. 135–140.

Aronson, J. L., T. J. Schmitt, R. C. Walter, M. Taieb, J. J. Tiercelin, D. C. Johanson, C. W. Naeser, and A. E. M. Nairn. 1977. New geochronologic and palaeomagnetic data for the hominid-bearing Hadar Formation of Ethiopia. Nature 267 (5609):323–327.

Aronson, J. L., and M. Taieb. 1981. Geology and paleogeography of the Hadar hominid site, Ethiopia. In Hominid sites: Their geologic settings, edited by G. Rapp, Jr. and C. F. Vondra, 165–195. Boulder, CO: Westview Press.

Asfaw, Berhane. 1983. A new hominid parietal from Bodo, Middle Awash Valley, Ethiopia. American Journal of Physical Anthropology 61(3):367–371.

———. 1987. The Belohdelie frontal: New evidence of early hominid cranial morphology from the Afar of Ethiopia. Journal of Human Evolution 16 (7–8):611–624.

Asfaw, B., Y. Beyene, G. Suwa, R. Walter, T. White, G. WoldeGabriel, and T. Yemane. 1992. The earliest Acheulean from Konso-Gardula. Nature 360:732–735.

Asfaw, B., T. White, O. Lovejoy, B. Latimer, S. Simpson, and G. Suwa. 1999. Australopithecus garhi: A new species of early hominid from Ethiopia. Science 284:629–635.

Awash in glory. June 1982. California Monthly.

Badoglio, Pietro. 1937. The war in Abyssinia. London: Methuen.

Baer, G. W. 1967. The Coming of the Italian–Ethiopian war. Cambridge, MA: Harvard Univ. Press.

Balk, Robert. 1953. Memorial to Hans Cloos (1886–1951). Proceedings Volume of the Geological Society of America, Annual Report for 1952, 87–94.

Barberi, F., S. Borsi, G. Ferrara, G. Marinelli, R. Santacroce, H. Tazieff, and J. Varet. 1972. Evolution of the Danakil depression (Afar, Ethiopia) in light of radiometric age determinations. Journal of Geology 80(6):720–729.

Barberi, F., G. Giglia, G. Marinelli, R. Santacroce, H. Tazieff, and J. Varet. 1971. Geological map of the Danakil Depression (Northern Afar—Ethiopia). Scale 1:500,000. France: Centre National de la Recherche Scientifique; Italy: Consiglio Nazionale delle Richerche. [See Varet, 1978.]

Barberi, F., G. Marinelli, R. Santacroce, H. Tazieff, and J. Varet. 1973. Geology of northern Afar (Ethiopia). Revue de géographie physique et de géologie dynamique 25(1): 443–490. [See Varet, 1978.]

Barberi, F., H. Tazieff, and J. Varet. 1971. Volcanism in the Afar Depression: Its tectonic and magmatic significance. Tectonophysics 15(1/2):19–29.

Barkeley, Richard. 1958. The road to Mayerling: Life and death of Crown Prince Rudolph of Austria. New York: St. Martin's Press.

Barker, A. J. 1968. The civilizing mission: The Italo-Ethiopian War 1935–6. London: Cassell.

Bauer, I. E., J. McMorrow, and D. W. Yalden. 1994. The historic ranges of three equid species in north-east Africa: A quantitative comparison of environmental tolerances. Journal of Biogeography 21:169–182.

Bell, Robert. 1992. Impure science: Fraud, compromise and political influence in scientific research. New York: Wiley.

Berggren, W. A., and J. A. Van Couvering. 1974. The late Neogene: Biostratigraphy, geochronology and paleoclimatology of the last 15 million years in marine and continental sequences. *Palaeogeography, Palaeoclimatology, Palaeoecology* 16(1/2): 1–216.

Bernstein, Carl, and Bob Woodward. 1974. *All the president's men.* New York: Simon and Schuster.

Beshah, T.-W., and J. W. Harbeson. 1978. Afar pastoralists in transition and the Ethiopian revolution. *Journal of African Studies* 5(3):249–267.

Black, R., W. H. Morton, D. C. Rex, and R. M. Schackleton. 1972. Sur la decouverte en Afar (Éthiopie) d'un granite hyperalcalin miocène: le massif de Limmo. *Comptes Rendus Hebdomadaires des Séances de l'Académie des Sciences, Serie D: Sciences Naturelles* 274(10):1453–1456.

Bliese, Loren. 1976. Afar. In *The Non-Semitic languages of Ethiopia,* edited by M. Lionel Bender, p. 133. East Lansing, MI: African Studies Center: Michigan State University.

Boaz, Noel T. 1993. *Quarry: Closing in on the missing link.* New York: Free Press.

Bonatti, Enrico. 1970. Submarine volcanoes in the Afar rift (Ethiopia). *Eos (American Geophysical Union, Transactions)—Abstracts Annual Meeting* 51(4):443.

Bonatti, Enrico, and Haroun Tazieff. 1970. Exposed guyot from the Afar Rift, Ethiopia. *Science* 168:1087–1089.

Bondestam, Lars. 1974. People and capitalism in the north-eastern lowlands of Ethiopia. *Journal of Modern African Studies* 12(3):423–439.

Bonnefille, R. 1983. Evidence for a cooler and drier climate in the Ethiopian uplands towards 2.5M yr ago. *Nature* 303 (5917):487–491.

Bonnefille, R., A. Vincens, and G. Buchet. 1987. Palynology, stratigraphy and palaeoenvironment of a Pliocene hominid site (2.9–3.3 m.y.) at Hadar, Ethiopia. *Palaeogeography, Palaeoclimatology, Palaeoecology* 60(3–4):249–281.

Booth, William. 1989. Change sought in rules for issuing science grants: Petition charges excessive secrecy. *Washington Post* (July 15), A12.

Bower, B. 1991. Fossil finds expand early hominid anatomy. *Science News* 139:182.

———. 1999. African fossils flesh out humanity's past. *Science News* 155:262.

Brandt, S. A. 1982. A late Quaternary cultural/environmental sequence from Lake Besaka, Southern Afar, Ethiopia. Ph.D. diss., University of California, Berkeley.

Brinckmann, J., and M. Kürsten. 1970. *Geological sketch map of the Danakil depression.* Scale 1:250,000. Hanover, Germany: Bundesanstalt für Bodenforschung.

Brown, Francis H. 1994. Development of Pliocene and Pleistocene chronology of the Turkana Basin, East Africa, and its relation to other sites. Chap. 15 in *Integrative paths to the past: Paleoanthropological advances in honor of F. Clark Howell,* edited by R. S. Corruccini and R. L. Ciochon, pp. 285–312. Englewood Cliffs, NJ: Prentice-Hall.

Brown, Monty. 1989. *Where giants trod: The saga of Kenya's desert lake.* London: Quiller Press.

Burton, Richard F. [1849] 1987. *First footsteps in East Africa or, An exploration of Harar.* New York: Dover Publications.

Butzer, Karl W. 1982. *Archaeology as human ecology: Method and theory for a contextual approach.* Cambridge: Cambridge Univ. Press.

Butzer, K. W., G. L. Isaac, J. L. Richardson, and C. Washbourn-Kamau. 1972. Radiocarbon dating of East African lake levels. *Science* 175:1069–1076.

Buxton, D. 1967. *Travels in Ethiopia.* New York: Praeger.

Callison, Lynn. 1976. For 2 SMU students, it was find of lifetime. *Dallas Times-Herald* (Nov. 28), 1, 2.

Cann, J. R., and F. J. Vine. 1966. An area on the crest of the Carlsberg Ridge: Petrology and magnetic survey. *Philosophical Transactions of the Royal Society of London* 259:198–217.

Casada, James Allen. 1968. John Hanning Speke: The Nile sources and controversy. Master's thesis, Virginia Polytechnic Institute, Blacksburg, VA.

Cecchi, Antonio. 1887. *Da Zeila alle frontiere del Caffa; Societa Geografica Italiana; spedizione Italiana nell Africa Equatoriale.* Rome: Loescher.

Cerulli, E. 1992. Ethiopia's relations with the Muslim world. Chap. 20 in *Africa from the seventh to the eleventh century,* edited by I. Hrbek. London: James Currey.

Chamberlain, Muriel E. 1974. *The scramble for Africa.* London: Longman.

Chavaillon, J., comp., 1970. Les civilisations Éthiopiennes de la préhistoire au moyen age. Massy, France: Centre National de la Recherche Scientifique.

Chavaillon, Jean. 1971. Prehistorical living-floors of Melka Kontouré in Ethiopia. Paper presented at the VIIth Pan-African Congress of Prehistory and Quaternary Studies, Addis Ababa.

———. 1976. Les habitats acheuléen de Melka-Kontouré. In *Proceedings of the VIIth (1971) Pan-African Congress of Prehistory and Quaternary Studies,* pp. 57–61. Addis Ababa: Ethiopian Antiquities Administration.

Chavaillon, Jean, and Maurice Taieb. 1968. Stratigraphie du Quaternaire de Melka Kontouré (vallée de l'Aouache, Éthiopie); premiers resultats. *Comptes Rendus Hebdomadaires des Séances de l'Académie des Sciences, Serie D: Sciences Naturelles* 266 (12):1210–1212.

Chavaillon, Jean, Claude Brahimi, and Yves Coppens. 1974. Première découverte d'Hominidé dans l'un des sites acheuléens de Melka Kontouré (Éthiopie). *Comptes Rendus Hebdomadaires des Séances de l'Académie des Sciences, Série D: Sciences Naturelles* 278: 3299–3302.

Chavaillon, N. 1976. Les habitats oldowayens de Melka Kontouré. In *Proceedings of the VIIth (1971) Pan-African Congress of Prehistory and Quaternary Studies,* pp. 63–66. Addis Ababa: Ethiopian Antiquities Administration.

Cherfas, J. 1983. Grave accusations for Lucy finders. *New Scientist* (Feb. 10), 390.

Christiansen, T. B., H.-U. Schaefer, and M. Schönfeld. 1975. Southern Afar and adjacent areas: Geology, petrology, geochemistry. In *Afar Depression of Ethiopia,* edited by A. Pilger and A. Rösler, pp. 259–277. Stuttgart: Schweizerbart'sche Verlagsbuchhandlung.

Churcher, C. S., and M. L. Richardson. 1978. Equidae. In *Evolution of African mammals,* edited by V. J. Maglio and H. B. S. Cooke. Cambridge, MA: Harvard Univ. Press.

Ciochon, Russell L. 1995. A remarkable lineage. Review of *Theropithecus: The rise and fall of a primate genus,* edited by Nina Jablonski. *Evolutionary Anthropology* 4:32–36.

Clapham, Christopher. 1969. *Haile-Selassie's government.* New York: Praeger.

Clark, J. D. 1969. *Kalambo Falls prehistoric site.* Vol. 1. London: Cambridge Univ. Press.

———. 1974. *Kalambo Falls prehistoric site.* Vol. 2. London: Cambridge Univ. Press.

———. 1970. *The prehistory of Africa.* New York: Praeger.

———. 1986. Glynn Isaac 1937–1985. *Antiquity* 60(228):55–56.

———. 1987. Fred Wendorf: A critical assessment of his career in and contribution to North African prehistory. Chap. 1 in *Prehistory of arid North Africa: Essays in honor of Fred Wendorf.* Dallas, TX: Southern Methodist Univ. Press.

———. 1988. The Middle Stone Age of East Africa and the beginnings of regional identity. *Journal of World Prehistory* 2(3):235–305.

———. 1989. Chap. 10 in *The pastmasters: Eleven modern pioneers of archaeology,* edited by G. Daniel and C. Chippindale. New York: Thames and Hudson.

———. 1995. Recent developments. *South African Archaeological Bulletin* 50:168.

Clark, J. D., B. Asfaw, G. Assefa, J. W. K. Harris, H. Kurashina, R. C. Walter, T. D. White, and M. A. Williams. 1984. Palaeoanthropological discoveries in the Middle Awash Valley, Ethiopia. *Nature* 307(5950):423–428.

Clark, J. D., J. de Heinzelin, K. D. Schick, W. K. Hart, O. White, G. WoldeGabriel, R. C. Walter, G. Suwa, B. Asfaw, E. Vrba, and Y. H.-Selassie. 1994. African *Homo erectus:* Old radiometric ages and young Oldowan assemblages in the Middle Awash Valley, Ethiopia. *Science* 264:1907–1910.

Clark, J. D., and J. W. K. Harris. 1985. Fire and its roles in early hominid lifeways. *African Archaeological Review* 3:3–27.

Clark, J. D., and H. Kurashina. 1979. Hominid occupation of the east-central highlands of Ethiopia in the Plio-Pleistocene. *Nature* 282(5734):33–39.

Clark, J. D., K. D. Williamson, J. M. Michels, and C. A. Marean. 1984. A Middle Stone Age occupation site at Proc Epic cave, Dire Dawa (east-central Ethiopia). *African Archaeological Review* 2:37–71.

Clay, Jason W., and Bonnie K. Holcomb. 1986. *Politics and the Ethiopian famine 1984–1985.* Cambridge, MA: Cultural Survival.

Cloos, Hans. 1939. Hebung-Spaltung-Vulkanismus. Elemente einer geometrischen Analyse indischer Grossformen. *Geologische Rundschau* 30:405–527.

———. 1942. Tektonische Bemerkungen über den Boden des Golfes von Aden. *Geologische Rundschau* 33:354–363.

———. 1954. *Conversation with the earth.* New York: Knopf.

Close, Angela E. 1987. Preface to *Prehistory of arid North Africa: Essays in honor of Fred Wendorf.* Dallas, TX: Southern Methodist Univ. Press.

Cobb, Jr., Charles E. 1996. Eritrea wins the peace. *National Geographic* 189(6): 82–105.

Cold war, commerce, and the Lucy trail. 1982. *New Scientist* (Dec. 2), 552.

Cole, Sonia. 1963. *The prehistory of East Africa.* New York: Macmillan.

Conroy, G. C., C. J. Jolly, D. Cramer, and J. E. Kalb. 1978. Newly discovered fossil hominid skull from the Afar Depression, Ethiopia. *Nature* 276(5683):67–70.

Cooke, H. B. S. 1976. Suidae from Pliocene-Pleistocene strata of the Rudolf basin. In *Earliest man and environments in the Lake Rudolf Basin,* edited by Y. Coppens, F. C. Howell, G. L. Isaac, and R. E. F. Leakey, pp. 251–263. Chicago: Univ. of Chicago Press.

———. 1978. Faunal evidence for the biotic setting of early African hominids. In *Early African hominids,* edited by C. J. Jolly, pp. 267–281. London: Duckworth.

———. 1978. Pliocene-Pleistocene Suidae from Hadar, Ethiopia. *Kirtlandia* 29:1–63.

———. 1986. *Changing perspectives on the age of man.* Raymond Dart Lecture 21. Johannesburg: Witwatersrand Univ. Press.

Cooke, H. B. S., J. W. K. Harris, and K. Harris. 1987. J. Desmond Clark: His career and contribution to prehistory. *Journal of Human Evolution* 16:549–581.

Cooke, H. B. S., and V. J. Maglio. 1972. Plio-Pleistocene stratigraphy in East Africa in relation to proboscidean and suid evolution. In *Calibration of hominoid evolution,* pp. 303–329. Edinburgh: Scottish Academic Press.

Coppens, Y. 1961. Decouverte d'un australopithecine dans le Villafranchian du Tchad. *Comptes rendus des séances de l'Académie des Sciences,* Paris 252:3851–3852.

———. 1972. Un nouveau proboscidean de Pliocene du Tchad, *Stegodibelodon schneideri* no. gen. nov. sp., et le phylum des Stegotetrabelodontinae. *Comptes rendus séances de l'Académie des Sciences,* Paris 274:2962–2965.

Coppens, Y., F. C. Howell, G. L. Isaac, and R. E. F. Leakey, eds. 1976. *Earliest man and environments in the Lake Rudolf basin: Stratigraphy, paleoecology, and evolution.* Chicago: Univ. of Chicago Press.

Cordes, Colleen. 1987. NSF apologizes to geologist who accused it of spreading CIA rumor. *Chronicle of Higher Education* 34(16):A17.

Corvinus, G. 1975. Palaeolithic remains at the Hadar in the Afar region. *Nature* 256(5517):468–471.

———. 1976. Prehistoric exploration at Hadar, Ethiopia. *Nature* 261(5561):571–572.

Corvinus, G., and H. Roche. 1980. Prehistoric exploration at Hadar in the Afar (Ethiopia) in 1973, 1974 and 1976. In *Proceedings (1977), VIIIth Pan-african Congress of Prehistory and Quaternary Studies,* edited by R. E. F. Leakey and B. A. Ogot, pp. 186–188. Nairobi: International Louis Leakey Memorial Institute for African Prehistory.

Culotta, Elizabeth. 1999. A new human ancestor? *Science* (April 23) 284:572–573.

Cross, Margaret W., and Vincent J. Maglio. 1975. *A bibliography of the fossil mammals of Africa, 1950–1972.* Princeton, NJ: Princeton Univ. Press.

Dal Piaz, Giambattista, Carlo Zanmatti, Giovanni Merla, Augusto Azzaroli, and Bruno Zanettin. 1973. In *Illustrazione Geologica Missione Geologica dell' Azienda Generale Italiani Petroli (A.G.I.P.) nella Dancalia Meriodionale e Sugli Altipiani Hararini.* Rome: Accademia Nazionale del Lincei.

Darnton, John. 1978. Addis Ababa's forces launch major drive on Eritrean rebels. *New York Times* (May 17), 9.

———. 1978. Ethiopia uses terror to control capital: 1,000 reported killed in drive on "counterrevolutionaries." *New York Times* (Feb. 10), A1, A9.

———. 1978. Ethiopia drives Somalia's forces from key areas. *New York Times* (Feb. 13), A1, A7.

Dart, Raymond A. 1959. *Adventures with the missing link.* New York: Viking.

Day, Dan. 1985. Drought halts anthropological dig in Ethiopia: Finder of fossil hominid called Lucy in Omaha. *Lincoln Nebraska Journal* (March 28), 18.

Day, M. 1977. *Guide to fossil man.* 3d ed. Chicago: Univ. of Chicago Press.

De Bono, Emilio. 1937. *Anno XIII: The conquest of Ethiopia.* London: Cresset Press Ltd.

de Contenson, H. 1990. Pre-Axumite Culture. Chap. 13 in *Ancient civilizations of Africa,* edited by G. Mokhtar. London: James Currey.

Degens, E. T., and D. A. Ross, eds. 1969. *Hot brines and recent heavy metal deposits in the Red Sea.* New York: Springer-Verlag.

de Heinzelin, J., J. D. Clark, T. White, W. Hart, P. Renne, G. WoldeGabriel, Y. Beyene, and E. Vrba. 1999. Environment and behavior of 2.5-million-year-old Bouri hominids. *Science* 284:625–628.

Deino, Alan L., and Richard Potts. 1990. Single-crystal ^{40}Ar/^{39}Ar dating of the Olorgesailie Formation, southern Kenya Rift. *Journal of Geophysical Research B: Solid Earth and Planets* 95(6):8453–8470.

Del Boca, Angelo. 1965. *The Ethiopian war, 1935–1941.* Translated by P. D. Cummins. Chicago: Univ. of Chicago Press.

Delson, E., and D. Dean. 1993. Are *Papio baringensis* R. Leakey, 1969, and *P. quadratirostris* Iwamoto, 1982, species of *Papio* or *Theropithecus?* Chap. 4 in *Theropithecus: The rise and fall of a primate genus,* edited by N. G. Jablonski, pp. 125–156. Cambridge: Cambridge Univ. Press.

deMenocal, P. B., and J. Bloemendal. 1995. Plio-Pleistocene climatic variability in subtropical Africa and the paleoenvironment of hominid evolution: A combined data-model approach. Chap. 19 in *Paleoclimate and evolution, with emphasis on human origins,* edited by E. S. Vrba, G. H. Denton, T. C. Partridge, and L. H. Burckle. New Haven, CT: Yale Univ. Press.

de Prorok, Byron. 1942. *Dead men do tell tales.* New York: Creative Age Press.

Diel, Louise. 1939. *Behold our new empire—Mussolini.* Translated by Kenneth Kirkness. London: Hurst and Blackett.

Dietz, Robert S. 1961. Continent and ocean basin evolution by spreading of the sea floor. *Nature* 190:854–857.

———. 1961. The spreading ocean floor. *Saturday Evening Post* (Oct.), 34–36.

Dietz, Robert S., and John C. Holden. 1970. Reconstruction of Pangaea: Breakup and dispersion of continents, Permian to present. *Journal of Geophysical Research* 75(26): 4939–4956.

Dilebo, Getahun. 1974. Emperor Menelik's Ethiopia, 1865–1916: National unification of Amhara communal domination. Ph.D. diss., Howard University, Washington, D.C.

Doe, Brian. 1971. *Southern Arabia.* London: Thames and Hudson.

Don't dig here. 1982. *The Economist* (Oct. 30), 48.

Doresse, Jean. 1959. *Ethiopia.* Translated by Elsa Coult. New York: Frederick Ungar.

Drake, C. L., and R. W. Girdler. 1964. A geophysical study of the Red Sea. *Geophysical Journal* 8(5):473–495.

Dudley, J. P. 1996. Mammoths, gomphotheres, and the Great American Faunal Interchange. Chap. 29 in *The Proboscidea: Evolution and palaeoecology of elephants and their relatives,* edited by J. Shoshani and P. Tassy. Oxford: Oxford Univ. Press.

Duff, P. McL. D., ed. 1993. *Holmes' principles of physical geology.* London: Chapman and Hall.

Early man fossil, stone tools discovered in Hararghe region. 1975. *Ethiopian Herald* (Sept. 23), 1.

Eck, G. G. 1993. *Theropithecus darti* from the Hadar Formation, Ethiopia. Chap. 2 in *Theropithecus: The rise and fall of a primate genus,* edited by N. G. Jablonski. Cambridge: Cambridge Univ. Press.

Emerson, Barbara. 1979. *Leopold II of the Belgians: King of colonialism.* New York: St. Martin's Press.

Emperor presents prizes to five award winners. 1972. *Ethiopian Herald* (Nov. 3), 1.

Estes, Richard D. 1993. *The safari companion: A guide to watching African mammals.* White River Junction, VT: Chelsea Green.

Ethiopia—man of the year. 1936. *Time* 27(1):13–18.

Ethiopia reports 2 fugitives slain. 1976. *New York Times* (Aug. 13), A3.

Ethiopia tells of 9 tries on Mengistu's life. 1978. *New York Times* (June 29), 4.

Ethiopia victims list. 1974. *Washington Post* (Nov. 25), A25.

Ethiopian deserters talk of massacres. 1977. *New York Times* (Dec. 18), 4.

Ethiopian fossil hominids. 1975. *Nature* (Jan. 24) 253:232–233.

Ethiopian head and 6 in capital reported slain. 1977. *New York Times* (Feb. 4), A1, A4.

Ethiopian officials executed. 1974. *Washington Post* (Nov. 24), A1, A23.

Ethiopian students attack U.S. and British centers. 1977. *New York Times* (Jan. 29), 4:3.

Falcon, N. L., I. G. Gass, R. W. Girdler, and A. S. Laughton. 1970. A discussion on the structure and evolution of the Red Sea and the nature of the Red Sea, Gulf of Aden and Ethiopian Rift junction. *Philosophical Transaction of the Royal Society of London* 267:1–417.

Faure, H., F. Gasse, C. Roubet, and M. Taieb. 1976. Les formations lacustres holocènes (argiles et diatomées) et l'industrie epipaléolithique de la région de Logghia. In *Proceedings of the VIIth (1971) Pan-African Congress of Prehistory and Quaternary Studies,* pp. 391–404. Addis Ababa: Ethiopian Antiquities Administration.

Feibel, C. S. 1992. [Letter]. Earliest *Homo* debate. *Nature* 358:289.

———. 1997. Debating the environmental factors in hominid evolution. *GSA Today* 7(3):1–7.

Feibel, C. S., F. H. Brown, and I. McDougall. 1989. Stratigraphic context of fossil hominids from the Omo Group deposits: Northern Turkana Basin, Kenya and Ethiopia. *American Journal of Physical Anthropology* 78:595–622.

Fernandez-Armesto, Felipe, ed. 1991. *The Times atlas of world exploration*. London: Times Books.

Ferrara, G., and R. Santacroce. 1980. Dama' Ale: A tholeiitic volcano in southern Afar (Ethiopia). *Geodynamic evolution of the Afro-Arabian rift system*, pp. 437–453. Rome: Accademia Nazionale dei Lincei.

Fighting erupts in Ethiopia. 1975. *Washington Post* (June 5), A16.

Fisher, R. V., and H.-U. Schmincke. 1984. *Pyroclastic rocks*. Berlin: Springer-Verlag.

Fitch, F. J., and J. A. Miller. 1970. Radioisotopic age determinations of Lake Rudolf artefact site. *Nature* 226(5242):226–228.

Fitzgerald, M. A. 1995. Rift valley: How the Getty millions stoked up a bitter feud in the African desert over man's origins. *Sunday Times* (March 19), section 3, 1–3.

Fleagle, John G. 1988. *Primate adaptation and evolution*. San Diego, CA: Academic Press.

Folk, R. L., and G. K. Hoops. 1982. An early iron-age layer of glass made from plants at Tel Yin'am, Israel. *Journal of Field Archaeology* 9:455–466.

Franchetti, Raimondo. 1930. *Nella dancalia Etiopica: Spedizione Italiana 1928–29*. Milan: A. Mondadori.

Gallagher, James P. 1971. A preliminary report on archaeological research near Lake Zuai. *Annales d'Éthiopie* 9:13–18.

———. 1976. Ethno-archaeology in south-central Ethiopia. In *Proceedings of the VIIth (1971) Pan-African Congress of Prehistory and Quaternary Studies*, pp. 391–404. Addis Ababa: Ethiopian Antiquities Administration.

———. 1977. Contemporary stone tools in Ethiopia: Implications for archaeology. *Journal of Field Archaeology* 4(4):407–414.

Gasse, Françoise. 1974. Les diatomées holocènes du bassin inférieur de l'Aouache (Dépression des Danakil, Éthiopie). Leur signification paléoecologique. *Internationale Revue der Gesamten Hydrobiologie* 59(1):123–146.

———. 1978. Sedimentary formations in the northern and southern Afar. In *Geology of Central and Southern Afar*, J. Varet, pp. 37–59. Paris: Centre National de la Recherche Scientifique.

Gasse, F., and P. Rognon. 1973. Le Quaternaire des bassins lacustres de l'Afar. *Numero special sur l'Afar, Revue de Géographie Physique et de Géologie Dynamique* 15(4):405–414.

Gasse, F., and F. A. Street. 1978. Late Quaternary lake-level fluctuations and environments of the northern rift valley and Afar region (Ethiopia and Djibouti). *Palaeogeography, Palaeoclimatology, Palaeoecology* 24(4):279–325.

Gebre-Mariam, Ayele. 1994. The alienation of land rights among the Afar in Ethiopia. *Nomadic Peoples* 34/35:137–146.

Gee, Henry. 1995. Uprooting the human family tree. *Nature* 373:15.

Georgi, J. 1962. Memories of Alfred Wegener. In *Continental drift*, edited by S. K. Runcorn, pp. 309–324. New York: Academic Press.

Ghedamu, Tefera. 1992. Pandemonium at the National Museum (English translation; original in Amharic). *IFOYTA* 1(10):6–9.

———. 1993. Causes of the pandemonium at the National Museum (English translation; original in Amharic). *IFOYTA* 1(12):18–19.

Gibbons, A. 1990. Anthropology goes back to Ethiopia. *Science* (Sept. 21) 249:1373.

———. 1991. First hominid finds from Ethiopia in a decade. *Science* (March 22) 251:428.

———. 1994. Clash with billionaire costs anthropology institute dearly. *Science* (May 27) 264:1247.

———. 1994. Lab custody fight in institute "divorce." *Science* (Aug. 12) 265:864.

———. 1994. [Letter]. Response to Johanson, Kimbel and Shea. *Science* (July 1) 265:13.

———. 1995. Claim-jumping charges ignite controversy at meeting. *Science* (April 14) 368:196–197.

Gigli, Giuseppe Cobolli. 1938. *Strade imperiali.* Milan: A. Mondadori.

Girdler, R. W. 1957. The relationship of the Red Sea to the East African rift system. *Geological Society of London Quarterly Journal* 114:79–105.

———. 1962. Initiation of continental drift. *Nature* 194:521–524.

———. 1984. The evolution of the Gulf of Aden and Red Sea in space and time. *Deep-Sea Research, Part A: Oceanographic Research Papers* 31(6–8A):747–762.

———. 1991. The Afro-Arabian rift system: An overview. *Tectonophysics* 197(2–4): 139–153.

Girdler, R. W., and P. Styles. 1974. Two stage Red Sea floor spreading. *Nature* 247(5435):7–11.

Glen, William. 1982. *The road to Jaramillo.* Stanford, CA: Stanford Univ. Press.

Glitzenstein, E. R., and J. E. Kalb. 1989. [Letter]. NSF peer review. The *Scientist* (Oct. 30), 3.

Gortani, M., and A. Bianchi. 1973. Itinerario Miesso-Gauani-Tiho carta topografica e carta geologica. Scala 1: 250,000. In *Atlante geologico.* Rome: Accademia Nazionale dei Lincei.

———. 1973. *Carta geologica della Dancalia meridionale e degli altipiani Hararini* (in quqattro fogli. Scala 1: 500,000. In *Atlante geologico.* Rome: Accademia Nazionale dei Lincei.

Goss, John. 1993. *The mapmaker's art.* Skokie, IL: Rand McNally.

Gray, T. 1980. Environmental reconstruction of the Hadar Formation (Afar, Ethiopia) Ph.D. diss., Case Western Reserve University, Cleveland, OH.

Greenfield, Richard. 1965. *Ethiopia: A new political history.* New York: Praeger.

Gregory, J. W. 1896. *The Great Rift Valley.* London: John Murray.

———. 1921. *The rift valleys and geology of East Africa.* London: Seeley Service.

Groves, Colin P. 1981. [Comment on Shipman, Bosler, and Davis.] *Current Anthropology* 22(3):265.

Haberland, E. 1992. The horn of Africa. Chap. 24 in *Africa from the sixteenth to the eighteenth century,* edited by B. A. Ogot. London: Heinemann.

Hall, C. M., R. C. Walter, J. A. Westgate, and D. York. 1984. Geochronology, stratigraphy and geochemistry of Cindery Tuff in Pliocene hominid-bearing sediments of the Middle Awash, Ethiopia. *Nature* 308(5954):26–31.

Harbeson, J. W. 1978. Territorial and development politics in the Horn of Africa: The Afar of the Awash valley. *African Affairs: Journal of the Royal African Society* 77(309):479–498.

———. 1988. *The Ethiopian transformation: The quest for the post-imperial state.* Boulder, CO: Westview Press.

Harland, W. B., R. L. Armstrong, A. V. Cox, L. E. Craig, A. G. Smith, and D. G. Smith. 1990. A geologic time scale. Cambridge: Cambridge Univ. Press.

Harris, J. M. 1978. Deinotherioidea and Barytherioidea. In *Evolution of African mammals,* edited by V. J. Maglio and H. B. S. Cooke, pp. 315–332. Cambridge, MA: Harvard Univ. Press.

Harris, J. M., and T. D. White. 1979. Evolution of the Plio-Pleistocene African Suidae. *Transactions of the American Philosophical Society* 69(2):1–128.

Harris, John W. K. 1983. Cultural beginnings: Plio-Pleistocene archaeological occurrences from the Afar, Ethiopia. *African Archaeological Review* 1:3–31.

Harrison, C. G. A., E. Bonatti, and L. Stieltjes. 1975. Tectonism of axial valleys in spreading centers: Data from the Afar Rift. In *Afar Depression of Ethiopia,* edited by A. Pilger and A. Rösler, pp. 178–198. Stuttgart: A. E. Schweizerbart'sche Verlagsbuchhandlung.

Harrison, James J. 1901. A journey from Zeila to Lake Rudolf. *Geographical Journal* 18:258–275.

Hess, H. H. 1962. History of ocean basins. In *Petrologic studies: A volume to honor A. F. Buddington,* edited by A. E. J. Engel, H. L. James, and B. F. Leonard, pp. 599–620. Boulder, CO: Geological Society of America.

Hess, Robert L. 1970. *Ethiopia: The modernization of autocracy.* Ithaca, NY: Cornell Univ. Press.

Heybroek, F. 1965. The Red Sea Miocene evaporite basin. In *Salt basins around Africa,* pp. 17–40. London: Institute of Petroleum and the Geological Society.

High Ethiopian casualties reported. 1977. *New York Times* (Aug. 24), 2:3.

Hill, A. 1994. Late Miocene and early Pliocene hominoids from Africa. Chap. 6 in *Integrative paths to the past: Paleoanthropological advances in honor of F. Clark Howell,* edited by R. S. Corrucini and R. L. Ciochon. Englewood Cliffs, NJ: Prentice-Hall.

Hill, A., S. Ward, A. Deino, G. Curtis, and R. Drake. 1992. Earliest *Homo. Nature* 355:719–722.

———. 1992. [Letter] Reply to Feibel. *Nature* 358:290.

Hill, Bob. 1955. "Cat (*Catha edulis* Forsk)." *Journal of Ethiopian Studies* 3(2):13–23.

Hill, M. N. 1966. A discussion concerning the floor of the northwest Indian Ocean. *Philosophical Transactions of the Royal Society of London* 259:135–278.

Hochschild, Adam. 1998. *King Leopold's ghost.* Boston: Houghton Mifflin.

Hofmann, C., V. Courtillot, G. Féraud, P. Rouchett, G. Yirgu, E. Ketefo, and R. Pik. 1997. Timing of the Ethiopian flood basalt event and implications for plume birth and global change. *Nature* 389(6653):838–841.

Holcomb, Bonnie K., and Sisai Ibssa. 1990. *The invention of Ethiopia.* Trenton, NJ: Red Sea Press.

Holden, C. 1981. The politics of palaeoanthropology. *Science* 213:737–740.

Holloway, Ralph L. 1970. [Letter]. *Man,* n.s., 5(3):518.

Holmes, Arthur. 1913. *The age of the earth.* Edinburgh: Nelson and Sons.

———. 1929. Radioactivity and earth movements. *Geological Society of Glasgow* 18:559–606.

———. 1944. *Principles of physical geology.* London: Nelson and Sons.

———. 1965. *Principles of physical geology.* 2d ed. London: English Language Book Society.

Hovey, Graham. 1978. Brzezinski asserts that Soviet general leads Ethiopia units. *New York Times* (Feb. 25), 1, 5.

Howell, F. C. 1968. Omo Research Expedition. *Nature* 219:567–572.

———. 1969. Remains of Hominidae from Pliocene/Pleistocene formations in the Lower Omo Basin, Ethiopia. *Nature* 223:1234–1239.

———. 1978. Hominidae. In *Evolution of African mammals,* edited by Vincent J. Maglio and H. B. S. Cooke, pp. 154–248. Cambridge, MA: Harvard Univ. Press.

Howell, F. C., P. Haesaerts, and J. de Heinzelin. 1987. Depositional environments, archeological occurrences and hominids from Members E and F of the Shungura Formation (Omo basin, Ethiopia). *Journal of Human Evolution* 16:665–700.

Human skull, tools discovered in Awash. 1977. *Ethiopian Herald* (Jan. 30), 1.

Huntingford, G. W. B. 1989. *The historical geography of Ethiopia: From the first century AD to 1704.* Edited by Richard Pankhurst. Oxford: British Academy.

Ibrahim, H. A. 1985. Politics and nationalism in north-east Africa, 1919–35. Chap. 23 in *Africa under colonial domination 1880–1935,* edited by A. Adu Boahen. London: Heinemann.

Imbrie, John, and Katherine Palmer Imbrie. 1979. *Ice ages: Solving the mystery.* Short Hills, NJ: Enslow Publishers.

Imperato, Pascal James. 1998. *Quest for the Jade Sea: Colonial competition around an East African lake.* Boulder, CO: Westview Press.

Irvine, T. N., ed. 1966. *The world rift system.* Geological Survey of Canada, Paper 66-14. Paper presented at symposium, The World Rift System, 4–5 September 1965, Ottawa, Canada.

Isaac, Glynn. 1977. *Olorgesailie: Studies of a middle Pleistocene lake basin in Kenya.* Chicago: Univ. of Chicago Press.

Isacks, Bryan, Jack Oliver, and Lynn R. Sykes. 1968. Seismology and the new global tectonics. *Journal of Geophysical Research* 73:5855–5899.

Jablonski, N. G. 1993. Evolution of the masticatory apparatus in *Theropithecus.* Chap. 11 in *Theropithecus: The rise and fall of a primate genus,* edited by N. G. Jablonski, pp. 299–330. Cambridge: Cambridge Univ. Press.

Jaeger, M. M., R. L. Bruggers, B. E. Johns, and W. A. Erickson. 1986. Evidence of itinerant breeding of the red-billed quelea *Quelea quelea* in the Ethiopian Rift Valley. *Ibis* 128:469–482.

James, H. L. 1973. Harry Hammond Hess: May 24, 1906–August 25, 1969. In Vol. 43 of *Biographical Memoirs,* pp. 109–125. New York: Columbia Univ. Press.

Jefferson, David J. 1995. This anthropologist has a style that is bone of contention. *Wall Street Journal* (Jan. 31), A1.

Jennings, Karla. 1985. The daybreak inquirey. *Express: The East Bay's Free Weekly* (Aug. 30) 7(46):38.

Jepsen, Darwin H., and Murray J. Athearn. 1964. General geology map of the Blue Nile River Basin, Ethiopia—Plate I. Scale 1: 1,000,000. In Vol. 1 of *Land and water resources of the Blue Nile basin, Ethiopia.* Washington, D.C.: U.S. Bureau of Reclamation.

Johanson, D. C. 1974. Progress report on [November 26] 1974 International Afar Research Expedition to Central Afar, Ethiopia supported by the National Science Foundation grant no. GS-39624. Cleveland, OH: Case Western Reserve University.

———. 1976. Ethiopia yields first "family of early man." *National Geographic* 150:788–811.

———. 1978. Our roots go deeper. In *Science Year: World Book Science Annual, 1979,* pp. 42–55. Chicago: Field Enterprises Educational Corp.

Johanson, D. C., and Maitland A. Edey. 1981. *Lucy: The beginnings of humankind.* New York: Simon and Schuster.

Johanson, D. C., and B. Edgar. 1996. *From Lucy to language.* New York: Simon and Schuster.

Johanson, D. C., W. H. Kimbel, and S. C. Shea. 1994. [Letter]. Institute of human origins: Separation issues. *Science* (July 1)265:14.

Johanson, D. C., W. H. Kimbel, and R. C. Walter. 1995. [Letter]. Fossil collecting. *Science* 268:1113.

Johanson, D. C., and J. Shreeve. 1989. *Lucy's child: The discovery of a human ancestor.* New York: William Morrow.

Johanson, D. C., and M. Taieb. 1976. Plio-Pleistocene hominid discoveries in Hadar, Ethiopia. *Nature* 260(5549):293–297.

———. 1978. Plio-Pleistocene hominid discoveries in Hadar, central Afar, Ethiopia. In *Early hominids in Africa,* edited by C. J. Jolly, pp. 29–44. New York: Duckworth.

Johanson, D. C., M. Taieb, and Y. Coppens. 1982. Pliocene hominids from the Hadar Formation, Ethiopia (1973–1977): Stratigraphic, chronologic, and paleoenvironmental contexts, with notes on hominid morphology and systematics. *American Journal of Physical Anthropology* 57(4):373–402.

Johanson, D. C., M. Taieb, B. T. Gray, and Y. Coppens. 1978. Geological framework of the Pliocene Hadar Formation (Afar, Ethiopia) with notes on palaeontology including hominids. In *Geological background to fossil man: Recent research in the Gregory Rift Valley, East Africa,* edited by W. W. Bishop, pp. 549–564. Edinburgh: Scottish Academic Press.

Johanson, D. C., and T. D. White. 1979. A systematic assessment of early African hominids. *Science* 202:321–330.

Johanson, D. C., T. D. White, and Y. Coppens. 1978. A new species of the genus *Australopithecus* (Primates: Hominidae) from the Pliocene of eastern Africa. *Kirtlandia* 28:1–14.

Johnson, Charles. [1844] 1972. *Travels in southern Abyssinia.* New York: Books for Libraries Press.

Jolly, C. J. 1970. The seed eaters: A new model of hominid differentiation based on a baboon analogy. *Man,* n.s., 5:5–26.

———. 1970. The seed eaters. [Letter]. *Man,* n.s., 5(3):518–519.

———. 1972. The classification and natural history of *Theropithecus (Simopithecus)* (Andrews, 1916) baboons of the African Plio-Pleistocene. *Bulletin of the British Museum (Natural History), Geology* 22:1–123.

Jolly, C. J., and F. Plog. 1976. *Physical anthropology and archeology.* New York: Knopf.

Jolly, C. J., and R. J. Ucko. 1971. The riddle of the sphinx-monkey. In *Man in Africa,* edited by M. Douglas and P. M. Kaberry, pp. 317–333. Garden City, NY: Anchor Books.

Jones, A. H. M., and Elizabeth Monroe. 1935. *A history of Ethiopia.* Oxford: Clarendon Press.

Jones, Roger. 1972. *The rescue of Emin Pasha: The story of Henry M. Stanley and the Emin Pasha Relief Expedition, 1887–1889.* London: Allison and Busby.

Jones, Steve, Robert Martin, and David Pilbeam, eds. 1992. *The Cambridge encyclopedia of human evolution.* Cambridge: Cambridge Univ. Press.

Kalb, J. E. 1973. Preliminary bibliography of the geology of Ethiopia. Addis Ababa: Ethiopian Geological Survey.

———. 1973. Preliminary map series for the Afar Depression. Addis Ababa: Ethiopian Geological Survey.

———. 1973. Recent geological research in Ethiopia. Addis Ababa: Ethiopian Geological Survey.

———. 1974. The potential for prehistory research in the Afar depression. *Ethiopian Herald* (Sept. 28).

———. 1974. Special report: Recent geologic research in Ethiopia. *Geology* 2(6):266.

———. 1974. Geological, paleontological and anthropological investigations in the Afar triple junction, Ethiopia. International Afar Research Expedition, Addis Ababa.

———. 1975. Preliminary map series for a scientific park in the Awash Valley, Ethiopia. RVRME report no. 1 for the Ethiopian Ministry of Culture, Addis Ababa.

———. 1976. Annual report 1975–1976. RVRME report no. 8 for the Ethiopian Ministry of Culture, Addis Ababa.

———. 1977. Annual report 1976–1977. RVRME report no. 23 for the Ethiopian Ministry of Culture, Addis Ababa.

———. 1978. Mio-Pleistocene deposits in the Middle Awash Valley, Afar Depression. *Sinet* 1(2):87–98.

———. 1986. A case study of the peer review process of the U.S. National Science Foundation—prehistory research in Ethiopia. In *Science Policy Study—research project selection,* pp. 631–638. Washington, D.C.: U.S. Government Printing Office.

———. 1986. The peer review process of the U.S. National Science Foundation. In *Science Policy Study—research project selection,* pp. 625–630. Washington, D.C.: U.S. Government Printing Office.

———. 1993. Refined stratigraphy of the hominid-bearing Awash Group, Middle Awash Valley, Afar Depression, Ethiopia. *Newsletters on Stratigraphy* 29(1):21–62.

———. 1995. Fossil elephantoids, Awash paleolake basins, and the Afar triple junction, Ethiopia. *Palaeogeography, Palaeoclimatology, Palaeoecology* 114(2–3):357–368.

Kalb, J. E., Dejene Aneme, Katema Kidane, Mohamed Santur, and Afifa Kechrid. 2000. *Bibliography of earth sciences of the Horn of Africa: Ethiopia, Eritrea, Somalia, and Djibouti.* Alexandria, VA: American Geological Institute.

Kalb, J. E., and D. J. Froehlich. 1995. Interrelationships of late Neogene elephantoids: New evidence from the Middle Awash Valley, Afar, Ethiopia. *Geobios* 28(6):727–736.

Kalb, J. E., D. J. Froehlich, and G. L. Bell. 1996. Phylogeny of African and Eurasian Elephantoidea of the late Neogene. Chap. 12A in *The Proboscidea: Evolution and palaeoecology of elephants and their relatives,* edited by J. Shoshani and P. Tassy. Oxford: Oxford Univ. Press.

———. 1996. Palaeobiogeography of late Neogene African and Eurasian Elephantoidea. Chap. 12B in *The Proboscidea: Evolution and palaeoecology of elephants and their relatives,* edited by J. Shoshani and P. Tassy. Oxford: Oxford Univ. Press.

Kalb, J. E., and E. R. Glitzenstein. 1989. Rejected applicant's petition says agency kept "secret" filing system. *The Scientist* (Sept. 18), 3:13.

Kalb, J. E., E. R. Glitzenstein, K. A. Meyer, and A. B. Morrison. 1989. Petition for rulemaking changes—submitted to the U.S. National Science Foundation.

Kalb, J. E., M. Jaegar, C. J. Jolly, and B. Kana. 1982. Preliminary geology, paleontology and paleoecology of a Sangoan site at Andalee, Middle Awash Valley, Ethiopia. *Journal of Archaeological Science* 9(4):349–363.

Kalb, J. E., C. J. Jolly, A. Mebrate, S. Tebedge, C. Smart, E. B. Oswald, D. Cramer, P. Whitehead, C. B. Wood, G. C. Conroy, T. Adefris, L. Sperling, and B. Kana. 1982. Fossil mammals and artefacts from the Middle Awash Valley, Ethiopia. *Nature* 298(5869):25–29.

Kalb, J. E., C. J. Jolly, E. B. Oswald, and P. Whitehead. 1984. Early hominid habitation in Ethiopia. *American Scientist* 72(2): 168–178.

Kalb, J. E., C. J. Jolly, S. Tebedge, A. Mebrate, C. Smart, E. B. Oswald, P. Whitehead, C. B. Wood, T. Adefris, and V. Rawn-Schatzinger. 1982. Vertebrate faunas from the Awash Group, Middle Awash Valley, Afar, Ethiopia. *Journal of Vertebrate Paleontology* 2(2):237–258.

Kalb, J. E., and A. Mebrate. 1993. Fossil elephantoids from the hominid-bearing Awash Group, Middle Awash Valley, Afar Depression, Ethiopia. *Transactions of the American Philosophical Society* 83(1):1–114.

Kalb, J. E., A. Mebrate, S. Tebedge, and D. Aneme. 1977. Catalog (revised) of fossil fauna collected by RVRME in the Awash Valley, May 1975–December 1976. RVRME report no. 20 for the Ethiopian Ministry of Culture, Addis Ababa.

Kalb, J. E., E. B. Oswald, A. Mebrate, S. Tebedge, and C. J. Jolly. 1982. Stratigraphy of the Awash Group, Middle Awash Valley, Afar, Ethiopia. *Newsletters on Stratigraphy* 11(3):95–127.

Kalb, J. E., E. B. Oswald, and H. P. Mosca. 1975. Spatial and temporal distribution of mammalian fossil and archaeological sites in the western Awash. RVRME report no. 3 for the Ethiopian Ministry of Culture, Addis Ababa.

Kalb, J. E., E. B. Oswald, S. Tebedge, A. Mebrate, E. Tola, and D. Peak. 1982. Geology and stratigraphy of Neogene deposits, Middle Awash Valley, Ethiopia. *Nature* 298(5869):17–25.

Kalb, J. E., and C. D. Peak. 1975. Documentation of fossil sites along the lower Awash Valley in the Afar region; summary of field research by the International Afar Research Expedition, 1971–1974. Report for the Ethiopian Antiquities Administration, Addis Ababa.

Kalb, J. E., C. B. Wood, C. Smart, E. B. Oswald, A. Mabrete, S. Tebedge, and P. Whitehead. 1980. Preliminary geology and palaeontology of the Bodo D'ar hominid site, Afar, Ethiopia. *Palaeogeography, Palaeoclimatology, Palaeoecology* 30(1–2):107–120.

Kaplan, I., M. Farber, B. Marvin, J. McLaughlin, H. D. Nelson, and D. Whitaker. 1971. *Area handbook for Ethiopia*. Washington, D.C.: Government Printing Office.

Kappelman, John. 1996. The evolution of body mass and relative brain size in fossil hominids. *Journal of Human Evolution* 30:243–276.

Kappelman, John, and John G. Fleagle. 1995. [Letter]. Age of early hominids. *Nature* 376:558.

Kapuscinski, R. 1978. *The emperor: Downfall of an autocrat*. Translated by W. R. Brand and K. Mroczkowska-Brand. San Diego: Harcourt.

Kaufman, Michael T. 1977. Ethiopian official is believed to have been executed. *New York Times* (Nov. 15), 6.

———. 1977. Surrounded by enemies, Ethiopia wars against itself. *New York Times* (Dec. 4), IV 3:1.

———. 1978. Somalis abandoning North Ogaden. *New York Times* (March 9), 3:4.

Kazmin, V. 1975. *Explanation of the geological map of Ethiopia*. Addis Ababa: Geological Survey of Ethiopia.

Keller, Edmond. 1991. *Revolutionary Ethiopia: From empire to people's republic*. Bloomington: Indiana Univ. Press.

Kent, Peter. 1978. Historical background: Early exploration in the East African Rift— the Gregory Valley Rift. In *Geological background to fossil man: Recent research in the Gregory Rift Valley, East Africa*, edited by W. W. Bishop, pp. 1–6. Edinburgh: Scottish Academic Press.

Kimbel, W. H., D. C. Johanson, and Y. Rak. 1994. The first skull and other new discoveries of *Australopithecus afarensis* at Hadar, Ethiopia. *Nature* 368:449–451.

———. 1997. Systematic assessment of a maxilla of *Homo* from Hadar, Ethiopia. *American Journal of Physical Anthropology* 103:235–262.

Kimbel, W. H., R. C. Walter, D. C. Johanson, K. E. Reed, J. L. Aronson, Z. Assefa, C. W. Marean, G. G. Eck, R. Bobe, E. Hovers, Y. Rak, C. Vondra, T. Yemane, D. York, Y. Chen, N. M. Evensen, and P. E. Smith. 1996. Late Pliocene *Homo* and Oldowan tools from the Hadar Formation (Kada Hadar Member), Ethiopia. *Journal of Human Evolution* 31:549–561.

King, B. C. 1966. Obituary notice [for Arthur Holmes]. *Proceedings of the Geological Society of London* 1635:196–201.

Kingdon, Jonathan. 1989. *East African mammals: An atlas of evolution in Africa.* Vol. 3, Part C, *Bovids.* Chicago: Univ. of Chicago Press.

Klemp, Egon. 1968. *Africa—on maps dating from the twelfth to the eighteenth century.* Leipzig: Edition Leipzig.

Klingel, H. 1971. The Somali wild ass. *Oryx: Journal of the Fauna Preservation Society* 11(2–3):110.

———. 1972. Social behavior of African Equidae. *Zoologica africana (South African Journal of Zoology)* 7(1):175–185.

Kloos, Helmut, and Aklilu Lemma. 1977. Bilharziasis in the Awash Valley III: Epidemiological studies in the Nura Era, Abadir, Melka Sadi and Amibara irrigation schemes. *Ethiopian Medical Journal* 15:161–168.

Kobishchanov, Yuri M. 1979. *Axum.* Translated by Lorraine T. Kapitanoff. University Park: Pennsylvania State Univ. Press.

Korn, David A. 1986. *Ethiopia, the United States and the Soviet Union.* Carbondale: Southern Illinois Univ. Press.

Krentz, Hartmut B. 1993. Postcranial anatomy of extant and extinct species of *Theropithecus.* Chap. 14 in *Theropithecus: The rise and fall of a primate genus,* edited by N. G. Jablonski, pp. 383–424. Cambridge: Cambridge Univ. Press.

Kreuzer, H., P. Miller, W. Harre, and H. Lenz. 1969. Radiometrische Altersbestimmungen an Gesteinen der Danakil Senke und ihrer Umgebung (Nordathiopien). *Geologisches Jahrbuch* 88:83–86.

Kunz, K., H. Kreuzer, and P. Mueller. 1975. Potassium–argon age determinations of the trap basalt of the south-eastern part of the Afar Rift. In *Afar Depression of Ethiopia,* Vol. 1, edited by A. Pilger and A. Rösler, pp. 370–374. Stuttgart: E. Schweizerbart'sche Verlagsbuchhandlung.

Larson, Jr., P. 1977. Matabaietu, an Oldowan site from the Afar, Ethiopia. *Nyame Akuma* 11:6–10.

Laughton, A. S., R. B. Whitmarsh, and M. T. Jones. 1970. The evolution of the Gulf of Aden. *Philosophical Transactions of the Royal Society of London* 267:227–266.

Leakey, L. S. B., J. F. Evernden, and G. H. Curtis. 1961. Age of Bed I, Olduvai Gorge, Tanganyika. *Nature* 191(4787):478–479.

Leakey, M. D. 1971. *Excavations in Beds I and II, 1960–1963.* Vol. 3, *Olduvai Gorge.* Cambridge: Cambridge Univ. Press.

———. 1978. Pliocene footprints at Laetoli, northern Tanzania. *Antiquity* 52(205):133.

———. 1984. *Disclosing the past.* Garden City, NY: Doubleday.

Leakey, M. D., R. L. Hay, G. H. Curtis, R. E. Drake, M. K. Jackes, and T. D. White. 1976. Fossil hominids from the Laetoli beds. *Nature* 262:460–466.

Leakey, M. G., C. S. Feibel, R. L. Bernor, J. M. Harris, T. E. Cerling, K. M. Stewart, G. W. Storrs, A. Walker, L. Werdelin, and A. J. Winkler. 1996. Lothagam: A record of faunal change in the late Miocene of East Africa. *Journal of Vertebrate Paleontology* 16(3):556–570.

Leakey, R. E. 1970. In search of man's past at Lake Rudolf. *National Geographic* 137(5):712–732.

Leakey, R. E., and R. Lewin. 1977. *Origins: What new discoveries reveal about the emergence of our species and its possible future.* New York: E. P. Dutton.

Lefort, Rene. 1981. *Ethiopia: An heretical revolution?* Translated by A. M. Berrett. London: Zed Press.

L. E. J. 1936. Review of *Desolate marches,* by L. M. Nesbitt, *The Geographical Journal* 87(1):84–85.

Levine, Donald N. 1974. *Greater Ethiopia: The evolution of a multiethnic society.* Chicago: Univ. of Chicago Press.

Lewin, Roger. 1981. Ethiopian stone tools are world's oldest. *Science* 211:806–807.

———. 1987. *Bones of contention: Controversies in the search for human origins.* New York: Simon and Schuster.

Lewis, Herbert S. 1965. *A Galla monarchy: Jimma Abba Jifar, Ethiopia 1830–1932.* Madison: Univ. of Wisconsin Press.

Lewis, I. M. 1955. *Peoples of the Horn of Africa: Somali, Afar and Saho.* Ethnographic Survey of Africa, North Eastern Africa, Part 1, edited by Daryll Forde. London: International African Institute.

"Lucy," crucial early human ancestor, finally gets a head. 1994. *Science* 264:34–35.

Macilwain, Colin. 1993. U.S. agencies urged to tighten up peer review. *Nature* 364(6437): 469.

Maglio, Vincent J. 1970. Early Elephantidae of Africa and a tentative correlation of African Plio-Pleistocene deposits. *Nature* 225(5230):328–332.

———. 1970. Four new species of Elephantidae from the Plio-Pleistocene of northwestern Kenya. *Breviora* 341:1–43.

———. 1971. Vertebrate faunas from the Kubi Algi, Koobi Fora and Ileret areas, East Rudolf, Kenya. *Nature* 231(5300):248–249.

———. 1972. Vertebrate faunas and chronology of hominid-bearing sediments east of Lake Rudolf, Kenya. *Nature* 289(5372):379–385.

———. 1973. Origin and evolution of the Elephantidae. *Transactions of the American Philosophical Society* 63(3):1–126.

Maknun, G. A., and R. J. Hayward. 1981. Tolo Hanfade's song of accusation: An Afar text. *Bulletin of the School of Oriental and African Studies, University of London* 44(2):327–333.

Marcus, Harold G. 1972. *The modern history of Ethiopia and the Horn of Africa: A select and annotated bibliography.* Stanford: Hoover Institution Press.

———. 1987. *Haile Sellassie I: The formative years, 1892–1936.* Berkeley: Univ. of California Press.

———. 1994. *A history of Ethiopia.* Berkeley: Univ. of California Press.

———. 1995. *The life and times of Menelik II: Ethiopia 1844–1913.* Lawrenceville, NJ: Red Sea Press.

———. 1995. *The politics of empire: Ethiopia, Great Britain and the United States, 1941–1974.* Lawrenceville, NJ: Red Sea Press.

Mariam, Mesfin Wolde. 1984. *Rural vulnerability to famine in Ethiopia, 1958–1977.* New Delhi: Vikas.

Marsh, Charmayne. 1976. Curious dealer paid expedition that found skull. *Dallas Morning News* (Nov. 28), 38A.

Marshall, Eliot. 1987. Gossip and peer review at NSF. *Science* 238(4833):1502.

———. 1989. NSF peer review under fire from Nader group. *Science* 245(4915):250.

———. 1990. Peer review under review. *Science* (June 15)248(4961):1307.

———. 1994. Congress finds little bias in system. *Science* 265:863.

Martelli, George. 1938. *Italy against the world.* New York: Harcourt.

Martin, R. D. 1990. *Primate origins and evolution: A phylogenetic reconstruction.* Princeton, NJ: Princeton Univ. Press.

———. 1993. Allometric aspects of skull morphology in *Theropithecus*. Chap. 10 in *Theropithecus: The rise and fall of a primate genus,* edited by N. G. Jablonski, pp. 273–298. Cambridge: Cambridge Univ. Press.

Matthews, D. H. 1963. A major fault scarp under the Arabian Sea, displacing the Carlsberg Ridge near Socotra. *Nature* 198:950–952.

———. 1966. The Owen Fracture Zone and the northern end of the Carlsberg Ridge. *Philosophical Transactions of the Royal Society of London* 159: 172–186.

Matthews, D. H., F. J. Vine, and J. R. Cann. 1965. Geology of an area of the Carlsberg Ridge, Indian Ocean. *Geological Society of America Bulletin* 76:675–682.

McCann, James. 1995. *People of the plow.* Madison: Univ. of Wisconsin Press.

McCullough, Jim. 1989. NSF official scoffs at allegations, asking: "What secret filing system?" *The Scientist* (Sept. 18), 13, 15.

McDougall, Ian, and Craig Feibel. 1999. Numerical age control for the Miocene-Pliocene succession at Lothagam, a hominid-bearing sequence in the northern Kenyan Rift. *Journal of the Geological Society, London* 156:731–745.

McGarity, Thomas O. 1992. Bias in awarding scientific grants: A modest proposal for an audit system. *Accountability in Research* 2:203–218.

———. 1994. Peer review in awarding federal grants in the arts and sciences. *High Technology Law Journal* 9(1):1–92.

McKenzie, D. P., and R. L. Parker. 1967. The North Pacific: An example of tectonics on a sphere. *Nature* 216:1276–1280.

McKenzie, D. P., D. Davies, and P. Molnar. 1970. Plate tectonics of the Red Sea and East Africa. *Nature* 226:243–248.

McLynn, Frank. 1989. *The making of an explorer.* Chelsea, MN: Scarborough House.

———. 1992. *Hearts of darkness: The European exploration of Africa.* New York: Carroll and Graf.

Mebrate, A. 1976. A primitive fossil elephant molar from northwest Hararghe; Rift Valley research mission in Ethiopia. In RVRME report no. 5 for the Ethiopian Ministry of Culture, Addis Ababa, pp. 1–16.

———. 1976. Fossil Proboscidea from the Middle Awash Valley; Rift Valley research mission in Ethiopia. In RVRME report no. 9 for the Ethiopian Ministry of Culture, Addis Ababa, pp. 1–15.

———. 1983. Late Miocene–middle Pleistocene proboscidean fossil remains from the Middle Awash Valley, Afar Depression, Ethiopia. Master's thesis, University of Kansas, Lawrence.

Mebrate, A., and J. E. Kalb. 1985. Anancinae (Proboscidea: Gomphotheriidae) from the Middle Awash Valley, Afar, Ethiopia. *Journal of Vertebrate Paleontology* 5:93–102.

Mekouria, Tekle Tsadik. 1990. Christian Axum. Chap. 16 in *Ancient civilizations of Africa,* edited by G. Mokhtar. London: James Currey.

Menard, H. W. 1986. *The ocean of truth.* Princeton, NJ: Princeton Univ. Press.

Merla, Giovanni, Ernesto Abbate, Augusto Azzaroli, Piero Bruni, Paolo Canuti, Milvio Fazuoli, Mario Sagre, and Paolo Tacconi. 1979. *A geological map of Ethiopia and Somalia and comment with a map of major landforms.* Scale 1:2,000,000. Firenze: Consiglio Nazionale delle Richerche Italia.

Mervis, Jeffrey. 1990. NSF makes it easier to appeal as it opens up review process. *The Scientist* (March 19), 4(6):2.

Michels, Joseph W. 1979. Axumite archaeology: An introductory essay. In Kobishchanov, Y. M., *Axum,* pp. 1–34. University Park: Pennsylvania State Univ. Press.

Mockler, Anthony. 1984. *Haile Selassie's war: The Italian–Ethiopian campaign, 1935–1941.* New York: Random House.

Mohr, Paul. 1962. *The geology of Ethiopia.* Asmara, Eritrea: II Poligrafico.
———. 1963. *Geological map of Horn of Africa.* Scale 1:2,000,000. London: Philip and Tacey.
———. 1967. Major volcano-tectonic lineament in the Ethiopian Rift System. *Nature* 213:664–665.
———. 1970. Notes on the Afar triple junction. *Eos (American Geophysical Union, Transactions)—Abstracts Annual Meeting* 51(4):420.
———. 1970. Plate tectonics of the Red Sea and East Africa. *Nature* 228:547–548.
Mohr, Paul, and Bruno Zanettin. 1988. The Ethiopian flood basalt province. In *Continental flood basalts,* edited by J. D. Macdougall, pp. 63–110. Dordrecht: Kluwer.
Moorehead, Alan. 1962. *The Blue Nile.* New York: Harper & Row.
Morgan, W. Jason. 1968. Rises, trenches, great faults and crustal blocks. *Journal of Geophysical Research* 73:1959–1982.
Morrell, Virginia. 1995. *Ancestral passions: The Leakey family and the quest for humankind's beginnings.* New York: Simon and Schuster.
Mortan, William, comp., 1972. Geological map of the Addis Ababa area. Scale 1: 25,000. Department of Geology, Addis Ababa University.
Mosca, Herb. 1975. Preliminary report on archeological surveys in the Middle Awash Valley, northwest Hararghe Province, Ethiopia. Report for the RVRME, Addis Ababa.
Mundy, P. J., and M. J. F. Jarvis. 1989. *Africa's feathered locust.* Harare, Zimbabwe: Baobab Books.
Munzinger, W. 1869. Narrative of a journey through the Afar country. *Journal of the Royal Geographical Society* 39:188–232.
Murphy, Kim. 1992. The Qat culture. *Austin American-Statesman,* 20 August, C20.
Murray, Neil. 1990. *The love of elephants.* Seacaucus, NJ: Chartwell Books.
Nahum, Fasil. 1997. *Constitution for a nation of nations.* Lawrenceville, NJ: Africa World Press.
National Portrait Gallery. 1996. *David Livingstone and the Victorian encounter with Africa.* London: National Portrait Gallery.
Nelson, H. D., ed. 1979. *Tunisia, a country study.* Washington, D.C.: U.S. Government Printing Office.
Nelson, H. D., and Irving Kaplan, eds. 1981. *Ethiopia, a country study.* Washington, D.C.: U.S. Government Printing Office.
Nesbitt, Ludovico M. 1930. *La Dancalia esplorata.* Firenze: R. Bemporad e Figlio.
Nesbitt, L. M. 1929. From south to north through Danakil. *The Geographical Journal* 73(6):529–539.
———. 1930. Danakil traversed from south to north in 1928. *The Geographical Journal* 64(4–6):298–315, 391–414, 545–557.
———. 1935. *Hell-hole of creation—the exploration of the Abyssinian Danakil.* New York: Knopf. [Also published in London in 1934 by Jonathan Cape under the title *Desert and forest—the exploration of the Abyssinian Danakil.*]
———. 1936. *Gold fever.* New York: Harcourt. [Reprinted under the name Lewis Mariano Nesbitt in 1974 by Arno Press, New York.]
New York Times. 1977. *New York Times index for year 1976.* New York: New York Times.
Newman, James L. 1995. *The peopling of Africa: A geographic interpretation.* New Haven, CT: Yale Univ. Press.
NSF erred on Privacy Act. 1990. *Science News* (June 9) 137(23):359.
Ofcansky, T. P., and L. Berry, eds. 1991. *Ethiopia, a country study.* Washington, D.C.: Government Printing Office.

Office of Inspector General. 1990. *Semiannual report to the Congress: Number 2, October 1, 1989–March 31, 1990.* Washington, D.C.: National Science Foundation.

Oldest human ancestors found in Ethiopia. 1994. *Ethiopian Herald* (Sept. 22), 1.

Oswald, E. B. 1975. Geologic and paleontologic investigations at the Geraru Site, central Afar. Report for the Ethiopian Antiquities Administration, Addis Ababa.

Patterson, Bryan, Anna K. Behrensmeyer, and William D. Sill. 1970. Geology and fauna of a new Pliocene locality in north-western Kenya. *Nature* 226(5249):918–921.

Patterson, J. H. [1907] 1947. *The man-eaters of Tsavo.* London: Macmillan.

Pavitt, Nigel. 1989. *Kenya: The first explorers.* London: Aurum Press.

Perham, Margery. 1969. *The government of Ethiopia.* London: Faber and Faber.

Petit, Charles. 1982. A fight for bones and money. *San Francisco Chronicle* (Nov. 15).

Perkins, Kenneth J. 1986. *Tunisia—crossroads of the Islamic and European worlds.* Boulder, CO: Westview Press.

Petrocchi, C. 1941. Il giacimento fossilifero di Sahabi. *Bollettino della Societa Geologica Italiana* 50:107–114.

———. 1943. Il giacimenta fossilifero di Sahabi. Parte 2, Paleontologia. *Collezione Scientifica e Documentaria dell'Africa Italiana, Ministero dell'Africa Italiana* 12:69–167.

Pettijohn, F. J. 1977. Memorial to Ernst Cloos, 1898–1974. In *Memorials,* vol. 6, 1–9. Boulder, CO: Geological Society of America.

Pilger, A., and A. Rösler., eds. 1975. *Afar Depression of Ethiopia.* Stuttgart: E. Schweizerbart'sche Verlagsbuchhandlung.

Post, R. January 13, 1984. The skeleton's in our closet, *University of Washington Daily,* Seattle.

Potts, Rick. 1996. *Humanity's descent: The consequences of ecological instability.* New York: William Morrow.

Powell-Cotton, P. H. G. 1902. *A sporting trip through Abyssinia: A narrative of a nine months' journey from the plains of the Hawash to the snows of Simien, with a description of the game, from elephant to ibex, and notes on the manners and customs of the natives.* London: Rowland Ward.

Public Citizen, Jon Ervin Kalb, petitioners. July 13, 1989. Petition for rulemaking. U.S. National Science Foundation. Eric Glitzenstein, Katherine A. Meyer, Alan B. Morrison, Public Citizen Litigation Group, attorneys for petitioners. Washington, D.C.

Putting our oldest ancestors in their proper place. 1994. *Science* 265:2011–2012.

Radosevich, S. C., G. J. Retallack, and M. Taieb. 1992. Reassessment of the paleoenvironment and preservation of hominid fossils from Hadar, Ethiopia. *American Journal of Physical Anthropology* 87:15–27.

Raloff, Janet. 1990. Revamping peer review. *Science News* 137:234–235.

Reader, John. 1998. *Africa: A biography of the continent.* New York: Knopf.

Renne, P. 1994. [Letter]. Institute of Human Origins breakup. *Science* 265:721–722.

Renne, P., R. Walker, K. L. Verosub, M. Sweitzer, and J. Aronson. 1993. New data from Hadar (Ethiopia) support orbitally tuned time scale to 3.3 Ma. *Geophysical Research Letters* 20(11):1067–1070.

Renne, P. R., G. WoldeGabriel, W. K. Hart, G. Heiken, and T. White. 1999. Chronostratigraphy of the Miocene-Pliocene Sagantole Formation, Middle Awash Valley, Afar rift, Ethiopia. *Geological Survey Bulletin* 111(6):869–885.

Rensberger, Boyce. 1975. Skeleton fossils linked to ancient "near man." *New York Times* (Feb. 15), 17.

———. 1987. NSF admits spreading spy rumor: Agency apologizes to grant applicant. *Washington Post* (Dec. 4), 1:225.

Rice, Edward. 1990. *Captain Sir Richard Francis Burton: The secret agent who made the pilgrimage to Mecca, discovered the Kama Sutra, and brought the Arabian Nights to the West.* New York: Charles Scribner's Sons.

Roberts, D. G. 1970. A discussion mainly concerning the contribution by Hutchinson and Baker. *Philosophical Transactions of the Royal Society of London* 267(1181): 399–405.

Rochet-D'Hericourt, C. 1841. *Voyage sur la côte orientale de la Mer Rouge, dans le pays d'Adel et le Royaume de Choa.* Paris: Arthus Bertrand.

Rognon, P. and F. Gasse. 1973. Dépots lacustres Quaternaires de la basse Vallée de l'Aouache (Afar, Éthiopie); leurs rapports avec la tectonique et le volcanisme sous-aquatique. *Travaux des Laboratoires des Sciences de la Terre, Série B* 11: 129–130.

———. 1973. Dépots lacustres Quaternaires de la basse Vallée de l'Aouache (Afar, Éthiopie); leurs rapports avec la tectonique et le volcanisme sous-aquatique. *Reveue de Géographie physique et de géologie dynamique* 15(2):295–316.

Rotberg, Robert I., ed. 1973. *Africa and its explorers: Motives, methods, and impact.* Cambridge, MA: Harvard Univ. Press.

Rothé, J. P. 1954. La zone seismique médiane Indo-Atlantique. *Proceedings of the Royal Society of London* A222:387–397.

Rubenson, S. 1978. *The survival of Ethiopian independence.* London: Heinemann Educational Books.

S. I. T. 1967. Obituary notice of Arthur Holmes. *Proceedings of the Geologists Association* 78(2):374–377.

Said, Ali. 1994. *Pastoralism and the state policies in mid-Awash valley: The case of the Afar, Ethiopia.* African Arid Lands, Working Paper Series. Uppsala, Sweden: Nordiska Afrikainstitutet.

Salt, Henry. [1814] 1967. *A voyage to Abyssinia.* London: Frank Cass.

Sampson, C. Garth. 1974. *The stone age of archaeology of southern Africa.* New York: Academic Press.

Saunders, Jeffrey, J. 1996. North American Mammutidae. Chap. 27 in *The Proboscidea: Evolution and palaeoecology of elephants and their relatives,* edited by J. Shoshani and P. Tassy. Oxford: Oxford Univ. Press.

Savage, R. J. G. 1978. Carnivora. In *Evolution of African mammals,* edited by V. J. Maglio and H. B. S. Cooke, pp. 249–267. Cambridge, MA: Harvard Univ. Press.

Scherman, David E., and Richard Wilcox. 1944. *Literary England: Photographs of places made memorable in English literature.* New York: Random House.

Schick, Kathy D., and Nicholas Toth. 1993. *Making silent stones speak: Human evolution and the dawn of technology.* New York: Simon and Schuster.

Schrenk, F., T. G. Bromage, C. G. Betzler, U. Ring, and Y. M. Juwayeyi. 1993. Oldest *Homo* and Pliocene biogeography of the Malawi Rift. *Nature* 365:833–836.

Schwab, Peter. 1985. *Ethiopia: Politics, economics and society.* London: Frances Pinter.

Scientific American. 1973. *Continents adrift.* San Francisco: W. H. Freeman.

Scientist says man originated in Ethiopia. 1971. *Ethiopian Herald* (December 6): 1.

Selassie, Bereket Habte. 1980. *Conflict and intervention in the Horn of Africa.* New York: Monthly Review Press.

Sellassie I, Haile. 1997. *My life and Ethiopia's progress.* Vol. 2. Chicago: Research Associates School Times Publications.

Sellassie, Sergew Hable. 1972. *Ancient and medieval Ethiopian history to 1270.* Addis Ababa: United Printers.

Semaw, S., P. Renne, J. W. K. Harris, C. S. Feibel, R. L. Bernor, N. Fesseha, and K. Mowbray. 1997. [Letters.] 2.5-million-year-old stone tools from Gona, Ethiopia. *Nature* 385:333–336.

Severin, Timothy. 1973. *The African adventure.* New York: E. P. Dutton.

Shehim, Kassim. 1982. The influence of Islam on the Afar. Ph.D. diss., University of Washington, Seattle.

———. 1985. Ethiopia, revolution, and the question of nationalities: The case of the Afar. *Journal of Modern African Studies* 23(2):331–348.

Shillington, Kevin. 1995. *History of Africa.* Rev. ed. New York: St. Martin's Press.

Shipman, Pat. 1975. Implications of drought for vertebrate fossil assemblages. *Nature* 257 (5528):667–668.

———. 1981. *Life history of a fossil: An introduction to taphonomy and paleoecology.* Cambridge, MA: Harvard Univ. Press.

Shipman, Pat, Wendy Bosler, and Karen Lee Davis. 1981. Butchering of giant geladas at an Acheulian site. *Current Anthropology* 22 (3):257–268.

Shoshani, J., and P. Tassy. 1996. Summary, conclusions, and a glimpse into the future. Chap. 34 in *The Proboscidea: Evolution and palaeoecology of elephants and their relatives,* edited by J. Shoshani and P. Tassy. Oxford: Oxford Univ. Press.

Shoshani, J., R. M. West, N. Court, R. J. G. Savage, and J. M. Harris. 1996. The earliest proboscideans: General plan, taxonomy, and palaeoecology. Chap. 8 in *The Proboscidea: Evolution and palaeoecology of elephants and their relatives,* edited by J. Shoshani and P. Tassy. Oxford: Oxford Univ. Press.

Sickenberg, O., and M. Schönfeld. 1975. The Chorora Formation—lower Pliocene limnical sediments in the southern Afar (Ethiopia). In *Afar Depression of Ethiopia,* edited by A. Pilger and A. Rösler, pp. 277–284. Stuttgart: E. Schweizerbart'sche Verlagsbuchhandlung.

Simons, E. L., and E. Delson. 1978. Cercopithecidae and Parapithecidae. In *Evolution of African mammals,* edited by V. J. Maglio and H. B. S. Cooke, pp. 100–119. Cambridge, MA: Harvard Univ. Press.

Simpson, Kieran, ed. 1983. H. B. S. Cooke. In *The Canadian Who's Who,* 231–232. Toronto: Univ. of Toronto Press.

Singleton, W. 1976. Archeology report for the Middle Awash Valley. RVRME report no. 16 for the Ethiopian Ministry of Culture, Addis Ababa.

———. 1976. Preliminary archaeological report on the Hounda Bodo hominid site. RVRME report no. 17 for the Ethiopian Ministry of Culture, Addis Ababa.

Singleton, W., and P. Larson Jr. 1976. Catalog of artifacts collected by the RVRME in the Middle Awash Valley. RVRME report no. 15 for the Ethiopian Ministry of Culture, Addis Ababa.

Skeleton may be 3 million years old: Most complete early man discovery in Africa. 1974. *Austin American-Statesman* (Dec. 23), 14.

Smart, C. 1976. The Lothagam 1 fauna: Its phylogenetic, ecological, and biogeographic significance. In *Earliest man and environments in the Lake Rudolf Basin,* edited by Y. Coppens, F. C. Howell, G. L. Isaac, and R. Leakey, pp. 361–369. Chicago: Univ. of Chicago Press.

Smart, C. 1976. Report of the Bovidae from Bodo, Middle Awash Valley, Ethiopia. Report for the RVRME, Addis Ababa.

———. 1976. Report of the preliminary paleoecology of Bodo. Report for the RVRME, Addis Ababa.

Smith, Jack. 1994. "[Letter]." *Science* (July 10) 265:15–16.

Sorenson, John. 1993. *Imagining Ethiopia: Struggles for history and identity in the horn of Africa.* New Brunswick, NJ: Rutgers Univ. Press.

Stanley, Steven M. 1986. *Earth and life through time.* New York: W. H. Freeman.

Steininger, F. F., W. A. Berggren, D. V. Kent, R. L. Bernor, S. Sen, and J. Agusti. 1996. Circum-Mediterranean Neogene (Miocene and Pliocene) marine-continental chronologic correlations of European mammal units. Chap. 2 in *The evolution of western Eurasian Neogene mammal faunas,* edited by R. L. Bernor, V. Fahlbusch, and H.-W. Mittmann. New York: Columbia Univ. Press.

Stern, J. T., and R. L. Susman. 1983. The locomotor anatomy of *Australopithecus afarensis. American Journal of Physical Anthropology* 60:279–317.

Stringer, C. B. 1992. Evolution of early humans. Part. 6.6 in *The Cambridge encyclopedia of human evolution,* edited by S. Jones, R. Martin, and D. Pilbeam. Cambridge: Cambridge Univ. Press.

Suess, Eduard. 1891. Di Bruche des ostlichen Afrika. *Beitraege zur geologischen Kenntnis oestlichen Afrikas. Denkschriften der Oesterreichischen Akademie der Wissenschaften. Math.-Phys. Klasse* 58:555.

———. 1909. *The face of the earth.* Translated by Hertha Sollas. Oxford: Clarendon Press.

Suit on rumor of tie to C.I.A. brings apology to geologist. 1987. *New York Times* (Dec. 5), 3:28.

Taieb, Maurice. 1969. Différents aspects du quaternaire de la vallée de l'Aouache (Éthiopie) *Comptes Rendus Hebdomadaires des Séances de l'Académie des Sciences, Série D: Sciences Naturelles* 269 (3):289–292.

———. 1969. Les études Françaises sur le Quaternaire d'Afrique, 209–210. *VIIe Congres International de l'INQUA,* Paris, 1969.

———. 1969. Melka Kontouré (Haute Vallée de l'Aouache, Éthiopie) stratigraphie du Quarternaire. *Palaeoecology of Africa* 4:60–61.

———. 1971. Aperçus sur les formations quaternaires et la néotectonique de la basse vallée de l'Aouache (Afar méridional, Éthiopie). *Bulletin de la Société Géologique de France* 2:63–65.

———. 1971. Compte-rendu des travaux de terrain Éthiopie et T.F.A.I. In *Rapport d'activité annuel du Laboratoire de Géologie du Quaternaire,* Oct. 1970–Dec. 1971, 14–20. Meudon-Bellevue: Centre National de la Recherche Scientifique.

———. 1971. Les dépots quaternaires sédimentaires de la vallée de l'Aouache (Éthiopie) et leurs relations avec la néotectonique cass ante du rift. *Quaternairia* 15:351–365.

———. 1972. Recherches sur le quaternaire continental de la vallée de l'Aouache (zone du rift), Éthiopie. *Palaeoecology of Africa and of the Surrounding Islands and Antarctica* 6:109–113.

———. 1974. Évolution quaternaire du bassin de l'Aouache. Ph.D. diss., Université de Paris.

———. 1976. Excursion de la vallée de l'Aouache (12–15 Decembre 1971). In *Proceedings of the VIIth (1971) Pan-African Congress of Prehistory and Quaternary Studies,* pp. 25–31. Addis Ababa: Ethiopian Antiquities Administration.

———. 1985. *Sur la terre des premiers hommes.* Paris: Éditions Robert Laffont.

Taieb, M., Y. Coppens, D. C. Johanson, and J. E. Kalb. 1972. Dépots sédimentaires et faunes du Plio-Pleistocène dans la basse vallée de l'Aouache (Afar central, Éthiopie). [With foldout map insert.] *Comptes-Rendus des Séances de l'Académie des Sciences, Série 2: Mécanique-Physique, Chimie, Sciences de l'Univers, Sciences de la Terre* 275(1):819–822.

Taieb, M., D. C. Johanson, and Y. Coppens. 1974. Découverte d'hominides plio-pleistocènes a l'Afar, Éthiopie, 3ᵉ campagne 1974. *Comptes Rendus Hebdomadaires des Séances de l'Académie des Sciences, Série D. Sciences Naturelles* 281 (18): 1297–1300.

Taieb, M., D. C. Johanson, Y. Coppens, and J. L. Aronson. 1976. Geological and palaeontological background of Hadar hominid site, Afar, Ethiopia. *Nature* 260 (5549): 289–293.

Taieb, M., D. C. Johanson, Y. Coppens, R. Bonnefille, and J. Kalb. 1974. Découverte d'hominides dans les séries plio-pleistocène d'Hadar (Bassin de l'Aouache; Afar, Éthiopie). *Comptes Rendus Hebdomadaires des Séances de l'Académie des Sciences, Série D. Sciences Naturelles* 279(9):735–738.

Taieb, M., D. C. Johanson, Y. Coppens, and J. J. Tiercelin. 1978. Expédition internationale de l'Afar, Éthiopie (4ᵉ et 5ᵉ campagnes 1975–1977): Chronostratigraphie des gisements à Hominides pliocènes d'Hadar et corrélations avec les sites préhistoriques du Kada Gona. *Comptes Rendus Hebdomadaires des Séances de l'Académie des Sciences, Série D: Sciences Naturelles,* 287 (5):459–461.

Tamrat, T. 1977. Ethiopia, the Red Sea and the Horn. Chap. 2 in *The Cambridge history of Africa, Vol. 3,* edited by Roland Oliver, pp. 134–164. Cambridge: Cambridge Univ. Press.

———. 1984. The Horn of Africa: The Solomonids in Ethiopia and the states of the Horn of Africa. Chap. 17 in *Africa from the twelfth to the sixteenth century,* edited by D. T. Niane. London: Heinemann.

Tassy, Pascal. 1996. Who is who among the Proboscidea? Chap. 6 in *The Proboscidea: Evolution and palaeoecology of elephants and their relatives,* edited by J. Shoshani and P. Tassy. Oxford: Oxford Univ. Press.

Tattersall, Ian. 1986. Species recognition in human paleontology. *Journal of Human Evolution* 15:165–175.

———. 1992. Species concepts and species identification in human evolution. *Journal of Human Evolution* 22:341–349.

———. 1995. *The fossil trail: How we know what we think we know about human evolution.* New York: Oxford Univ. Press.

Tazieff, H. 1970. The Afar Triangle. *Scientific American* 222 (2):32–40.

Tazieff, H., J. Varet, F. Barberi, and G. Giglia. 1972. Tectonic significance of the Afar (or Danakil) depression. *Nature* 235 (5334):144–147.

Tebedge, Sleshi. 1975. Fossil Suidae in northwest Hararghe. RVRME report no. 4 for the Ethiopian Ministry of Culture, Addis Ababa.

———. 1975. Genus *Sus* (Suidae) from northwest Hararghe. RVRME report no. 6 for the Ethiopian Ministry of Culture, Addis Ababa.

———. 1980. Fossil Suidae from the Middle Awash Valley, Afar Depression, Ethiopia. Master's thesis, University of Texas, Austin.

The forbidden desert of Danakil. 1973. Produced by Aubrey Buxton for Anglia Television.

The unknown famine: A report on famine in Ethiopia. 1973. Produced and directed by Jonathon Dimbleby for Thames Television.

Thesiger, Wilfred. 1935. The Awash River and the Aussa Sultanate. *Geographical Journal of London* 85:1–23.

———. 1987. *The life of my choice.* New York: Norton.

Thesiger, W., and M. Meynell. 1935. On the collection of birds from Danakil, Abyssinia. *The Ibis* 5 (4):774–807.

Thompson, Virginia, and Richard Adloff. 1968. *Djibouti and the Horn of Africa*. Stanford: Stanford Univ. Press.

Thomson, Blair. 1975. *Ethiopia, the country that cut off its head: A diary of the revolution*. London: Robson Books.

Throngs in Addis Ababa condemn regime's foes. 1977. *New York Times* (Jan. 31), 4.

Tiercelin, J. J. 1984. The Pliocene Hadar Formation: Afar Depression of Ethiopia. In *Sedimentation in the African rifts*, 221–240. London: Geological Society of London.

Tiercelin, J. J., and J. Michaux. 1977. Un exemple de remplissage Mio-Pliocène de rift dans l'Est Africain: La formation de Ch'orora, Éthiopie; Stratigraphie et Mammiferes fossiles. *Réunion Annuelle des Sciences de la Terre* 5:449.

Tiercelin, J. J., J. Michaux, and Y. Bandet. 1979. Le Miocène supérieur du sud de la dépression de l'Afar, Éthiopie: Sediments, faunes, âges isotopiques. *Bulletin de la Société Géologique de France* 21(3):255–258.

Tiruneh, Andargachew. 1993. *The Ethiopian revolution, 1974–1987: A transformation from an aristocratic to a totalitarian autocracy*. Cambridge: Cambridge Univ. Press.

Trimingham, J. Spencer. 1965. *Islam in Ethiopia*. London: Frank Cass.

Turnbull, William D. 1980. Bryan Patterson, 1909–1979. *Field Museum of Natural History Bulletin* 51(2):11–13.

Tuttle, R, H., 1981. Paleoanthropology without inhibitions. *Science* 212:798.

United Kingdom. Ministry of Defense. 1968. *Aiscia*, series 1404, sheet 789-B, edition 2-GSGS. Scale 1: 500,000. London: United Kingdom, Ministry of Defense.

United Kingdom, War Office. [1908] 1922. *Abyssinia*, map no. 2319. Scale 1: 3,000,000. In Vol. 2, *Handbook of Ethiopia, 1922*. London: United Kingdom, War Office.

———. 1946. Geographical section, *Abbe*, map no. 4355, sheet NC 37/3. Scale 1: 500,000. London: United Kingdom, War Office.

———. 1946. *Lechemt*, map no. 1754, sheet NC37/4, series Y401,. Scale 1: 500,000. London: United Kingdom, War Office.

———. 1946. *Soddu*, sheet NB37/1. Scale 1: 500,000. London: United Kingdom, War Office.

United Nations Development Fund. 1973. *Geology, geochemistry and hydrology of hot springs of the East African rift system within Ethiopia*. New York: U.N. Development Programme.

United Nations Food and Agricultural Organization. 1965. *Report on survey of the Awash River Basin*. 5 vols. Rome: United Nations Special Fund.

U.S. Army. 1972. *Serdo, Ethiopia; French Territory of Afars and Issas: Joint operations graphic (ground), Ethiopia*, sheet NC 37-4, series 1501 (edition 1). Scale 1: 250,000. Washington D.C.: United States Army Topographic Command.

U.S. Bureau of Reclamation [Office of International Affairs], Department of Interior. 1964. *Land and water resources of the Blue Nile Basin, Ethiopia*. 4 vols. Washington, D.C.: U.S. Bureau of Reclamation.

U.S. District Court for the District of Columbia. (Dec. 31) 1986. Jon Ervin Kalb, plaintiff, v. National Science Foundation and United States of America, defendants. Complaint for damages and injunctive relief. Civ. No. 86-3557.

U.S. District Court for the District of Columbia. (Dec 8) 1987. Jon Ervin Kalb, plaintiff, v. National Science Foundation et al., Defendants. Stipulation of settlement. Civ. No. 86-3557 (SSH).

U.S. General Accounting Office. 1991. *Peer review: Compliance with the Privacy Act and Federal Advisory Committee Act*. Washington, D.C.: General Accounting Office.

————. 1994. *Peer review: Reforms needed to ensure fairness in federal agency grant selection.* Washington, D.C.: General Accounting Office.

U.S. House Committee on International Relations. 1978. *War in the horn of Africa: A firsthand report on the challenges for United States policy. Report of a factfinding mission to Egypt, Sudan, Ethiopia, Somalia, and Kenya,* December 12 to 22, 1977. 95th Cong., 2d sess. Washington D.C.: U.S. Government Printing Office.

————. 1986. Committee on Science and Technology. *Science policy study—research project selection: Hearings before the Task Force on Science Policy.* 99th Cong., 2d sess., 8–10 April.

Varet, J. 1978, comp., *Geological map of central and southern Afar (Ethiopia and F.T.A.I).* Scale 1:500,000. France: Centre National de la Recherche Scientifique; Italy: Consiglio Nazionale delle Richerche. [See Barberi et al., 1971, 1973.]

————. 1978. *Geology of central and southern Afar (Ethiopian and Djibouti Republic).* Paris: Centre National de la Recherche Scientifique. [See Barberi et al., 1971, 1973.]

Variable but singular. 1994. *Nature* 368:399.

Verin, P. 1990. Madagascar. Chap. 28 in *Ancient civilizations of Africa,* edited by G. Mokhtar. London: James Currey.

Vinassa de Regny, P. 1924. *Dancalia.* [With geologic map, scale 1:500,000.] Rome: Alfieri and Lacroix.

Vine, Fred J. 1966. Spreading of the ocean floor: New evidence. *Science* 154:1405–1415.

Vine, F. J., and D. H. Matthews. 1963. Magnetic anomalies over oceanic ridges. *Nature* 199:947–949.

Vine, Fred J., and J. Tuzo Wilson. 1965. Magnetic anomalies over a young oceanic ridge off Vancouver Island. *Science* 150:485–489.

Vivó, Raul Valdés. 1978. *Ethiopia's revolution.* New York: International Publishers.

Vrba, E. S. 1995. The fossil record of African antelopes (Mammalia, Bovidae) in relation to human evolution and paleoclimate. Chap. 27 in *Paleoclimate and evolution, with emphasis on human origins,* edited by E. S. Vrba, G. H. Denton, T. C. Partridge, and L. H. Burckle. New Haven, CT: Yale Univ. Press.

Wallace, Alfred Russel. [1876] 1962. *The geographical distribution of animals.* Vol. 1. New York: Hafner.

Walter, R. C., P. C. Manega, R. L. Hay, R. E. Drake, and G. H. Curtis. 1991. Laser-fusion [40]Ar/[39]Ar dating of Bed I, Olduvai Gorge, Tanzania. *Nature* 354:145.

Waugh, Evelyn. 1936. *Waugh in Abyssinia.* London: Longmans.

Wegener, Alfred. 1915. *Die Entstehung der Kontinente und Ozeane.* Braunschweig: Friedr. Vieweg und Sohn.

————. [1929] 1966. *The origin of continents and oceans.* New York: Dover Publications.

Wegener, Kurt. 1966. Alfred Wegener. In *The origin of continents and oceans.* New York: Dover Publications.

Wendorf, Fred, and Romuald Schild. 1974. A *Middle Stone Age sequence from the central rift valley, Ethiopia.* Wroclaw: Polska Adademii Nauk [Warsaw: Polish Academy of Sciences].

White, Tim. 1985. *Acheulian man in Ethiopia's Middle Awash Valley: The implications of cutmarks on the Bodo cranium.* Netherlands: Joh. Enschede en Zonen Haarlem.

————. 1986. Cut marks on the Bodo cranium: A case of prehistoric defleshing. *American Journal of Physical Anthropology* 69(4):503–509.

————. 1992. *Prehistoric cannibalism at Mancos 5MTUMR-2346.* Princeton, NJ: Princeton Univ. Press.

———. 1993. For sure there is pandemonium at the National Museum (English translation; in Amharic). *IFOYTA* 1(11):20–22.

White, T. D., and J. M. Harris. 1977. Suid evolution and correlation of African hominid localities. *Science* 198(4312):13–21.

White, T. D., G. Suwa, and B. Asfaw. 1995. *Australopithecus ramidus,* a new species of early hominid from Aramis, Ethiopia. *Nature* 371(6495):306–312.

———. 1995. Corrigendum: *Australopithecus ramidus,* a new species of early hominid from Aramis, Ethiopia. *Nature* 375:88.

White, T. D., G. Suwa, W. K. Hart, R. C. Walter, G. WoldeGabriel, J. de Heinzelin, J. D. Clark, B. Asfaw, and E. Vrba. 1993. New discoveries of *Australopithecus* at Maka in Ethiopia. *Nature* 366(6452):261–265.

Whitehead, Paul F. 1976. Summary of primate fossils collected by the RVRME during the fall, 1976 field season. Report for the RVRME, Addis Ababa.

———. 1982. Hominid discovery in the Awash Valley. *Explorer's Journal* (September), 122–125.

Who's Who in the World, 1976–1977. 1976. 3d ed. Chicago: Marquis Who's Who Inc.

Whybrow, Peter J. 1984. Geological and faunal evidence from Arabia for mammal "migrations" between Asia and Africa during the Miocene. *Courier Forschungsinstitut Senckenberg* 69:189–198.

———. 1992. Land movements and species dispersal. In *The Cambridge encyclopedia of human evolution,* edited by S. Jones, R. Martin, and D. Pilbeam, pp. 169–173. Cambridge: Cambridge Univ. Press.

Wilford, John Noble. 1994. New fossils take science close to dawn of humans. *New York Times* (Sept. 22), A1, A11.

———. 1995. Human ancestors' tools are found in Africa. *New York Times* (April 25), B7, B9.

———. 1996. 2.3-million-year-old jaw extends human family. *New York Times* (Nov. 19), A1, B7.

Willis, Delta. 1989. *The hominid gang: Behind the scenes in the search for human origins.* New York: Viking.

Wilson, J. Tuzo, 1965. A new class of faults and their bearing on continental drift. *Nature* 207:343–347.

Winid, B. 1981. Comments on the development of the Awash Valley, Ethiopia. In *River basin planning: Theory and practice,* edited by S. K. Saha and C. J. Barrow, pp. 147–165. New York: Wiley.

Wiseman, J. D. H., and R. B. S. Sewell. 1937. The floor of the Arabian Sea. *Geological Magazine* 74:219–230.

WoldeGabriel, G., G. Heiken, W. K. Hart, P. Renne, and T. White. 1996. Silicic tephra in the middle Awash Valley, Ethiopia: Implications for regional stratigraphic correlations. *Geological Society of America Annual Meeting Abstracts* 28(6):71.

WoldeGabriel, G., P. Renne, T. D. White, G. Suwa, J. de Heinzelin, W. K. Hart, and G. Heiken. 1995. Age of early hominids. [Letter]. *Nature* 376:559.

WoldeGabriel, G., T. D. White, G. Suwa, P. Renne, J. de Heinzelin, W. K. Hart, and G. Heiken. 1994. Ecological and temporal placement of early Pliocene hominids at Aramis, Ethiopia. *Nature* 371(6495):330–333.

Wolpoff, Milford H. 1996. *Human evolution: 1996–1997 edition.* New York: McGraw-Hill.

Wood, Bernard. 1993. Rift on the record. *Nature* 365:789–790.

———. 1997. The oldest whodunnit in the world. *Nature* 385:292–293.

Wood, C. B. 1976. A detailed report on a personal reconnaissance investigation of Miocene to Recent sediments and faunas in the Awash River area of the Afar Triangle, Ethiopia. Report for the Museum of Comparative Zoology, Harvard University, Cambridge, MA.

————. Hounda Bodo site maps. RVRME report no. 18 for the Ethiopian Ministry of Culture, Addis Ababa.

Wylde, Augustus B. 1901. *Modern Abyssinia.* London: Methuen.

Young, Patrick. 1974. Meet an ancestor, age 3 million. *The National Observer* (Feb. 16).

Zanettin, Bruno. 1992. Evolution of the Ethiopian volcanic province. *Atti della Accademia Nazionale dei Lincei, Memorie Lincee Scienze Fisiche e Naturali, Series 1* 6:153–181.

Index

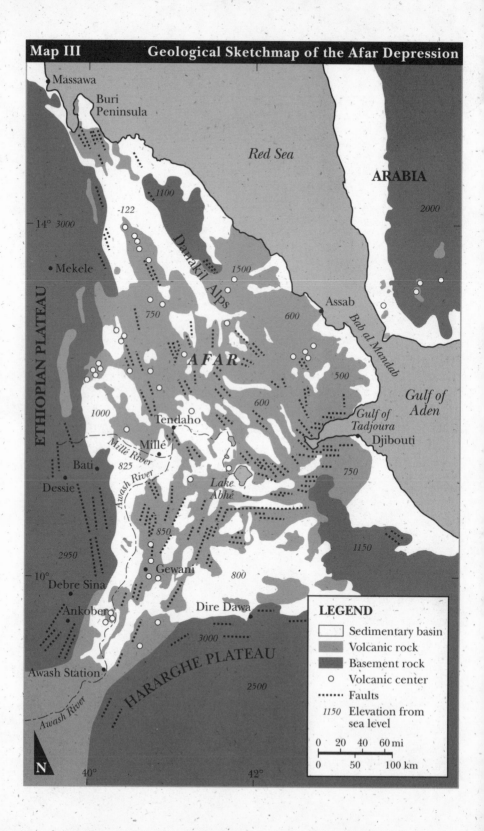

Map III　Geological Sketchmap of the Afar Depression

Massawa

Buri
Peninsula

Red Sea

ARABIA

2000

1100

−122

−14° 3000

Danakil Alps

1500

• Mekele

Assab

Bab al Mandab

600

750

AFAR

500

Gulf of
Aden

1000

600

Tendaho

Gulf of
Tadjoura

Millé

Millé River

Djibouti

• Bati

825

750

Dessie

Awash River

Lake
Abhé

2950

850

− 10° Debre Sina

Gewani

800

1150

Ankober

Dire Dawa

3000

Awash Station

HARARGHE PLATEAU

2500

Awash River

ETHIOPIAN PLATEAU

LEGEND

☐ Sedimentary basin

▨ Volcanic rock

▧ Basement rock

○ Volcanic center

⋯⋯ Faults

1150 Elevation from
sea level

0　20　40　60 mi

0　　50　　100 km

N

40°　　　　42°